T0317533

Magnetic Memory
Technology

Magnetic Memory Technology

Spin-Transfer-Torque MRAM and Beyond

Denny D. Tang
Tang Consultancy

Chi-Feng Pai
National Taiwan University

Library of Congress Cataloging-in-Publication Data
Names: Tang, Denny D., author. | Pai, Chi-Feng, author. | John Wiley &
    Sons, Inc., publisher.
Title: Magnetic memory technology : spin-transfer-torque MRAM and beyond /
    Denny D. Tang, Chi-Feng Pai.
Description: Hoboken : Wiley-IEEE Press, [2021] | Includes bibliographical
    references and index.
Identifiers: LCCN 2020033767 (print) | LCCN 2020033768 (ebook) | ISBN
    9781119562238 (cloth) | ISBN 9781119562221 (adobe pdf) | ISBN
    9781119562283 (epub)
Subjects: LCSH: Magnetic memory (Computers). | Spintronics. | Nonvolatile
    random-access memory.
Classification: LCC TK7895.M3 T365 2021 (print) | LCC TK7895.M3 (ebook) |
    DDC 621.39/763–dc23
LC record available at https://lccn.loc.gov/2020033767
LC ebook record available at https://lccn.loc.gov/2020033768

Cover Design: Wiley
Cover Image: Fractal image courtesy of Denny D. Tang, (inset image) IEEE

MIX
Paper from
responsible sources
FSC
www.fsc.org   FSC® C013604

Contents

Preface *xi*
Author Biographies *xiv*
List of Cited Tables and Figures *xvi*

1   **Basic Electromagnetism** *1*
1.1   Introduction *1*
1.2   Magnetic Force, Pole, Field, and Dipole *1*
1.3   Magnetic Dipole Moment, Torque, and Energy *3*
1.4   Magnetic Flux and Magnetic Induction *5*
1.5   Ampère's Circuital Law, Biot-Savart Law, and Magnetic Field from Magnetic Material *6*
1.5.1   Ampère's Circuital Law *6*
1.5.2   Biot-Savart's Law *8*
1.5.3   Magnetic Field from Magnetic Material *10*
1.6   Equations, cgs-SI Unit Conversion Tables *11*
    Homework *13*
    References *17*

2   **Magnetism and Magnetic Materials** *19*
2.1   Introduction *19*
2.2   Origin of Magnetization *19*
2.2.1   From Ampère to Einstein *19*
2.2.2   Precession *21*
2.2.3   Electron Spin *22*
2.2.4   Spin-Orbit Interaction *24*
2.2.5   Hund's Rules *25*
2.3   Classification of Magnetisms *28*
2.3.1   Diamagnetism *30*
2.3.2   Paramagnetism *30*

2.3.3   Ferromagnetism   *34*

2.3.4   Antiferromagnetism   *37*

2.3.5   Ferrimagnetism   *40*

2.4   Exchange Interactions   *42*

2.4.1   Direct Exchange   *43*

2.4.2   Indirect Exchange: Superexchange   *45*

2.4.3   Indirect Exchange: RKKY Interaction   *46*

2.4.4   Dzyaloshinskii-Moriya Interaction (DMI)   *48*

2.5   Magnetization in Magnetic Metals and Oxides   *49*

2.5.1   Slater-Pauling Curve   *49*

2.5.2   Rigid Band Model   *50*

2.5.3   Iron Oxides and Iron Garnets   *51*

2.6   Phenomenology of Magnetic Anisotropy   *51*

2.6.1   Uniaxial Anisotropy   *52*

2.6.2   Cubic Anisotropy   *53*

2.7   Origins of Magnetic Anisotropy   *54*

2.7.1   Shape Anisotropy   *55*

2.7.2   Magnetocrystalline Anisotropy (MCA)   *56*

2.7.3   Perpendicular Magnetic Anisotropy (PMA)   *57*

2.8   Magnetic Domain and Domain Walls   *57*

2.8.1   Domain Wall   *58*

2.8.2   Single Domain and Superparamagnetism   *59*

Homework   *60*

References   *64*

3   **Magnetic Thin Films**   *67*

3.1   Introduction   *67*

3.2   Magnetic Thin Film Growth   *67*

3.2.1   Sputter Deposition   *68*

3.2.2   Molecular Beam Epitaxy (MBE)   *71*

3.3   Magnetic Thin Film Characterization   *72*

3.3.1   Vibrating-Sample Magnetometer (VSM)   *73*

3.3.2   Magneto-Optical Kerr Effect (MOKE)   *74*

References   *76*

4   **Magnetoresistance Effects**   *77*

4.1   Introduction   *77*

4.2   Anisotropic Magnetoresistance (AMR)   *78*

4.3   Giant Magnetoresistance (GMR)   *79*

4.4   Tunneling Magnetoresistance (TMR)   *81*

4.5   Contemporary MTJ Designs and Characterization   *84*
4.5.1   Perpendicular MTJ (p-MTJ)   *85*
4.5.2   Fully Functional p-MTJ   *85*
4.5.3   CIPT Approach for TMR Characterization   *87*
        Homework   *89*
        References   *89*

5   Magnetization Switching and Field MRAMs   *93*
5.1   Introduction   *93*
5.2   Magnetization Reversible Rotation and Irreversible Switching Under
        External Field   *93*
5.2.1   Magnetization Rotation Under an External Field in the Hard Axis
        Direction   *94*
5.2.2   Magnetization Rotation and Switching Under an external Field
        in the Easy Axis Direction   *95*
5.2.3   Magnetization Rotation and Switching Under Two Orthogonal
        External Fields   *96*
5.2.4   Magnetization Behavior of a Synthetic Anti-ferromagnetic Film
        Stack   *97*
5.3   Field MRAMs   *99*
5.3.1   MTJ of Field MRAM   *100*
5.3.2   Half-Select Bit Disturbance Issue   *101*
        Homework   *102*
        References   *103*

6   Spin Current and Spin Dynamics   *105*
6.1   Introduction to Hall Effects   *105*
6.1.1   Ordinary Hall Effect   *105*
6.1.2   Anomalous Hall Effect and Spin Hall Effect   *106*
6.2   Spin Current   *109*
6.2.1   Electron Spin Polarization in NM/FM/NM Film Stack   *109*
6.2.2   Spin Current Injection, Diffusion, and Inverse Spin Hall Effect   *111*
6.2.3   Generalized Carrier and Spin Current Drift-Diffusion Equation   *114*
6.3   Spin Dynamics   *116*
6.3.1   Landau-Lifshitz and Landau-Lifshitz-Gilbert Equations of
        Motion   *116*
6.3.2   Ferromagnetic Resonance   *118*
6.3.3   Spin Pumping and Effective Damping in FM/NM Film Stack   *120*
6.3.4   FM/NM/FM Coupling Through Spin Current   *122*
6.4   Interaction Between Polarized Conduction Electrons and Local
        Magnetization   *124*

6.4.1   Electron Spin Torque Transfer to Local Magnetic Magnetization   *124*
6.4.2   Macrospin Model   *125*
6.4.3   Spin-Torque Transfer in a Spin Valve   *127*
6.4.3.1 Switching Threshold Current Density   *128*
6.4.3.2 Switching Time   *129*
6.4.4   Spin-Torque Transfer Switching in Magnetic Tunnel Junction   *131*
6.4.5   Spin-Torque Ferromagnetic Resonance and Torkance   *133*
6.5   Spin Current Interaction with Domain Wall   *134*
6.5.1   Domain Wall Motion under Spin Current   *135*
6.5.2   Threshold Current Density   *137*
         Homework   *138*
         References   *144*

7   Spin-Torque-Transfer (STT) MRAM Engineering   *151*
7.1   Introduction   *151*
7.2   Thermal Stability Energy and Switching Energy   *152*
7.3   STT Switching Properties   *154*
7.3.1   Switching Probability and Write Error Rate (WER)   *156*
7.3.2   Switching Current in Precessional Regime   *160*
7.3.3   Switching Delay of an STT-MRAM Cell   *161*
7.3.4   Read Disturb Rate   *161*
7.3.5   Switching Under a Magnetic Field – Phase Diagram   *162*
7.3.6   MTJ Switching Abnormality   *164*
7.3.6.1 Magnetic Back-Hopping   *164*
7.3.6.2 Bifurcation Switching (Ballooning in WER)   *165*
7.3.6.3 Domain Mediated Magnetization Reversal   *166*
7.4   The Integrity of MTJ Tunnel Barrier   *166*
7.4.1   MgO Degradation Model   *167*
7.5   Data Retention   *169*
7.5.1   Retention Determination Based on Bit Switching Probability   *169*
7.5.2   Energy Barrier Determination Based on Aiding Field   *170*
7.5.3   Energy Barrier Extraction with Retention Bake at Chip Level   *171*
7.5.4   Data Retention Fail at the Chip Level   *173*
7.6   The Cell Design Considerations and Scaling   *173*
7.6.1   STT-MRAM Bit Cell and Array   *174*
7.6.2   CMOS Options   *174*
7.6.3   Cell Switching Efficiency   *176*
7.6.4   Cell Design Considerations   *177*
7.6.4.1 WRITE Current and Cell Size   *178*
7.6.4.2 READ Access Performance and *RA* Product of MTJ   *178*

7.6.4.3   READ and WRITE Voltage Margins   *178*
7.6.4.4   Stray Field Control for Perpendicular MTJ   *179*
7.6.4.5   Suppress Stochastic Switching Time Variation Ideas   *181*
7.6.5   The Scaling of MTJ for Memory   *182*
7.6.5.1   In-Plane MTJ   *183*
7.6.5.2   Out-of-Plane (Perpendicular) MTJ   *184*
7.7   MTJ SPICE Models   *188*
7.7.1   Basic MTJ Equivalent Circuit Model for Circuit Design Simulation   *188*
7.7.2   MTJ SPICE Circuit Model with Embedded Macrospin Calculator   *189*
7.8   Test Chip, Test, and Chip-Level Weak Bit Screening   *191*
7.8.1   Read Marginal Bits   *192*
7.8.2   Write Marginal Bits   *193*
7.8.3   Short Retention Bits   *193*
7.8.4   Low Endurance Bits   *194*
        Homework   *195*
        References   *197*

8   Advanced Switching MRAM Modes   *205*
8.1   Introduction   *205*
8.2   Current-Induced-Domain-Wall Motion (CIDM) Memory   *206*
8.2.1   Single-Bit Cell   *207*
8.2.2   Multibit Cell: Racetrack   *209*
8.3   Spin-Orbit Torque (SOT) Memory   *211*
8.3.1   Spin Orbit Torque (SOT) MRAM Cells   *211*
8.3.1.1   In-Plane SOT Cell   *212*
8.3.1.2   Perpendicular SOT Cell   *218*
8.3.2   Materials Choice for SOT-MRAM Cell   *219*
8.3.2.1   Transition Metals and their Alloys   *219*
8.3.2.2   Emergent Materials Systems   *221*
8.3.2.3   Benchmarking of SOT Switching Efficiency   *222*
8.4   Magneto-Electric Effect and Voltage-Control Magnetic Anisotropy
        (VCMA) MRAM   *224*
8.4.1   Magneto-Electric Effects   *224*
8.4.2   VCMA-Assisted MRAMs   *227*
8.4.2.1   VCMA-Assisted Field-MRAM   *227*
8.4.2.2   VCMA-Assisted Multi-bit-Word SOT-MRAM   *229*
8.4.2.3   VCMA-Assisted Precession-Toggle MRAM   *229*
8.5   Relative Merit of Advanced Switching Mode MRAMs   *231*
        Homework   *233*
        References   *233*

9   **MRAM Applications and Production** *241*
9.1   Introduction *241*
9.2   Intrinsic Characteristics and Product Attributes of Emerging Nonvolatile
        Memories *242*
9.2.1   Intrinsic Properties *243*
9.2.2   Product Attributes *244*
9.3   Memory Landscape and MRAM Opportunity *247*
9.3.1   MRAM as Embedded Memory in Logic Chips *248*
9.3.1.1 Integration Issues of Embedded MRAM *248*
9.3.1.2 MRAM as Embedded Flash in Microcontroller *249*
9.3.1.3 Embedded MRAM Cell Size *250*
9.3.1.4 MRAM as Cache Memory in Processor *250*
9.3.1.5 Improvement of Access Latency *251*
9.3.2   High-Density Discrete MRAM *254*
9.3.2.1 Technology Status *254*
9.3.2.2 Ideal CMOS Technology for High-Density MRAM *256*
9.3.2.3 Improvement to Endurance and Write Error Rate with Error Buffer in
        Chip Architecture *258*
9.3.3   Applications and Market Opportunity of MRAM *258*
9.3.3.1 Battery-Backed DRAM Applications *260*
9.3.3.2 Internet of Things (IoT) and Cybersecurity Applications *261*
9.3.3.3 Applications to In-Memory Computing, and Artificial
        Intelligence (AI) *264*
9.3.3.4 MRAM-Based Memory-Driven Computer *265*
9.4   MRAM Production *266*
9.4.1   MRAM Production Ecosystem *266*
9.4.2   MRAM Product History *267*
9.4.2.1 First-Generation MRAM – Field MRAM (Also Called Toggle
        MRAM) *268*
9.4.2.2 The Second-Generation MRAM – STT-MRAM *269*
9.4.2.3 The Potential Third-Generation MRAM – SOT MRAM *270*
        Homework *271*
        References *271*

Appendix A   Retention Bake (Including Two-Way Flip) *277*
Appendix B   Memory Functionality-Based Scaling *279*
**Appendix C   High-Bandwidth Design Considerations for
               STT-MRAM** *299*

Index *323*

Preface

In early 1900s, quantum mechanics successfully explained that the discrete light emission spectrum is associated with the discrete electron energy of an atom. Schrödinger further confirmed that three integer quantum numbers are sufficient to describe the Hydrogen emission spectrum. In 1925, Ralph Kronig, a young Ph. D. student who studied emission spectrum at Columbia University came to the conclusion that he needed to add a new quantum number to explain the spectrum of certain materials. The quantum number is $m_s = \pm 1/2$ and is associated with electron *spin*. This hypothesis caused debates in the science community, since it was entirely derived from experimental observations, which lacks a theoretical base. One of Kronig's thesis advisers, Wolfgang Pauli, commented, "It is indeed a very clever, but of course has nothing to do with reality." Shortly after, two graduate students, George Uhlrenbeck and Samuel Gousmit of Leiden University, Netherlands, also came up with the same idea and wrote a note to their thesis adviser, Paul Ehrenfest. Ehrenfest sent the note for publication. The two students felt uncomfortable and wanted to retract, but Ehrenfest said, "You are both young enough to be able to afford a stupidity." In 1928, Paul Dirac included the concept of relativity into the Schrödinger equation, and the fourth quantum number fell out naturally and established the theoretical base of electron *spin*. The discoveries of these scientists (see Figure P.1) revolutionized not only fundamental physics but also electronic industry in the future.

The properties of electron spin were not fully commercially exploited in the next 60 years, until the days of the development of giant magnetoresistance (GMR) sensor in the 1980s by the hard disk drive (HDD) industry. The introduction of GMR sensors accelerated the recording density growth rate to two times every year. *Spintronics*, a name created by putting *spin* and *electronics* into a single word, became a hot subject. Cheap HDD data storage devices began to become ubiquitous and made a significant impact on people's lives. Today, storing data in HDD is cheaper than on a piece of paper. Nonetheless, GMR is based on the

Pauli

Uhlenbeck

Goutsmit

Fermi

Kronig

Ehrenfest

Ann Arbor 1928

Ehrenfest's students, Leiden 1924. Left to right: Gerhard Heinrich Dieke, Samuel Abraham Goudsmit, Jan Tinbergen, Paul Ehrenfest, Ralph Kronig, and Enrico Fermi

Dirac

Figure P.1 Stars of the discovery of electron spin.

John Slonczewski, 1928-2019   Luc Berger, 1933-.

Figure P.2 J.C. Slonczsewski and Luc Berger were the two people who proposed the spin torque exchange.

spin-dependent transport properties of electrons only. The *spin torque* exchange properties were not exploited.

The torque exchange between electrons was first recognized by John Slonczewski and also independently by Luc Berger (see Figure P.2). The best description of the spin torque exchange of electrons can be found in the patent issued to John Slonczewski in 1997:

> It is a fundamental fact that the macroscopic magnetization intensity of a magnet such as iron arises from the cooperative mutual alignment of

elementary magnetic moments carried by electrons. An electron is little more than a mass particle carrying an electrostatic charge, which spins at a constant rate, like a planet about its axis. The electric current of this spin induces a surrounding magnetic field distribution resembling that, which surrounds the Earth. Thus, each electron is effectively a miniscule permanent magnet....

... The exchange interaction is that force, arising quantum-mechanically from electrostatic interactions between spinning electrons, which causes this mutual alignment ... Not only does it couple the bound spins of a ferro-magnet to each other, but it also couples the spins of moving electrons, such as those partaking in current flow, to these bound electrons.

We believe that electron *spin* will make a big impact on people's lives in the twenty-first century as much as the electron *charge* did in the twentieth century. At the end of twentieth century, scientists and engineers had made remarkable progress in the research of spin torque exchange. The torque exchange between electrons, itinerary and local, is fully exploited. Electrons are now treated not only as charge-carrying particles but also tiny magnets. This little magnet is not easily observable, since the net magnetic torque of two opposite spins cancel, unless one can filter off one spin and keep the other. Through the exchange, they can pass on the magnetic moment in an efficient manner, much more efficient than an external magnetic field. We believe that the most important product entry is spin-transfer torque magnetic memory (STT-MRAM). STT-MRAM has moved out of laboratory and is the only fast read/write nonvolatile memory in production today. The technology is still very young, and not as mature as the existing volatile dynamic memory (DRAM) or static memory (SRAM). Since its data latency is close to SRAM and DRAM, we believe its potential is tremendous. The success of this technology will offer storage at the speed of a data processor, redefines the memory architecture, and drastically lowers the power dissipation of computers.

This book was written to inspire students and professionals to push the frontier of spintronics, to exploit its potentials further, and to help it make more of an impact on our lives.

Denny D. Tang and Chi-Feng Pai

Author Biographies

Denny D. Tang

Dr. Denny D. Tang received his PhD degree in Electrical Engineering from The University of Michigan in 1975. He then joined IBM T.J. Watson Research Center, Yorktown, New York, where he conducted research in Silicon technology and managed a bipolar transistor team. He was elected as IEEE Fellow for his work in bipolar scaling. In 1990, he transferred to IBM Almaden Research Center in San Jose, California, where he conducted research in magnetic recording and then managed the read channel design group and later the write head magnetics group. His group demonstrated the MRAM concept at IEDM in 1995. In 2001, he joined Taiwan Semiconductor Manufacturing Company (TSMC) in Hsinchu, Taiwan, where he managed an exploratory Si research group. In 2003, he received a multi-year grant from the Taiwan government to start MRAM research. In 2008, he joined MagIC, Milpitas, CA, as VP of product engineering to develop and manufacture MRAM products. He is the author of *Magnetic Memory: Fundamentals and Technology* (Cambridge University Press, 2010) and has been granted more than 70 US patents. He is an IEEE Live Fellow, a Fellow of Industrial Technology Research Institute (ITRI) and a Fellow of TSMC Academy.

Chi-Feng Pai

Dr. Chi-Feng Pai received his PhD degree in Applied Physics from Cornell University in 2015. His research on the giant spin Hall effect in various materials systems led to the invention of spin-orbit torque MRAM. He then joined the Department of Materials Science and Engineering at Massachusetts Institute of Technology as a postdoctoral research associate. From 2016 to 2020, he served as assistant professor at the Department of Materials Science and Engineering of National Taiwan University (NTU). He is currently an associate professor at NTU and consulting research fellow at Industrial Technology Research Institute (ITRI), Taiwan. He was the recipient of the Young Researcher Award of Asian Union of Magnetic Society (AUMS) in 2016, Young Researcher Fellowship of Ministry of Science and Technology (MOST, Taiwan) in 2019, and Young Researcher Award of Taiwan Semiconductor Industry Association (TSIA) in 2020.

List of Cited Tables and Figures

Chapter 2

Table/Figure Number	Source
Figure 2.8	J. M. D. Coey, Magnetism and Magnetic Materials (Cambridge University Press, Cambridge, UK, 2010).
Figure 2.10	S. Blundell, Magnetism in condensed matter (Oxford University Press, Oxford; New York, 2001), Oxford master series in condensed matter physics.
Figure 2.13	C. M. Hurd, Varieties of Magnetic Order in Solids, Contemp Phys 23, 469 (1982)
Figure 2.15	R. C. O'Handley, Modern Magnetic Materials: Principles and Applications (Wiley, New York, 2000)
Figure 2.17	S. S. P. Parkin and D. Mauri, Spin engineering: Direct determination of the Ruderman-Kittel-Kasuya-Yosida far-field range function in ruthenium, Phys. Rev. B 44, 7131 (1991)
Figure 2.18	A. Fert, V. Cros, and J. Sampaio, Skyrmions on the track, Nat. Nanotechnol. 8, 152 (2013).
Figure 2.19	R. C. O'Handley, Modern Magnetic Materials: Principles and Applications (Wiley, New York, 2000)
Figure 2.20	R. C. O'Handley, Modern Magnetic Materials: Principles and Applications (Wiley, New York, 2000)
Figure 2.22	R. C. O'Handley, Modern Magnetic Materials: Principles and Applications (Wiley, New York, 2000)
Figure 2.23	R. C. O'Handley, Modern Magnetic Materials: Principles and Applications (Wiley, New York, 2000)
Figure 2.27	T. Liu, J. W. Cai, and L. Sun, Large enhanced perpendicular magnetic anisotropy in CoFeB/MgO system with the typical Ta buffer replaced by an Hf layer, AIP Adv. 2, 032151 (2012)
Figure 2.30	R. C. O'Handley, Modern Magnetic Materials: Principles and Applications (Wiley, New York, 2000)

Chapter 3

Table/Figure Number	Source
Figure 3.2.4	Applied Materials Inc. www.appliedmaterials.com/products/MRAM.

Chapter 4

Table/Figure Number	Source
Figure 4.2.1	W. Gil, D. Görlitz, M. Horisberger, and J. Kötzler, Magnetoresistance anisotropy of polycrystalline cobalt films: Geometrical-size and domain effects, Phys. Rev. B 72, 134401 (2005).
Figure 4.3.2	1. M. N. Baibich, J. M. Broto, A. Fert, F. N. Vandau, F. Petroff, P. Eitenne, G. Creuzet, A. Friederich, and J. Chazelas, Giant Magnetoresistance of (001)Fe/(001) Cr Magnetic Superlattices, Phys. Rev. Lett. 61, 2472 (1988).
	2. G. Binasch, P. Grunberg, F. Saurenbach, and W. Zinn, Enhanced Magnetoresistance in Layered Magnetic-Structures with Antiferromagnetic Interlayer Exchange, Phys. Rev. B 39, 4828 (1989).
Figure 4.4.1	S. A. Wolf, D. D. Awschalom, R. A. Buhrman, J. M. Daughton, S. von Molnar, M. L. Roukes, A. Y. Chtchelkanova, and D. M. Treger, Spintronics: A spin-based electronics vision for the future, Science 294, 1488 (2001)
Figure 4.4.2	1. J. S. Moodera, L. R. Kinder, T. M. Wong, and R. Meservey, Large Magnetoresistance at Room Temperature in Ferromagnetic Thin Film Tunnel Junctions, Phys. Rev. Lett. 74, 3273 (1995).
	2. T. Miyazaki and N. Tezuka, Giant magnetic tunneling effect in Fe/Al2O3/Fe junction, Journal of Magnetism and Magnetic Materials 139, L231 (1995).
Figure 4.4.3	1. S. Yuasa, T. Nagahama, A. Fukushima, Y. Suzuki, and K. Ando, Giant room-temperature magnetoresistance in single-crystal Fe/MgO/Fe magnetic tunnel junctions, Nat. Mater. 3, 868 (2004).
	2. S. S. P. Parkin, C. Kaiser, A. Panchula, P. M. Rice, B. Hughes, M.

(Continued)

Table/Figure Number	Source
Figure 4.5.2	Samant, and S. H. Yang, Giant tunnelling magnetoresistance at room temperature with MgO (100) tunnel barriers, Nat. Mater. 3, 862 (2004). S. Ikeda, K. Miura, H. Yamamoto, K. Mizunuma, H. D. Gan, M. Endo, S. Kanai, J. Hayakawa, F. Matsukura, and H. Ohno, A perpendicular-anisotropy CoFeB-MgO magnetic tunnel junction, Nat. Mater. 9, 721 (2010).
Figure 4.5.3	S. Ikeda, H. Sato, H. Honjo, E. C. I. Enobio, S. Ishikawa, M. Yamanouchi, S. Fukami, S. Kanai, F. Matsukura, T. Endoh, and H. Ohno, in 2014 IEEE International Electron Devices Meeting2014), pp. 33.2.1.
Figure 4.5.6	D. C. Worledge and P. L. Trouilloud, Magnetoresistance measurement of unpatterned magnetic tunnel junction wafers by current-in-plane tunneling, Appl. Phys. Lett. 83, 84 (2003).

Chapter 6

Table/Figure Number	Source
Figure 6.3	Based on Y. K. Kato, R. C. Myers, A. C. Gossard, and D. D. Awschalom, "Observation of the spin Hall effect in semiconductors", Science Express 1105514 (2004)
Figure 6.5	M.I. Dyakonov, Spin Hall Effects,International Journal of Modern Physics B, 23(12n13), 2556–2565. ©2009, World Scientific Publishing Company
Figure 6.7	Feng et. al. Prospects of spintronics based on 2D materials. WIREs Comput Mol Sci, 7: e1313. © 2017 John Wiley & Sons
Figure 6.8	Based on T. Kimura, J. Hamrle, and Y. Otani, "Spin-dependent boundary resistance in the lateral spin-valve structure", Appl. Physics Lett. VOLUME 85, NUMBER 16 18 OCTOBER (2004)
Figure 6.9	Based on S. O. Valenzuela and M. Tinkham, "Spin-polarized tunneling in room-temperature mesoscopic spin valves", APPLIED PHYSICS LETTERS VOLUME 85, NUMBER 24 13 DECEMBER 2004
Figure 6.14	H. Lee, L. Wen, M. Pathak, P. Janssen, P. LeClair, C. Alexander and T. Mewes, "Spin pumping in Co56Fe24B20 multilayer systems", Journal of Physics D: Applied Physics, Volume 41, Number 21. © 2008 IOP Publishing

Table/Figure Number	Source
Figure 6.15	Bret Heinrich, Yaroslav Tserkovnyak, GeorgWoltersdorf, Arne Brataas, Radovan Urban, and Gerrit E.W. Bauer "Dynamic Exchange Coupling in Magnetic Bilayers", Physical review Lett. 90, NUMBER 18. 187601. © 2003 American Physical Society
Figure 6.21	Modified from Chen Wang, Yong-Tao Cui, Jordan A. Katine, Robert A. Buhrman and Daniel C. Ralph1, "Time-resolved measurement of spin-transfer-driven ferromagnetic resonance and spin torque in magnetic tunnel junctions", NATURE PHYSICS, VOL 7, p.496, JUNE 2011
Figure 6.23	Based on S. Maekawa, editor. Concepts in spin electronics, Oxford: Oxford University Press current in very thin metallic films," J. Appl. Phys. 55, 1954 (1984)

Chapter 7

Table/Figure Number	Source
Figure 7.3	From W. H. Butler, Tim Mewes, Claudia K. A. Mewes, P. B. Visscher, William H. Rippard, Stephen E. Russek, and Ranko Heindl, "Switching Distributions for Perpendicular Spin-Torque Devices Within the Macrospin Approximation," IEEE Trans. On Magentics, VOL. 48, NO. 12, p.4684, DECEMBER 2012. © 2012 IEEE.
Figure 7.4	From W. H. Butler, Tim Mewes, Claudia K. A. Mewes, P. B. Visscher, William H. Rippard, Stephen E. Russek, and Ranko Heindl, "Switching Distributions for Perpendicular Spin-Torque Devices Within the Macrospin Approximation," IEEE Trans. On Magentics, VOL. 48, NO. 12, p.4684, DECEMBER 2012. © 2012 IEEE.
Figure 7.6	From Guenole Jan, et.al, "Achieving Sub-ns switching of STT-MRAM for future embedded LLC applications through improvement of nucleation and propagation switching mechanisms", Symposium on VLSI Technology Digest of Technical Papers, (2016). © 2016 IEEE.
Figure 7.10	From W. H. Butler, Tim Mewes, Claudia K. A. Mewes, P. B. Visscher, William H. Rippard, Stephen E. Russek, and Ranko Heindl, "Switching Distributions for Perpendicular Spin-Torque Devices Within the Macrospin Approximation," IEEE Trans. On Magentics, VOL. 48, NO. 12, p.4684, DECEMBER 2012. © 2012 IEEE.
Figure 7.11	From Y. Higo, K. Yamane, K. Ohba, H. Narisawa, K. Bessho, M. Hosomi, and H. Kano, "Thermal activation effect on spin transfer switching in magnetic tunnel junctions", APPLIED PHYSICS LETTERS 87, 082502 2005. © 2005 AIP Publishing.

(*Continued*)

Table/Figure Number	Source
Figure 7.12	From Sun, J. Z. et al. High-bias backhopping in nanosecond time-domain spin-torque switches of MgO-based magnetic tunnel junctions. J. Appl. Phys. 105, 07D109 (2009). © 2009 AIP Publishing.
Figure 7.13	From T. Min, Q. Chen, R. Beach, G. Jan, C. Horng, W. Kula, T. Torng, R. Tong, T. Zhong, D. Tang, P. Wang, M. Chen, J.Z. Sun, J. K. Debrosse, D. C. Worledge, T. M. Maffitt, W. J. Gallagher, "A Study of Write Margin of Spin Torque Transfer Magnetic Random Access Memory Integrated with CMOS Technology", Joint MMM-Intermag Conference paper, AA-05, (2009). © 2009 IEEE.
Figure 7.14	From Guenole Jan, et.al, "Achieving Sub-ns switching of STT-MRAM for future embedded LLC applications through improvement of nucleation and propagation switching mechanisms", Symposium on VLSI Technology Digest of Technical Papers, (2016). © 2016 IEEE.
Figure 7.15	From R. Carboni, S. Ambrogio, W. Chen, M. Siddik, J. Harms, A. Lyle, W. Kula, G. Sandhu, and D. Ielmini, "Understanding cycling endurance in perpendicular spin-transfer torque (p-STT) magnetic memory", Dig. of IEDM, p.516, (2016). © 2016 IEEE.
Figure 7.16	From Guenole Jan, Yu-Jen Wang, Takahiro Moriyama, Yuan-Jen Lee, Mark Lin, Tom Zhong, Ru-Ying Tong, Terry Torng, and Po-Kang Wang, "High Spin Torque Efficiency of Magnetic Tunnel Junctions with MgO/CoFeB/MgO Free Layer," Applied Physics Express, 5, (2012) 093008.
Figure 7.17	From Guenole Jan, Yu-Jen Wang, Takahiro Moriyama, Yuan-Jen Lee, Mark Lin, Tom Zhong, Ru-Ying Tong, Terry Torng, and Po-Kang Wang, "High Spin Torque Efficiency of Magnetic Tunnel Junctions with MgO/CoFeB/MgO Free Layer," Applied Physics Express, 5, (2012) 093008.
Figure 7.19	From Private communication, Courtesy to Industrial Technology Research Institute
Figure 7.22	From Luc Thomas, Guenole Jan, Santiago Serrano-Guisan, Huanlong Liu, Jian Zhu, Yuan-Jen Lee, Son Le, Jodi Iwata-Harms, Ru-Ying Tong, Sahil Patel, Vignesh Sundar, Dongna Shen, Yi Yang, Renren He, Jesmin Haq, Zhongjian Teng, Vinh Lam, Paul Liu, Yu-Jen Wang, Tom Zhong, Hideaki Fukuzawa, and PoKang Wang, "STT-MRAM devices with low damping and moment optimized for LLC applications at 0x nodes", IEEE Dig. Of IEDM 2018, paper 27.03. © 2018 IEEE.
Figure 7.23	From private communication, courtesy to Applied Material
Figure 7.25	From Sheng-Huang Huang, Ding-Yeong Wang, Kuei-Hung Shen, Cheng-Wei Chien, Keng-Ming Kuo, Shan-Yi Yang, Yung-Hung Wang, "Impact of Stray Field on the Switching Properties of Perpendicular MTJ for Scaled MRAM", IEEE Dig. Of IEDM, p. 29.2.1-4 (2012). © 2012 IEEE.
Figure 7.26	From Sheng-Huang Huang, Ding-Yeong Wang, Kuei-Hung Shen, Cheng-Wei Chien, Keng-Ming Kuo, Shan-Yi Yang, Yung-Hung Wang, "Impact of Stray Field on the Switching Properties of Perpendicular MTJ for Scaled MRAM", IEEE Dig. Of IEDM, p. 29.2.1-4 (2012). © 2012 IEEE.

Table/Figure Number	Source
Figure 7.27	From Sheng-Huang Huang, Ding-Yeong Wang, Kuei-Hung Shen, Cheng-Wei Chien, Keng-Ming Kuo, Shan-Yi Yang, Yung-Hung Wang, "Impact of Stray Field on the Switching Properties of Perpendicular MTJ for Scaled MRAM", IEEE Dig. Of IEDM, p. 29.2.1-4 (2012). © 2012 IEEE.
Figure 7.28	From Thibaut Devolder, "Scalability of Magnetic Random Access Memory based on an In-Plane Magnetized Free Layer", Appl. Phys. Express, 4, 2011 (093001).
Figure 7.29	From S. Ikeda, H. Sato, H. Honjo, E. C. I. Enobio, S. Ishikawa, M. Yamanouchi, S. Fukami, S. Kanai, F. Matsukura, T. Endoh and H. Ohno, "Perpendicular-anisotropy CoFeB-MgO based magnetic tunnel junctions scaling down to 1X nm"., IEEE Digest of IEDM paper 32.2 , p.796, (2014). © 2014 IEEE.
Figure 7.30	Reproduced from 62. Luc Thomas, Guenole Jan, Jian Zhu, Huanlong Liu, Yuan-Jen Lee, Son Le, Ru-Ying Tong, Keyu Pi, Yu-Jen Lee, Dongna Shen, Renren He, Jesmin Haq, Jeffrey Teng, Vinh Lam, Kenlin Huang, Tom Zhong, Terry Torng, and Po-Kang Wang, "Perpendicular spin transfer torque magnetic random access memories with high spin torque efficiency and thermal stability for embedded applications (invited)," Journal of Applied Physics 115, 172615 (2014), with the permission of AIP Publishing.
Figure 7.31	From Guenole Jan, Yu-Jen Wang, Takahiro Moriyama, Yuan-Jen Lee, Mark Lin, Tom Zhong, Ru-Ying Tong, Terry Torng, and Po-Kang Wang, "High Spin Torque Efficiency of Magnetic Tunnel Junctions with MgO/CoFeB/MgO Free Layer," Applied Physics Express, 5, (2012) 093008 & S. Ikeda, H. Sato, H. Honjo, E. C. I. Enobio, S. Ishikawa, M. Yamanouchi, S. Fukami, S. Kanai, F. Matsukura, T. Endoh and H. Ohno, "Perpendicular-anisotropy CoFeB-MgO based magnetic tunnel junctions scaling down to 1X nm"., IEEE Digest of IEDM paper 32.2 , p.796, (2014)
Figure 7.33	http://www.eecg.utoronto.ca/~ali/mram.html. © IEEE
Table 7.1	From Guenole Jan, Yu-Jen Wang, Takahiro Moriyama, Yuan-Jen Lee, Mark Lin, Tom Zhong, Ru-Ying Tong, Terry Torng, and Po-Kang Wang, "High Spin Torque Efficiency of Magnetic Tunnel Junctions with MgO/CoFeB/MgO Free Layer," Applied Physics Express, 5, (2012) 093008.

Chapter 8

Table/Figure Number	Source
Figure 8.2.2	From S. Fukami, T. Suzuki, K Nakahara, N. Ohashima, Y. Ozaki, "low-current Perpendicular domain wall motion cell for scalable high-speed MRAM", VLSI Symp. Tech., 230-1, 2009. © 2009 IEEE.
Figure 8.2.4	Based on See-hun Yang, Kwang-Su Ryu, Stuart Parkin, "Domain-wall velocities of up to 750 m s−1 driven by exchange-coupling torque in synthetic antiferromagnets", Nature Nanotechnology, PUBLISHED ONLINE: 23 FEBRUARY 2015 I DOI: 10.1038/NNANO.2014.324. © 2015 Springer Nature.
Figure 8.3.1	From D.Y. Wang, et al., "A statistical study of the reliability of SOT MRAM cell structures by thermal baking," Abstract of IEDM 2019, MRAM Poster paper. © 2019 IEEE.
Figure 8.3.2	From Yao-Jen Chang, et.al, IEDM, MRAM Poster, No.7 (2018)
Figure 8.3.5	From Rahaman Sk A., et.al., IEEE IEDM MRAM Poster Abstract, paper 24 (2018) & D.Y. Wang, et.al., "A statistical study of the reliability of SOT MRAM cell structures by thermal baking," Abstract of IEDM 2019, MRAM Poster paper.
Figure 8.3.6	Data Summarized from A. Hoffmann, Spin Hall Effects in Metals, IEEE Trans. Magn. 49, 5172 (2013).
Figure 8.3.7	1. Adapted from Wiki 2. Modified from Y. L. Chen, J. G. Analytis, J. H. Chu, Z. K. Liu, S. K. Mo, X. L. Qi, H. J. Zhang, D. H. Lu, X. Dai, Z. Fang, S. C. Zhang, I. R. Fisher, Z. Hussain, and Z. X. Shen, Experimental Realization of a Three-Dimensional Topological Insulator, Bi2Te3, Science 325, 178 (2009). 3. James G. Analytis, Jiun-Haw Chu, Yulin Chen, Felipe Corredor, Ross D. McDonald, Z. X. Shen, and Ian R. Fisher, "Bulk Fermi surface coexistence with Dirac surface state in Bi2Se3 : A comparison of photoemission and Shubnikov–de Haas measurements", Phys. Rev. B 81, 205407 (2010)
Figure 8.4.1	Modified from Maruyama, T. et al. "Large voltage-induced magnetic anisotropy change in a few atomic layers of iron." Nature Nanotech. 4, 158–161 (2009). © 2009 Springer Nature.
Figure 8.4.2	From Zhenchao Wen, Hiroaki Sukegawa, Takeshi Seki, Takahide Kubota, Koki Takanashi, and Seiji Mitani, "Voltage control of magnetic anisotropy in epitaxial Ru/Co2FeAl/MgO heterostructures "https://arxiv.org/pdf/1611.02827. Licensed under CC BY 4.0
Figure 8.4.5	From H. Yoda, N. Shimomura, Y. Ohsawa, S. Shirotori, Y. Kato, T. Inokuchi, Y. Kamiguchi, B., Altansargai, Y. Saito, K. Koi, H. Sugiyama,

Table/Figure Number	Source
Figure 8.4.6	S. Oikawa, M. Shimizu, M. Ishikawa, K. Ikegami, and A. Kurobe, "Voltage-Control Spintronics Memory (VoCSM) Having Potentials of Ultra-Low Energy-consumption and High-Density", IEEE Dig. Of IEDM 2018, paper 27.6. © 2018 IEEE. Modified from Shiota, Y. et al. "Induction of coherent magnetization switching in a few atomic layers of FeCo using voltage pulses." Nature Mater. 11, 39–43 (2012). © 2012 Springer Nature.

Chapter 9

Table/Figure Number	Source
Figure 9.6	From Luc Thomas, Guenole Jan, Jian Zhu, Huanlong Liu, Yuan-Jen Lee, Son Le, Ru-Ying Tong, Keyu Pi, Yu-Jen Wang, Dongna Shen, Renren He, Jesmin Haq, Jeffrey Teng, Vinh Lam, Kenlin Huang, Tom Zhong, Terry Torng, and Po-Kang Wang, "Perpendicular spin transfer torque magnetic random access memories with high spin torque efficiency and thermal stability for embedded applications", (invited) J. of Appl. Phys. 115, 172615 (2014). © 2014 AIP Publishing.
Figure 9.7	From IEEE 2018 Global MRAM innovation Forum, Bernard Dieny (2018). © 2018 IEEE.
Figure 9.9	Modified from H. Noguchi et al., "Highly Reliable and Low-Power Nonvolatile Cache Memory with Advanced Perpendicular STT-MRAM for High-Performance CPU", IEEE Symp. VLSI Circuits Dig. Tech. Papers, June 2014. © 2014 IEEE.
Figure 9.10	From S.-W. Chung, T. Kishi, J.W. Park, M. Yoshikawa, K. S. Park, T. Nagase, K. Sunouchi, H. Kanaya, G.C. Kim, K. Noma, M. S. Lee, A. Yamamoto, K. M. Rho, K. Tsuchida, S. J. Chung, J. Y. Yi, H. S. Kim, Y.S. Chun, H. Oyamatsu, and S. J. Hong, "4Gb bit density STT-MRAM using perpendicular MTJ realized with compact cell structure", IEEE IEDM 2016 Technical Digest, 27.1, 2016. © 2016 IEEE.
Figure 9.17	From Yu-Sheng Chen , Ding-Yeong Wang, Yu-Chen Hsin, Kai-Yu Lee, Guan-Long Chen, Shan-Yi Yang, Hsin-Han Lee, Yao-Jen Chang, I-Jung Wang, Pei-Hua Wang, Chih-I Wu, and D. D. Tang, "On the Hardware Implementation of MRAM Physically Unclonable Function", IEEE Transactions on Electron Devices (Volume: 64, Issue: 11, Nov. 2017). © 2017 IEEE.

(Continued)

Table/Figure Number	Source
Figure 9.18	From Yu-Sheng Chen , Ding-Yeong Wang, Yu-Chen Hsin, Kai-Yu Lee, Guan-Long Chen, Shan-Yi Yang, Hsin-Han Lee, Yao-Jen Chang, I-Jung Wang, Pei-Hua Wang, Chih-I Wu, and D. D. Tang, "On the Hardware Implementation of MRAM Physically Unclonable Function", IEEE Transactions on Electron Devices (Volume: 64, Issue: 11, Nov. 2017). © 2017 IEEE.
Figure 9.19	From Baohua Sun, Daniel Liu, Leo Yu, Jay Li, Helen Liu, Wenhan Zhang, Terry Torng, "MRAM Co-designed Processing-in-Memory CNN Accelerator for Mobile and IoT Applications" arXiv:1811.12179v1 [eess.SP] 26 Nov 2018

1

Basic Electromagnetism

1.1 Introduction

This chapter introduces basic electromagnetism. Starting from the simple attractive (or repelling) force between magnets, we define magnetic field, dipole moment, torque, and magnetic energy and its equivalence to current. At the end, we give a few methods to calculate magnetic field from electric current and from magnetic pole on material surface.

1.2 Magnetic Force, Pole, Field, and Dipole

Since electrostatic phenomenon was studied earlier than magnetostatic one, magnetic phenomena were described in analogy to electrical phenomena. Magnetic pole was defined as the source of magnetic field and force. Magnetic poles exert magnetic force to each other, like electric charges. In centimeter-gram-second (cgs) units, the force is proportional to the strength of the magnetic poles, defined as

$$F = \frac{p_1 p_2}{r^2},\tag{1.1}$$

where r is the distance between two poles, the unit is cm, the force is F, and the unit is dyne. So far, the unit of the poles are not defined; however, the dimension of a pole is $\sqrt{F \cdot r^2}$. When the distance between the poles is 1 cm, one unit of p_1 exerts a 1 dyne force to a unit of p_2, and vice versa. Like in Coulomb's law, the force can be described as a magnetic field H, produced by pole p_1 and exerts on p_2. Thus,

$$F = \left(\frac{p_1}{r^2}\right)p_2 = H\,p_2,\tag{1.2}$$

Magnetic Memory Technology: Spin-Transfer-Torque MRAM and Beyond,
First Edition. Denny D. Tang and Chi-Feng Pai.
© 2021 The Institute of Electrical and Electronics Engineers, Inc.
Published 2021 by John Wiley & Sons, Inc.

where H is defined as

$$H = \frac{p_1}{r^2}. \tag{1.3}$$

Thus, a magnetic field H of unit strength exerts a force of 1 dyne onto 1 unit of magnetic pole. The unit of the magnetic field in cgs unit is oersted (Oe). To get a feeling about the strength of magnetic field, the magnetic field at the end of a magnetic bar on the classroom whiteboard can be as high as 200–1000 Oe, while the earth magnetic field is smaller than 0.5 Oe.

In 1820, H.C. Oersted discovered that a compass needle could be deflected when electric current passes through a wire nearby the compass. That is the first time electricity is linked to magnetic phenomenon. Subsequent works by André-Marie Ampère established the basis of modern electromagnetism. He established the relation between magnetic dipole and a circulating current in a conductor loop around an axis. The direction of the dipole is along the axis of the loop, which is orthogonal to the loop plane. Figure 1.1b illustrates the relation between the dipole and the current loop. The polarity is dictated by the direction of the current. Reversing the direction of the electric current changes the polarity of the dipole. Thus, the magnetic dipole is another form of electric current, or moving electric charge.

Although both electric field and magnetic field originate from electric charges, the difference is that the magnetic field must come from moving electric charges or

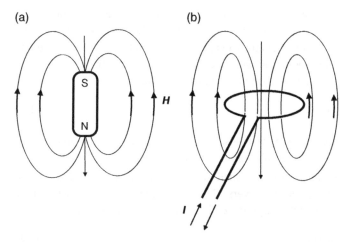

Figure 1.1 (a) Magnetic field lines from a magnet, (b) magnetic field from a circular current loop.

electric current and not from a stationary electric charge. A stationary charge emits only an electric field.

The moving electric charge concept explains well the origin of magnetic pole or magnetic moment at the time. However, this description was later proven incorrect when the electron spin is considered. We will discuss this topic in the next chapter of this book.

1.3 Magnetic Dipole Moment, Torque, and Energy

Although in the early days, magnetic pole was considered as the counterpart of electric charge, there is a major difference. In any magnetic materials, its magnetic poles always come in pairs: the north and south poles. A single monopole has not been found in nature. The positive and negative poles show up in the same time and form a dipole. For example, a bar magnet always has a north pole in one end and a south pole in the other. If one cuts the bar magnet into two, one gets two bar magnets, each with a north pole and a south pole. Magnetic field lines emit from one pole, diverge into the surroundings, and then converge and return into the other pole of the magnet. Figure 1.1a shows the field lines around a magnet.

A magnetic field applies a torque to a magnetic dipole. The dipole gains in angular momentum and rotates. For a bar magnet positioned at an angle ϕ to a uniform magnetic field, H, as shown in Figure 1.2a, the two forces on the pair of poles are

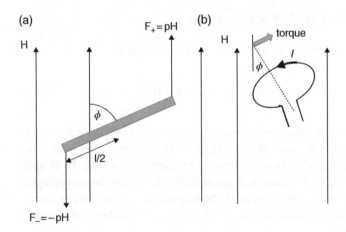

Figure 1.2 Magnetic field exerts a torque on moment (a) on moment of a bar magnet, (b) on a current coil.

$F_+ = + pH$ and $F_- = - pH$. The two forces are equal but have opposite direction. So, the moment acting on the magnet, which is the force times the perpendicular distance from the center of the mass, is

$$pH \sin \phi \, (l/2) + pH \sin \phi \, (l/2) = pH \, l \, \sin \phi = mH \, \sin \phi, \tag{1.4}$$

where $m = pl$, the product of the pole strength and the length of the magnet, is the amplitude of magnetic *moment*. Magnetic moment is a vector, pointing to a direction normal to the plane of the magnet and the magnetic field. One cgs unit of magnetic moment is the angular moment exerted on a magnet when it is perpendicular to a uniform field of 1 Oe. The cgs unit of magnetic moment is emu (electromagnetic unit).

Since the magnetic dipole moment is equivalent to a current loop, which can be quantified by loop area A and a current I in the loop, then the magnetic dipole moment is defined as

$$m = IAn, \tag{1.5}$$

where n is a vector normal to the plane of the current loop. Figure 1.2b illustrates that a torque is exerted by the H field on the coil. The current coil acquires angular momentum from the magnetic field. In the Système International d'Unités (SI) units, the magnetic moment is measured in Am^2.

The magnetostatic energy of a magnetic dipole in the presence of a magnetic field is defined to be zero when the dipole is perpendicular to a magnetic field. So, the work done in turning through an angle ϕ against the field is

$$\delta E = 2(pH \, \sin \phi)(l/2) \, d\phi = mH \, \sin \phi \, d\phi,$$

and the energy of a dipole at an angle ϕ to a magnetic field is

$$E = \int_{\pi/2}^{\phi} mH \, \sin \, \phi \, d\phi = - mH \, \cos \, \phi = - m \cdot H. \tag{1.6}$$

This expression for the energy of a magnetic dipole in a magnetic field is in cgs units. In Eq. (1.6), the unit of E is erg, of m is emu, and of H is Oe. The energy described in Eq. (1.6) is also known as magnetostatic energy. In SI units the energy is $E = - \mu_0 m \cdot H$, where μ_0 is permeability in vacuum. When the dipole moment, m, is in the same direction as H, the magnetostatic energy is the lowest.

The torque exerted on a dipole moment is the gradient of the dipole energy with respect to angle ϕ, or

$$\Gamma = dE/d\phi = mH \, \sin \phi. \tag{1.7}$$

In the H field, a torque is exerted onto the dipole in the direction that lowers the dipole energy, and the unit is expressed in erg/rad. When m and H are parallel, or $\phi = 0$, the energy is at a minimum, and the torque is zero. The torque is maximum when $\phi = \pi/2$.

1.4 Magnetic Flux and Magnetic Induction

Magnetic flux Φ is defined as the integrated strength of a normal component of magnetic field lines crossing an area, or

$$\Phi = \int (H \cdot n)\, dA, \tag{1.8}$$

where n is the unit vector normal to the plane of the cross-sectional area, A. In cgs unit, the flux is expressed in Oersted·cm^2.

The magnetic flux is an important parameter in electric motor and generator design. The time-varying flux induces an electric current in any conductor, which it intersects. Electromotive force (emf) ε is equal to the rate of change of the flux linked with the conductor:

$$\varepsilon = -\frac{d\Phi}{dt}. \tag{1.9}$$

This equation is Faraday's law of electromagnetic induction. The electromotive force provides the potential difference that drives electric current in a conductor. The minus sign indicates that the induced current sets up a time-varying magnetic field that acts against the change in the magnetic flux. This is known as Lenz's law. The units in Eq. (1.9) as expressed in SI units are flux in *Weber (Wb)*, time in *second (s)*, and an electromotive force in *Volt (V)*.

When a magnetic field, H, is applied to a material, the response of the material to H is called magnetic induction, B. The relationship between B and H is a property of the material. In material, B is not necessarily a linear function of H and the equation relating B and H is (in cgs units: gauss and Oe, respectively)

$$B = H + 4\pi M \tag{1.10}$$

where M is a function of H and saturates to a fixed value at high H. M is called the magnetization of the medium, and its saturated value is called the saturated magnetization, which is typically denoted as M_s. The magnetization is defined to be the magnetic moment per unit volume:

$$M = m/V \qquad (1.11)$$

M is a macroscopic property of the material and depends both on the individual magnetic moments of the constituent ions, atoms, and molecules, and on how these dipole moments interact with each other. The cgs unit of magnetization M is emu/cm^3, and that of magnetic induction B is gauss. Notice that although M is in cgs unit of emu/cm^3, the unit of $4\pi M$ in Eq. (1.10) is gauss. It means that 1 Oe of magnetic field induces 1 G of magnetic induction in vacuum. In vacuum, $M = 0$, and M only exists inside material. Consider a material with magnetization M. At the boundary of vacuum and such material, $H = 4\pi M$, since B must be continuous, according to Eq. (1.10). In a vacuum, $M = 0$, and $B = H$.

In SI units, the relation between B, H, and M is expressed as

$$B = \mu_0(H + M), \qquad (1.12)$$

where μ_0 is the vacuum permeability. The unit of M is obviously the same as that of H (A/m), and those of μ_0 are Wb/A·m, also known as henry/m. So, the unit of B is Wb/m^2, or tesla (T). Note that $\mu_0 = 4\pi \cdot 10^{-7}$ Wb/A·m and 1 G $= 10^{-4}$ T. Also note that the magnetic induction, B, is therefore the density of flux, Φ, inside the medium.

1.5 Ampère's Circuital Law, Biot-Savart Law, and Magnetic Field from Magnetic Material

This section describes the calculation of magnetic field from electrical current elements. For current flowing in a special path, Ampère's circuital law can provide a closed-form solution. For current flowing in a more complex geometry, the magnetic field at a position can be handled by Biot-Savart's law.

1.5.1 Ampère's Circuital Law

After Ampère discovered that magnetic field H is induced by current I, he further quantitatively established their relation as

$$\oint H \cdot dl = 4\pi \cdot 10^{-4} I, \qquad (1.13)$$

In Eq. (1.13), dl is the segment length of an arbitrary closed loop where the integration is performed, and I is the current within the closed loop and in cgs units; H

is in Oe, dl is in cm, and I is in mA. To calculate the magnetic field around the surface of a thin film, in which current flows uniformly in the film, it is handy to apply Eq. (1.13) with different units such that

$$\oint \boldsymbol{H} \cdot d\boldsymbol{l} = 4\pi \cdot I, \tag{1.13a}$$

where dl is in micrometers.

This law is simple in concept and particularly useful in computing the field generated by the current in a long conductor and conducting thin film. This field is called the Oersted field and is associated with a current flow.

We discuss two unit systems of magnetic field next. There has been two complementary ways of developing the theory and definitions of magnetism. As a result, there are two sets of units for magnetic field and magnetic pole (or magnetic moment). Their definitions are similar but not entirely identical. The major difference lies in how the magnetic field is defined inside the material. The cgs units are used for studying physics, such as the origin of magnetic pole and the magnetic properties in material. The SI units are frequently used for obtaining magnetic field strength from a current element. Engineers working on electromagnetic wave, electric motors, etc., like to adopt SI units. However, for magnetism and spintronics communities, cgs units are still popular. This book will use both sets of units interchangeably, depending on whichever makes more sense and is in line with relevant journal publications.

In SI unit, Ampère's law is given as

$$\oint \boldsymbol{H} \cdot d\boldsymbol{l} = I \quad \text{(in SI units : } H \text{ in A/m, } dl \text{ in m, and } I \text{ in A).} \tag{1.14}$$

Comparing Eq. (1.13) and Eq. (1.14), one finds that a magnetic field of 1 (Oe) $= 1000/4\pi$ (A/m) ~ 80 (A/m).

Example 1   Field Around a Conducting Wire

The magnetic field lines go around a current carrying wire in closed circles, as illustrated in Figure 1.3. At a distance r_0 from the conductor, the magnitude of the field H is constant. This makes the line integral of Ampère's law straightforward. It's simply

$$\oint \boldsymbol{H} \cdot d\boldsymbol{l} = 2\pi r H = I,$$

and so the field H is

$$H = \frac{I}{2\pi r_0}.$$

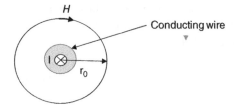

H

Conducting wire

$I \otimes$

r_0

Figure 1.3 Magnetic field around a conducting wire carrying a current I. The magnetic field at a distance r from the wire is $H = I/2\pi r_0$.

1.5.2 Biot-Savart's Law

An equivalent statement to the Ampère's circuital law (which is sometimes easier to use for certain systems) is given by the Biot-Savart law. The Biot-Savart law states that the fraction of a field δH is attributed by a current I flowing in an elemental length, δl, of a conductor,

$$\delta H = \frac{1}{4\pi r^2} I \delta l \times n \text{ (in SI unit)}, \tag{1.15}$$

where r is the radial distance from the current element, and n is a unit vector along the radial direction from the current element to the point where the magnetic field is measured. Notice that the direction of vector δH is orthogonal to the plane formed by $I\delta l$ and n, as a result of the vector operation x of two vectors $I\delta l$ and n, and the amplitude of $|I\delta l|\sin\theta$, where θ is the angle between vectors δl and n.

Example 2 Field from a Current in a Wire Loop
The magnetic field at the center of the loop plane as shown in Figure 1.3 is calculated by the Biot-Savart law as follows:

The radius of the loop is r_0. H can be in either the positive or the negative z-direction, depending on the current direction, and only in the z-direction. The vector sum is simplified into a scalar sum. On the loop plane, $z = 0$. So, $|H| = H_o$. H_o is the integral of field contributed by each segment dl of the loop and

$$H_o = 2\pi r_o \left[\frac{1}{4\pi r_o^2} I \right] = \frac{I}{2r_o} \text{ (SI unit)} \tag{1.16}$$

and

$$H = H_o \hat{n}_z,$$

where $\hat{n}_z$ is the unit vector in the z-direction.

Biot-Savart's law allows one to calculate the magnetic field at an arbitrary position away from the current element. Let us move one step forward and examine the H field at an arbitrary position (xa, ya, za) from a current loop on the $z = 0$ plane with center of the loop at $(0, 0, 0)$ and radius $r0$.

Start from Biot-Savart's law,

$$\delta H(xa, ya, za) = \frac{I}{4\pi r^2}\, \delta l \times \hat{r}.$$

The current element δl is located at $x0 = r0\cos\phi, y0 = r0\sin\phi, z0 = 0$, where ϕ is azimuth angle looking from the z-axis toward the current loop element. And the vector of the current element from the observation point (xa, ya, za) to the current element at $(x0, y0, 0)$ is

$$\hat{r} = \frac{(xa - x0)\hat{x} + (ya - y0)\hat{y} + (za - z0)\hat{z}}{r}. \tag{1.17}$$

The distance is

$$r^2 = (xa - x0)^2 + (ya - y0)^2 + (za - z0)^2 = (xa - r0\cos\phi)^2 + (ya - r0\sin\phi)^2 + za^2. \tag{1.18}$$

The current element vector is

$$\delta l = r0\delta\phi\,(-\sin\phi\,\hat{x} + \cos\phi\,\hat{y}) \tag{1.19}$$

Thus,

$$\delta l \times \hat{r} = \begin{vmatrix} \hat{x} & \hat{y} & \hat{z} \\ -\sin\phi & \cos\phi & 0 \\ xa - x0 & ya - y0 & za - z0 \end{vmatrix} \frac{r0}{r}\,\delta\phi$$

$$= \begin{vmatrix} \hat{x} & \hat{y} & \hat{z} \\ -\sin\phi & \cos\phi & 0 \\ xa - r0\cos\phi & ya - r0\sin\phi & za \end{vmatrix} \frac{r0}{r}\,\delta\phi.$$

Thus,

$$\delta l \times \hat{r} = \{(za\cos\phi)\hat{x} - (-za\sin\phi)\hat{y} + [(-\sin\phi)(ya - r0\sin\phi) \\ - (\cos\phi)(xa - r0\cos\phi)]\hat{z},\} \frac{r0}{r}\,\delta\phi. \tag{1.20}$$

The field at observation point (xa, ya, za) is the integration of the field element δH of the loop,

$$H(xa, ya, za) = \frac{I}{4\pi}\oint \frac{1}{r^2}\,\delta l \times \hat{r}.$$

For an arbitrary observation point (xa, ya, za), r^2 is not a constant; thus, a closed-form solution is not readily available. One may need to conduct numerical integration. Homework 1.Q3–1.Q6 will give the reader a good sense of the position dependence of the demagnetization field from current loops.

1.5.3 Magnetic Field from Magnetic Material

Both Ampère's law and Biot-Savart's law connect magnetic field and electrical current in a vacuum [1]. Biot-Savart's law is also readily to be modified to calculate the field distribution from a magnetic material, such as a bar magnet (a ferromagnetic material) or a paramagnetic bar under external magnetic field [2]. Magnetic dipoles exist on the surface for both cases. In other words, there are magnetic dipoles or magnetic moment m, which is equivalent to a current loop, Eq. (1.5). The moment is related to the saturated magnetization M_s by Eq. (1.11). Thus, one can rewrite Biot-Savart's law as

$$\delta H = \frac{1}{4\pi r^2} M_s \delta A \times \hat{n}, \tag{1.21}$$

where the magnetization M_s is in unit of A/m, r is in m (in SI unit) from pole to the point of measurement, A is the a vector surface area, and $\hat{n}$ follows the definition of Biot-Savart's law. As long as there are magnetic poles due to the termination of magnetization on material surface, there is magnetic field. The field exists both inside and outside of the material. Inside the magnet, the field is predominantly in opposite direction of the magnetization. Thus, the field is called the demagnetization field, or *demag* field in short.

Furthermore, the demag field strength inside a material depends on the geometry of the material. The field is not uniform and is position dependent (Figure 1.4). When an external field is applied to a magnetic material, the field inside is the superposition of the external field and the demag field.

For a given geometric shape material, the position-dependent demag field is averaged over position inside the material and can be expressed as $\overline{H_{demag}} = NM_s = \left(N_x\hat{x} + N_y\hat{y} + N_z\hat{z}\right)M_s$, where N is called the demag factor. The demag factor N of a cylindrical magnet is shown in Figure 1.5. For a thin disk, length $<<$ radius, the dominant demag field is in the thin (z) direction.

The demag field has served as a stabilizing mechanism of magnetic state of field magnetic memory (field MRAM) cells. It is referred to as *shape* anisotropy. This subject will be repeatedly discussed in later chapters. Ref. [3] presents a Green's function approach to dealing with the complex geometric analysis of the demag field calculation. Those interested in calculating the demag field are referred to that paper.

Figure 1.4 Comparison of magnetic field (flux density) **B**, demagnetizing field **H** and magnetization **M** inside and outside a cylindrical bar magnet. The right side is the north pole, and the left side is the south pole. (https://en.wikipedia.org/wiki/Demagnetizing_field).

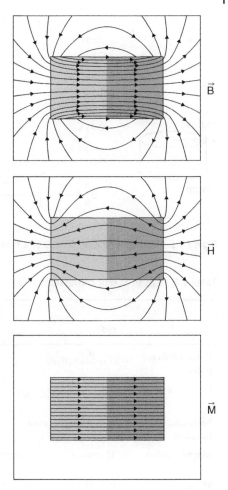

1.6 Equations, cgs-SI Unit Conversion Tables

As mentioned earlier, while cgs units are used in physics and magnetic material study, SI units are used when engineers investigate energy conversion in electric motor/generator as well as in electromagnetic wave propagation. For convenience, this book will follow the trend and use cgs unit, since most of the journal material in thin magnetic films and devices is written in cgs units. Table 1.1 gives the conversion table. Notice that two parameters, *susceptibility* and *permeability* of material, have not been discussed. The concepts are straightforward and are used mainly in the field of electromagnetic wave study, frequently in SI units. They are listed at the bottom of the table (see more discussion in Ref. [4]).

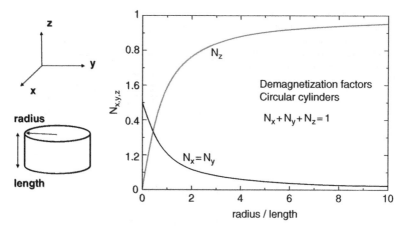

Figure 1.5 Demagnetization factors of a thin cylindrical disk.

Table 1.1 Cgs to SI unit conversion table.

	cgs	conversion factor**	SI
Force between poles	$F = \dfrac{p_1 p_2}{r^2}$ (dyne)	10^{-5}	$F = \dfrac{1}{4\pi\mu_0}\dfrac{p_1 p_2}{r^2}$ (Newton)
Field of a pole	$H = \dfrac{p}{r^2}$ (Oersted, Oe)	$10^3/4\pi$	$H = \dfrac{1}{4\pi\mu_0}\dfrac{P}{r^2}$ (Ampère/m)
Magnetic moment	$m = p\cdot$length (emu)	10^{-3}	$m = A\,I$ (Ampère-m^2)
Magnetization	$M = $ m/volume (emu/cm^3)	10^3	$M = $ m/volume (Ampère/m)
Magnetic induction	$\boldsymbol{B} = \boldsymbol{H} + 4\pi\,\boldsymbol{M}$ (Gauss, G)	10^{-4}	$\boldsymbol{B} = \mu_0\,(\boldsymbol{H} + \boldsymbol{M})$ (Tesla)
Energy of a dipole	$E = -\,\boldsymbol{m}\cdot\boldsymbol{H}$ (erg) *	10^{-7}	$E = -\mu_0\,\boldsymbol{m}\cdot\boldsymbol{H}$ (Joule) *
Susceptibility	$\chi = \dfrac{M}{H}$ (emu/(cm^3-Oe))	4π	$\chi = \dfrac{M}{H}$ (dimensionless)
Permeability	$\mu = \dfrac{B}{H} = (1 + 4\pi\chi)$ (Gauss/Oe)	$4\pi\times10^{-7}$	$\mu = \frac{B}{H} = \mu_0(1 + \chi)$ (Henry/m)

* Here "." between two vectors indicates vector products that results in scalar quantity. In x, y, z coordinate system, $\boldsymbol{m}\cdot\boldsymbol{H} = |m||H|\cos\phi = m_x H_x + m_y H_y + m_z H_z$, where ϕ is angle between vectors $\boldsymbol{m}$ and $\boldsymbol{H}$. This is the magnetostatic energy.
** For example, magnetization M of 1 emu/cm^3 is converted to $1*10^3$ ampere/m. Therefore, 1 emu/cm^3 = 1000 A/m.

Homework

Q1.1 A current I_0 is passing through a straight, infinitely long thin stripe conductor. The film thickness is t, and the width is L. The current density in the conductor is uniform. Calculate the magnetic field H on the surface of the conductor (Figure 1.Q1). Do this based on Ampère's law.

A1.1 Assuming that H is uniform on the top surface of the conducting thin conductor, Ampère's law states that $H \cong I/(2L + 2t)$. Since $L \gg t$,

$H \cong I/(2L)$ in SI units and $H \cong 4\pi I/(2L) = 2\pi I/L$ in cgs units.

Figure 1.Q1 Current induced magnetic field over a very large conducting film.

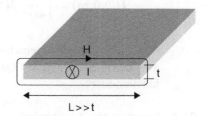

$$L \gg t$$

Q1.2 (a) On the package of a commercial NdFeB (NIB) permanent magnet, the label says the magnet has a strength of magnetic field around 3500 G. What is the magnetic induction B for this magnet in terms of SI units? (b) Assuming that B is solely coming from the magnetization M of the NIB magnet, what is the magnetization M for this magnet in terms of SI units?

A1.2 (a) $3500\,\text{G} = 0.35\,\text{T}$.
(b) $B = \mu_0(H + M) = \mu_0 M$; therefore, $M = B/\mu_0 = 0.35\,\text{T}/(4\pi \times 10^{-7}\,\text{T} \cdot \text{m/A}) \approx 2.8 \times 10^5\,\text{A/m}$.

Q1.3 Calculate the H field at a position $(0, 0, za)$ from a current loop on $z = 0$ plane with center of the loop at $(0, 0, 0)$ and radius $r0$.

A1.3 Start from Biot-Savart's law:

$$\delta H(xa, ya, za) = \frac{I}{4\pi r^2}\,\delta l(x0, y0, 0) \times \hat{r}.$$

The current element is located at $x0 = r0 \cos\phi$, $y0 = r0 \sin\phi$, $z0 = 0$. Observation point P1$(0, 0, za)$, a point at the center and above the loop, where $r^2 = (r0^2 + za^2)$ is a constant, and $r0/r$ is also a constant for a given za, Eq. (1.20), becomes

$$\delta l \times \hat{r} = \frac{r0}{r} \{ (za \cos\phi)\hat{x} - (-za \sin\phi)\hat{y}$$
$$+ [(-\sin\phi)(-r0\sin\phi) - (\cos\phi)(-r0\cos\phi)]\hat{z}\} \, \delta\phi.$$

Integrate over the loop, and the first two terms drop off, so one gets

$$H(0,0,za) = \oint \frac{I}{4\pi r^2} \, \delta l \times \hat{r} = \frac{I}{4\pi r^2} \oint \delta l \times \hat{r} = \frac{I \, r0^2}{2(r0^2 + za^2)^{3/2}} \hat{z}.$$

Only the z-component remains; the x- and y- component are equal to zero due to symmetry. Figure 1.Q3 shows the normalized field ($Hz \, r0/I$) as a function $za/r0$.

Q1.4 Based on the A1.3 analysis, calculate the field at point $(xa, 0, 0)$ for an in-plane current loop with radius $r0$ and center at $(0, 0, 0)$.

A1.4 From Eq. (1.20), $(xa, ya, za) = (xa, 0, 0)$, and let $xa/r0 = q$, one obtains

$$\delta l \times \hat{r} = \frac{r0^2}{r}(1 - q\cos\phi)\hat{z} \, \delta\phi.$$

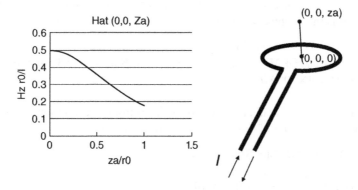

Figure 1.Q3 H-field at the center axis of a coil; only H_z exists. Its value decreases as the observation point $(0, 0, za)$ is move away from the coil. R0 is radius of the coil.

and

$$r^2 = (xa - r0\cos\phi)^2 + (-r0\sin\phi)^2$$
$$= xa^2 - 2r0\,xa\cos\phi + r0^2 = q^2 - 2q\cos\phi + 1$$

and

$$\oint \frac{1}{r^2}\,\delta l \times \hat{r} = \frac{1}{4\pi r0}\oint \frac{1 - q\cos\phi}{(1 - 2q\cos\phi + q^2)^{1.5}}\,\delta\phi\hat{z}.$$

Figure 1.Q4 shows a numerical calculation of $H\,r0\,/\,I$ as a function of $xa\,/\,r0$. H is mostly confined in the coil. The outside field is in reverse direction.

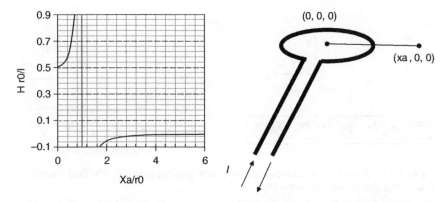

Figure 1.Q4 Calculated H/I versus $xa/r0$. xa is the distance from center of the coil. Outside of the coil when $xa > r0$, where field is in reverse direction and diminishes quickly from the edge of the coil (numerically integrated by dividing the coil into 100 segments).

Q1.5 A rectangular current loop with L/W = 1 : 7. What is the field distribution along the L direction along the center of the loop?

A1.5 Such a current loop can be partitioned into seven square loops in series along the long axis in the longitudinal direction. We can approximate a square loop by circular loop. This is the so-called 1D approximation analysis.

In A1.4, we have calculated the field outside of a circular loop. The field in each loop is a superposition of the field itself and its neighbors, i.e. summing H at $xa = 0$ and neighbors at $xa/r0 = 2, 4, 6, 8, 10$, which happens to be in opposite direction. Let H_{total} (n) designate the total field at the center of the loop in position n (total 7 loops), and let $H(a,b)$ designate the H field of loop a and observed at a distance $2b*r0$ away.

$$H_{total}(1) = H(1,0) + H(2,-1) + H(3,-2) + H(4,-3)... + H(7,-6)$$
$$H_{total}(2) = H(2,0) + H(1,1) + H(3,-1) + H(4,-2)... + H(7,-5)$$

...

$$H_{total}(7) = H(7,0) + H(6,1) + H(5,2) + H(4,3)... + H(1,6)$$

The answer is shown in Figure 1.Q5b. Students are encouraged to analyze the field distribution with an array of 2D current loops approximation.

Figure 1 Q5 shows the calculated field distribution long the long side. The field is nonuniform, and is stronger at the edge due to less canceling from neighbors. This is a 1D approximation of the field distribution of large current loop.

(a)

(b)

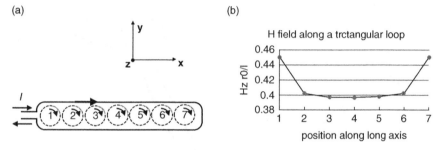

Figure 1.Q5 shows the calculated field distribution long the long side. The field is stronger at the edge due to less canceling from neighbors.

Q1.6 Consider a cylindrical magnetic bar with magnetization M_s and radius $r0$ and length $L = 2r0$. Calculate the field at the center plane along $(xa, 0, -r0)$, halfway between the top and bottom ends of the bar, as illustrated in Figure 1.4.

A1.6 Let the top surface be loop1 and the bottom loop be loop2. The two loop currents are in opposite sense. At $ya = 0$, $za = -r0$, and let $q = xa/r0$. Eq. (1.20) becomes

$$\delta l \times \hat{r} = \{ (r0\cos\phi)\hat{x} - (-r0\sin\phi)\hat{y} + [(-\sin\phi)(-r0\sin\phi)$$
$$- (\cos\phi)(xa - r0\cos\phi)]\hat{z} \} \frac{r0}{r} \delta\phi$$
$$= \frac{r0^2}{r} \{(\cos\phi)\hat{x} - (-\sin\phi)\hat{y} + [(-\sin\phi)(-\sin\phi)$$
$$- (\cos\phi)(q - \cos\phi)]\hat{z}\}\delta\phi$$
$$= \frac{r0^2}{r} \{(\cos\phi)\hat{x} - (-\sin\phi)\hat{y} + [1 - q(\cos\phi)]\hat{z}\}\delta\phi.$$

Eq. (1.19) becomes

$$r^2 = (xa - x0)^2 + (ya - y0)^2 + (za - z0)^2 = \left[(q - \cos\phi)^2 + (-\sin\phi)^2 + 1 \right] r0^2$$

$$= \left[q^2 - 2q\,\cos\phi + 2 \right] r0^2$$

$$\oint \frac{1}{4\pi r^2}\,\delta l \times \hat{r} = \frac{1}{4\pi} \oint \frac{r0^2\{ (\cos\phi)\hat{x} - (-\sin\phi)\hat{y} + [1 - q(\cos\phi)]\hat{z} \}}{r^3}$$

$$= \frac{1}{4\pi\,r0} \oint \frac{\{ (\cos\phi)\hat{x} - (-\sin\phi)\hat{y} + [1 - q(\cos\phi)]\hat{z} \}\,\delta\phi}{\left[q^2 - 2q\,\cos\phi + 2 \right]^{1.5}} \ .$$

Integrate over $\phi = 0$ to 2π. The bottom equivalent coil adds the same amount to the field, and one gets Figure 1.Q6.

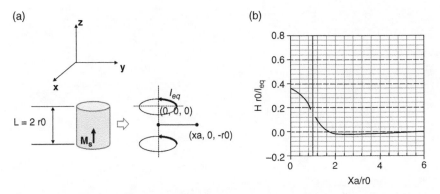

Figure 1.Q6   (a) A cylindrical magnet bar, with radius = $r0$, center at (0, 0, 0) and length $L = 2\ r0$. (b) H field along (xa, 0, $-r0$). H field along (xa, 0, $-r0$).

References

1 Collin, R.E. (1960). *Field Theory of Guided Waves*. Section 1.1. McGraw-Hill.

2 Nicola, S. (2003). *Magnetic Materials Fundamentals and Device Applications*. Cambridge, The Great Britain: Cambridge University Press.

3 Wysin, G.M. (2015). *Magnetic Excitations and Geometric Confinement, Theory and simulations*, Chapter 3, Demagnetization effects in thin magnets, 3-1–3-38. IOP Publishing Ltd.

4 Brown, W.E. Jr. (Jan 1984). Tutorial paper on dimensions and units. *IEEE Transactions on Magnetics* 20 (1): 112–117.

2
Magnetism and Magnetic Materials

2.1 Introduction

In this chapter we will start our journey of modern magnetism from the beginning of the twentieth century, a remarkable age of scientific breakthroughs that led to the birth of quantum mechanics and special relativity. Although magnetism, and its interplay with electricity, has long been discovered and studied both theoretically and experimentally by numerous well-known scientists, including Faraday, Ampère, Maxwell, Oersted, etc., providing a satisfactory interpretation of its microscopic origin in materials (condensed matter) was however difficult. A modern understanding of magnetism is based on the concept of spin, which is a purely quantum mechanical property with no classical analogy. To discuss magnetism in a contemporary perspective, therefore, we should start from quantum mechanics. Rather than simply going through formula derivations and jumping into the modern-yet-hindsight spin-based interpretation of microscopic origin of magnetism in matters, instead, we will briefly discuss the history of its development.

2.2 Origin of Magnetization

Although magnetism is a macroscopic phenomenon, its origin is microscopic. Many different models have been proposed to explain the formation of magnetization in solids. In this section, we will see some interesting models proposed by Ampère and Einstein. We will also see that classical physics is not sufficient to explain the existence of magnetism; therefore, quantum concepts need to come into play.

2.2.1 From Ampère to Einstein

It was proposed by Ampère that, soon after Oersted's discovery of current-induced magnetic field, the source of magnetism in materials should also stem from some

Magnetic Memory Technology: Spin-Transfer-Torque MRAM and Beyond,
First Edition. Denny D. Tang and Chi-Feng Pai.

certain types of molecular currents owing without resistance in solid matters. This argument, if proved to be true, then can reconcile the seemingly different origins of magnetic field coming from a permanent magnet and that from an electric current flowing in a wire. Ampère's molecular current theory attracted Albert Einstein's attention due to his personal interests in both electromagnetic theory and the concept of atomic structure. In 1915, Einstein and de Haas performed a series of measurements, later coined as the Einstein-de Haas experiment [1], to elucidate the connection between magnetic moment and the orbiting electron that gives rise to this so-called molecular current in matters. They suggested that a circularly orbiting electron will generate a magnetic moment via

$$\boldsymbol{\mu} = I\mathbf{A}, \tag{2.1}$$

where I is the current induced by the orbiting electron and $\mathbf{A}$ is the orbiting area (vector form, pointing along plane normal). By using $I = ef$, where e is the charge of electron and f the orbiting frequency, the previous equation can be written as

$$\boldsymbol{\mu} = ef\mathbf{A}. \tag{2.2}$$

On the other hand, the angular momentum of this orbiting electron with mass m_e is

$$\mathbf{L} = 2m_e f\mathbf{A}. \tag{2.3}$$

Therefore, the relation between the magnetic moment $\boldsymbol{\mu}$ and the angular momentum $\mathbf{L}$ of an orbiting electron can be naively written as

$$\boldsymbol{\mu} = \frac{e}{2m_e}\mathbf{L} = \gamma\mathbf{L}. \tag{2.4}$$

This equation indicates that the magnetic moment or the magnetization in materials is manifested by the angular momentum of orbiting electrons, with a *gyromagnetic ratio* of $\gamma = e/2m_e$. More important, if we consider the conservation of angular momentum, any change in magnetization should also accompany a change in angular momentum, which will lead to a variation in rotational motion. These concepts are summarized in Figure 2.1.

Followed by the confidence in previous equations, Einstein and de Haas designed and carried out experiments on the mechanical rotation of an hanging iron rod upon magnetization reversal along its long axis, with the aid of a solenoid-generated external field. Their conclusion from the measurements was that the observed rotational motion of the iron rod induced by magnetization reversal is consistent with their theoretical prediction. That is, in modern language, the measured $\gamma = e/2m_e$. However, this then-expected classical value of electron gyromagnetic ratio $e/2m_e$ is actually a factor of 2 off from the now-expected quantum value ($\gamma \approx e/m_e$ for ferromagnetic iron rod)! In fact, many later experiments of the similar type indicated that γ should be $\approx e/m_e$, and therefore the origin of magnetic moment in these ferromagnetic objects

Figure 2.1 Magnetic moment and angular momentum: Einstein-de Haas experiment.

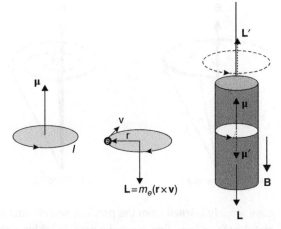

$$L = m_e(\mathbf{r} \times \mathbf{v})$$

cannot be simply explained by the orbital motion of electrons [2]. A purely quantum mechanical concept called electron *spin* (see Section 2.2.3) has to be taken into account to explain this anomaly.

2.2.2 Precession

Now we can go back to discuss, in more details, the dynamics of a magnetic moment **μ** under a uniform magnetic field **B** that we intentionally left out in the previous chapter. The **B**-induced torque on the moment **μ** will change the angular momentum **L** over time; therefore,

$$\frac{d\mathbf{L}}{dt} = \tau = \mu \times \mathbf{B} = \gamma \mathbf{L} \times \mathbf{B} \tag{2.5}$$

$$\Rightarrow \begin{cases} \mathbf{L} \cdot (d\mathbf{L}/dt) = \gamma \mathbf{L} \cdot (\mathbf{L} \times \mathbf{B}) = 0 \\ \mathbf{B} \cdot (d\mathbf{L}/dt) = \gamma \mathbf{B} \cdot (\mathbf{L} \times \mathbf{B}) = 0 \end{cases} \Rightarrow \begin{cases} \mathbf{L} \cdot d\mathbf{L} = 0 \\ \mathbf{B} \cdot d\mathbf{L} = 0 \end{cases}. \tag{2.6}$$

These relations indicate that the magnetic moment is going to *precess* along the axis of the applied magnetic field, as shown in Figure 2.2. By adopting a general solution of $\Delta L(t) = \gamma L B \sin \theta \, \Delta t = L \sin \theta \, (\gamma B) \Delta t$, the angular frequency of precession can be expressed as

$$\omega = \gamma B = \frac{eB}{2m_e}, \tag{2.7}$$

which is called the *Larmor precession (angular) frequency*. For example, if we apply a magnetic field of $B \sim 1$ Tesla, then the magnetic moment will precess around this field with a Larmor frequency of $f = \omega/2\pi \sim 14$ GHz. Also note that the case that we discuss here is by considering the scenario that the magnetic moment is solely originated from the orbital angular momentum of electron; therefore, $\gamma = e/2m_e$. We

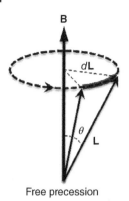

Free precession   With damping

Figure 2.2 Precession of a magnetic moment in a uniform magnetic field. We only focus on the free precession mode here, but will discuss the case with damping in Chapter 6.

have already learned from the previous section that for the case of ferromagnetic metal such as iron, this is not the case, in which $\gamma \approx e/m_e$.

It is also important to note that, although here we see a precession dynamics of magnetic moment around the direction of the applied field, in reality the magnetic moments in solids do tend to *align* with respect to the magnetic field. The precession dynamics will be damped down by magnetic damping, which will be discussed in Chapter 6.

2.2.3 Electron Spin

Since the orbital angular momentum alone cannot explain the observed phenomenon, such as the Einstein-de Hass effect, it means that some new physical properties need to be introduced. In this case, this is what we called *spin* or *spin angular momentum*. The idea of spin did not come out of thin air. This unique theoretical concept of spinning electron was proposed by Dutch physicists Goudsmit and Uhlenbeck [3] to explain the anomalous Zeeman effect, which is the unexpected splitting of spectral lines in the presence of an external magnetic field (the famous Stern-Gerlach experiment performed in 1922). From their theory, the electrons should possess intrinsic angular momentum of $\pm\hbar/2$, i.e. the spin quantum number $s = 1/2$. In contemporary quantum mechanics, we typically express these relations in terms of operator algebra with the so-called *bra-ket* notation:

$$\mathbf{S}^2|s, m_s\rangle = s(s + 1)\hbar^2|s, m_s\rangle = \frac{3}{4}\hbar^2|s, m_s\rangle \tag{2.8}$$

$$S_z|s, m_s\rangle = m_s\hbar|s, m_s\rangle = \pm\frac{1}{2}\hbar|s, m_s\rangle, \tag{2.9}$$

where $|s, m_s\rangle$ represents the quantum state of electron spin (s: spin quantum number, m_s: secondary spin quantum number ranges from $-s, -s + 1,...0, 1,..., s-1, s$).

$\mathbf{S}^2$ and S_z are the quantum operators for the magnitude and the z-direction projection (applied magnetic field direction) of the spin angular momentum, respectively. Also note that similar operators exist for the quantum state of orbital angular momentum $|l, m_l\rangle$:

$$\mathbf{L}^2|l, m_l\rangle = l(l + 1)\hbar^2|l, m_l\rangle \tag{2.10}$$

$$L_z|l, m_l\rangle = m_l\hbar|l, m_l\rangle. \tag{2.11}$$

With the concept of intrinsic electron spin angular momentum in mind, we can now go back to the relation between magnetic moment and angular momentum. Recall that when we first introduced the connection between magnetic moment and orbital angular momentum, $\boldsymbol{\mu} = e/2m_e\mathbf{L} = \gamma\mathbf{L}$. In present day, since we know that there will be possible angular momentum contributions from both *orbital* and *spin* components, a more general expression of μ-L relation is

$$\boldsymbol{\mu} = g_l\frac{e}{2m_e}\mathbf{L} = \gamma_l\mathbf{L}, \tag{2.12}$$

where $g_l = 1$ is the Landé g-factor for orbital contribution. This g-factor depends on the nature of particle of interest. For electrons, with spin-1/2, the g-factor is actually $g_e = 2$ based on Dirac's theory (quantum electrodynamics later provided an even more precise value of $g_e = 2.002319$ ...). So, the magnetic moment $\boldsymbol{\mu}$ from the electron spin angular momentum $\mathbf{S}$ should be written as

$$\boldsymbol{\mu} = g_e\frac{e}{2m_e}\mathbf{S} \approx \frac{e}{m_e}\mathbf{S} = \gamma_e\mathbf{S}. \tag{2.13}$$

Note that since $g_e = 2$ is roughly two times the value of $g_l = 1$, the corresponding Larmor frequency (as well as the gyromagnetic ratio) will also be doubled, ~28 GHz/Tesla ($\gamma_e \approx e/m_e$). This indicates that if the magnetic moment of, for instance, ferromagnetic iron, originated from electron spins, then $\gamma_e \approx e/m_e$ rather than $e/2m_e$, which is more consistent with the experimental results from Einstein-de Haas and other related gyromagnetic ratio measurements.

Furthermore, if we consider only the spin contribution of magnetic moment, the expectation value of magnetic moment along a specific direction, for instance the z-axis (along which the magnetic field is applied), can be expressed as

$$\langle\mu_z\rangle \approx \frac{e}{m_e}\langle s, m_s|S_z|s, m_s\rangle = \pm\frac{e\hbar}{2m_e} = \pm\mu_B, \tag{2.14}$$

where μ_B is called the *Bohr magneton*, which is a quantity that we will often adopt as a measure of magnetization (per formula unit) in materials. It is then useful to estimate the magnitude of the Bohr magneton:

$$\mu_B = 9.274 \times 10^{-24}\,\text{A}\cdot\text{m}^2. \tag{2.15}$$

For example, typical ferromagnetic elements Fe, Co, and Ni in their metallic forms have magnetic moments (per formula unit) of 2.22 μ_B, 1.715 μ_B, and 0.605 μ_B, respectively.

2.2.4 Spin-Orbit Interaction

Now we have a better picture of what's going on in the atom: Electrons are not just orbiting around the nucleus but also spinning at the same time. This indicates that the electron has both orbital **L** and spin **S** contributions. In fact, on many occasions, we have to consider the sum of both orbital and spin angular momentum to determine the overall magnetic moment. But how do these two angular momenta interact with each other?

A typical approach to deal with the spin-orbit coupling in an atom is to consider the rest frame of the electron, i.e. the nucleus orbiting around the electron, as shown in Figure 2.3. In this scenario, the orbiting nucleus generates a magnetic field **B** acting upon the magnetic moment **μ** of the electron. From the theory of electromagnetism,

$$\mathbf{B} = -\frac{1}{c^2}(\mathbf{v} \times \mathbf{E}),\tag{2.16}$$

where **v** and **E** are the velocity of the electron and the electric field the electron experiences, respectively. This can be further expressed as

$$\mathbf{B} = -\frac{1}{m_e c^2}(\mathbf{p} \times \mathbf{r})|E/r| = \frac{1}{m_e r c^2}\left(\frac{\partial V}{\partial r}\right)\mathbf{L},\tag{2.17}$$

where $\mathbf{p} = m_e\mathbf{v}$, $\mathbf{E} = |E/r|\mathbf{r}$, and $\mathbf{L} = \mathbf{r} \times \mathbf{p}$. The Hamiltonian (energy operator) due to the spin-orbit coupling then can be expressed as

$$\hat{H}_{SO} = -\boldsymbol{\mu}_s \cdot \mathbf{B} = -g_e \frac{e}{2m_e^2 c^2 r}\left(\frac{\partial V}{\partial r}\right)\mathbf{S} \cdot \mathbf{L}.\tag{2.18}$$

Using $g_e \approx 2$ and $U = -eV$ (potential energy), the previous relation can be rewritten as

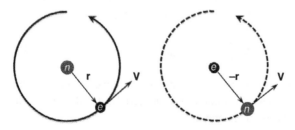

Figure 2.3 Illustration of a hydrogen atom under the rest frame of nucleus and of electron.

$$\hat{H}_{SO} \approx \frac{1}{m_e^2 c^2 r} \left(\frac{\partial U}{\partial r} \right) \mathbf{S} \cdot \mathbf{L}. \qquad (2.19)$$

However, a relativistic correction called *Thomas precession* needs to be considered here. This will give rise to a factor of $1/2$ difference from the original expression; therefore,

$$\hat{H}_{SO} \approx \frac{1}{2 m_e^2 c^2 r} \left(\frac{\partial U}{\partial r} \right) \mathbf{S} \cdot \mathbf{L}. \qquad (2.20)$$

The energy shift due to spin-orbit interaction then can be estimated by calculating the expectation value of this Hamiltonian,

$$\Delta E_{SO} = \langle \hat{H}_{SO} \rangle \approx \frac{1}{2 m_e^2 c^2} \left\langle \frac{1}{r} \frac{\partial U}{\partial r} \right\rangle \langle \mathbf{S} \cdot \mathbf{L} \rangle \propto Z^4, \qquad (2.21)$$

where Z is the atomic number of the atom of interest. Generally speaking, the larger the atomic number, the greater the possible spin-orbit interaction ($\propto Z^4$). Also note that spin-orbit interaction is essential to various magnetic and magneto transport properties, such as magnetocrystalline anisotropy (MCA), anisotropic magnetoresistance (AMR), and the spin Hall effect (SHE). We will discuss these properties in later chapters.

2.2.5 Hund's Rules

It is now clear that for an isolated atom that can be described by a simple atomic model, the magnetic dipole moment consists of two contributions: orbital angular momentum $\mathbf{L}$ and spin angular momentum $\mathbf{S}$. However, due to spin-orbit interaction, these two contributions can interact and give rise to a total angular momentum $\mathbf{J} = \mathbf{L} + \mathbf{S}$. The ground state of such isolated atom, which can be described by quantum numbers J, L, and S, then will determine the resulting atomic magnetic moment. Note that the corresponding eigenvalues for operators $\mathbf{J}^2$, $\mathbf{L}^2$, and $\mathbf{S}^2$ acting on this ground state will be $J(J+1)\hbar^2$, $L(L+1)\hbar^2$, and $S(S+1)\hbar^2$, respectively. Capital letters J, L, and S are used for quantum numbers to distinguish those from single electron atoms (j, l, s). The values of J, L, and S for the ground state are typically determined by the Hund's rules, listed here:

1) The spin quantum number S must be maximized without violating the Pauli exclusion principle to minimize the Coulomb interaction among electrons.
2) The orbital quantum number L must be maximized, in line with the first Hund's rule, to minimize the Coulomb interaction by having the electrons orbiting along the same direction.

3) If the shell is less than half full, then the total angular momentum quantum number $J = |L - S|$. If the shell is more than half full, then $J = |L + S|$. This rule is related to spin-orbit interaction and works well for rare-earth ions but fails for transition metal ions due to the crystal field effect and orbital quenching from the crystal environment.

Under this scheme, the total magnetic dipole moment is the combination of both orbital and spin angular momentum contributions:

$$\boldsymbol{\mu} = g_l \frac{e}{2m_e}\mathbf{L} + g_e \frac{e}{2m_e}\mathbf{S} \approx \frac{e}{2m_e}(\mathbf{L} + 2\mathbf{S}) = \frac{\mu_B}{\hbar}(\mathbf{L} + 2\mathbf{S}). \tag{2.22}$$

The expectation value of the magnitude of $\boldsymbol{\mu}$, typically expressed as μ_{eff} (effective magnetic moment), is related to the total angular momentum quantum number J through

$$\mu_{\text{eff}} = \langle \boldsymbol{\mu} \rangle = g_J \mu_B \sqrt{J(J + 1)}, \tag{2.23}$$

where g_J is the total Landé g-factor:

$$g_J = \frac{3}{2} + \frac{S(S + 1) - L(L + 1)}{2J(J + 1)}. \tag{2.24}$$

One can easily check that when $S = 0$ ($L = 0$), the Landé g-factor reduces to $g_l = 1$ ($g_s = 2$). The expectation value of $\boldsymbol{\mu}$ projected onto the z-axis (the direction of the applied magnetic field), on the other hand, is

$$\mu_z = g_J \mu_B m_J, \tag{2.25}$$

where m_J is the secondary magnetic quantum number of the total angular momentum, which ranges from $-J$, $-J+1$,... to $J-1$, J ($2J + 1$ possible values). Both μ_{eff} and μ_z are important in determining the magnetic properties of materials. Later we shall see that μ_{eff} is related to magnetic susceptibility χ, while the maximum value of μ_z corresponds to saturation magnetization M_s.

For isolated ions with $4f$ electrons (so-called the rare-earth family), the predictions from Hund's rules agree well with experimental measurements, as shown in Table 2.1. One can see that when the ground state values of S, L, and J are determined, then the estimated $\mu_{\text{eff}} = g_J \mu_B \sqrt{J(J + 1)}$ is close to the experimental $\mu_{\text{eff}}^{\text{exp}}$. However, Hund's rules fail for predicting the effective moments of $3d$ transition metal ions, as shown in Table 2.2. This is due to the crystal field effect and the resulting orbital quenching (therefore, $L = 0$), which will be discussed in Section 2.7.2.

Table 2.1 Predicted (from Hund's rules) and experimental values of effective moment per atom (μ_{eff}) in units of Bohr magneton (μ_B) for $4f$ rare earth ions.

$4f^n$	Ion	S	L	J	g_J	μ_z^{max}	μ_{eff}	μ_{eff}^{exp}
1	Ce^{3+}	1/2	3	5/2	6/7	2.14	2.54	2.5
2	Pr^{3+}	1	5	4	4/5	3.20	3.58	3.5
3	Nd^{3+}	3/2	6	9/2	8/11	3.27	3.52	3.4
4	Pm^{3+}	2	6	4	3/5	2.40	2.68	–
5	Sm^{3+}	5/2	5	5/2	2/7	0.71	0.85	1.7
6	Eu^{3+}	3	3	0	0	0	0	3.4
7	Gd^{3+}	7/2	0	7/2	2	7.0	7.94	8.9
8	Tb^{3+}	3	3	6	3/2	9.0	9.72	9.8
9	Dy^{3+}	5/2	5	15/2	4/3	10.0	10.65	10.6
10	Ho^{3+}	2	6	8	5/4	10.0	10.61	10.4
11	Er^{3+}	3/2	6	15/2	6/5	9.0	9.58	9.5
12	Tm^{3+}	1	5	6	7/6	7.0	7.56	7.6
13	Yb^{3+}	1/2	3	7/2	8/7	4.0	4.53	4.5

Table 2.2 Predicted (from Hund's rules) and experimental values of effective moment per atom (μ_{eff}) in units of Bohr magneton (μ_B) for 3d transition metal ions. Note that Hund's rules fail for 3d transition metal ions due to orbital quenching, and therefore $\mu_{eff} = g_s \mu_B \sqrt{S(S+1)}$.

$3d^n$	Ion	S	L	J	g_J	$\mu_{eff} = g_J \mu_B \sqrt{J(J+1)}$	$\mu_{eff} = g_s \mu_B \sqrt{S(S+1)}$	μ_{eff}^{exp}
1	Ti^{3+}, V^{4+}	1/2	2	3/2	4/5	1.55	1.73	1.7
2	Ti^{2+}, V^{3+}	1	3	2	2/3	1.63	2.83	2.8
3	V^{2+}, Cr^{3+}	3/2	3	3/2	2/5	0.78	3.87	3.8
4	Cr^{2+}, Mn^{3+}	2	2	0	0	0	4.90	4.9
5	Mn^{2+}, Fe^{3+}	5/2	0	5/2	2	5.92	5.92	5.9
6	Fe^{2+}, Co^{3+}	2	2	4	3/2	6.71	4.90	5.4
7	Co^{2+}, Ni^{3+}	3/2	3	9/2	4/3	6.63	3.87	4.8
8	Ni^{2+}	1	3	4	5/4	5.59	2.83	3.2
9	Cu^{2+}	1/2	2	5/2	6/5	3.55	1.73	1.9

2.3 Classification of Magnetisms

In previous sections we discussed the formation of magnetic moments (μ or μ_{eff}) from individual atoms or ions. However, magnetization ordering in materials should be considered as a collective behavior of all those individual moments within the material of interest. There exists different types of magnetization ordering in materials. Depending on their responses to the applied magnetic field, these magnetic properties can be roughly categorized into diamagnetism, paramagnetism, ferromagnetism, antiferromagnetism, and ferrimagnetism. Diamagnetic and paramagnetic materials correspond to matters with negative ($\chi < 0$) and positive ($\chi > 0$) susceptibilities, respectively. However, when the external magnetic field is absent, there will be no spontaneous macroscopic magnetization in these two types of materials. In contrast, spontaneous magnetization can be found in ferromagnetic and ferrimagnetic materials (typically with $\chi \gg 1$), as shown in Figure 2.4. Antiferromagnetic materials have spontaneous magnetic ordering, but the contributions from different sublattices cancel out with each other, which leads to a zero macroscopic magnetization. The rich variety of susceptibility of different types of materials can be found in Table 2.3.

Microscopically speaking, magnetic moments in diamagnetic and paramagnetic materials are independent from each other, and their responses to an external magnetic field can be derived from a simple Hamiltonian based on the atomic model that will be addressed in this section, whereas the magnetic moments in ferromagnetic, antiferromagnetic, and ferrimagnetic materials are coupled to each other through exchange interactions, which will be discussed in a much later

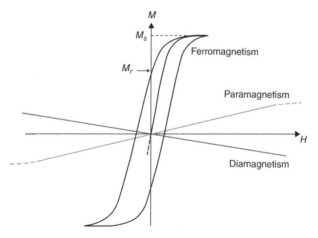

Figure 2.4 *M-H* plots of different types of materials. Note that the slope of *M-H* plot corresponds to magnetic susceptibility χ. M_s and M_r stand for the saturation magnetization and the remanence of a ferromagnetic material, respectively. The dashed line sections for paramagnetism stand for saturation of magnetization along the field direction.

Table 2.3 Magnetic properties of some materials. Note that most of the diamagnetic and paramagnetic susceptibilities are around the order of 10^{-6} (SI unit, unitless), whereas ferromagnetic susceptibilities are much greater than one.

Material	χ	Type of magnetism
Bi	-165×10^{-6}	Diamagnetic
Au	-34×10^{-6}	Diamagnetic
Ag	-24×10^{-6}	Diamagnetic
Cu	-9.7×10^{-6}	Diamagnetic
Si	-4.1×10^{-6}	Diamagnetic
Al	21×10^{-6}	Paramagnetic
W	78×10^{-6}	Paramagnetic
Pt	264×10^{-6}	Paramagnetic
Ni	600	Ferromagnetic
Fe	200 000	Ferromagnetic

section. For a Z-electron atom, without considering any exchange interactions from other atoms, the Hamiltonian of such system can be written as

$$\hat{H}_0 = \sum_{j=1}^{Z} \left(\frac{\mathbf{p}_j^2}{2m_e} + V_j \right), \tag{2.26}$$

where $\mathbf{p}_j$ and V_j stand for momentum operator and potential energy of the jth electron, respectively. Now, if this atom is experiencing an external magnetic field $\mathbf{B}$, then the momentum operator can be modified into $\mathbf{p}_j + e\mathbf{A}$, where $\mathbf{A}$ is the magnetic vector potential that satisfies $\mathbf{B} = \nabla \times \mathbf{A}$. For example, we can use $\mathbf{A} = (\mathbf{B} \times \mathbf{r})/2$. Substituting $\mathbf{p}_j$ with $\mathbf{p}_j + e\mathbf{A}$, the new Hamiltonian is

$$\hat{H} = \sum_{j=1}^{Z} \left[\frac{\left(\mathbf{p}_j + e\mathbf{A} \right)^2}{2m_e} + V_j \right] + \frac{\mu_B g_s}{\hbar} \mathbf{B} \cdot \mathbf{S}. \tag{2.27}$$

Here we also added the contribution of the Zeeman energy from the electron spin. Note that the angular momentum contribution is already included in the $(\mathbf{p}_j + e\mathbf{A})^2$ term. Also note that by definition the momentum operator is $\mathbf{p} = -i\hbar\nabla$ and the orbital angular momentum operator is $\mathbf{L} = \mathbf{r} \times \mathbf{p}$. Expanding the previous equation and with some simplifications, the Hamiltonian can be further organized into

$$\hat{H} = \hat{H}_0 + \frac{\mu_B}{\hbar} (\mathbf{L} + g_s \mathbf{S}) \cdot \mathbf{B} + \frac{e^2}{8m_e} \sum_{j=1}^{Z} \left(\mathbf{B} \times \mathbf{r}_j \right)^2. \tag{2.28}$$

The second term on the right side of Eq. (2.29) corresponds to the paramagnetic response, whereas the last term corresponds to the diamagnetic response of an isolated Z-electron ion/atom. It is interesting to note that if the d or f shell is completely filled such that $S = L = J = 0$, then the paramagnetic term vanishes (valid to first-order perturbation[1]), and only the diamagnetic term remains. For example, the diamagnetic materials listed in Table 2.3 such as Bi, Ag, Au, and Cu all have completely filled d or f shells with no unpaired electrons. In contrast, unpaired electrons and incompletely filled d or f shells can lead to paramagnetic response. Also note that if unpaired electrons exist, then the paramagnetic term will be dominating over the diamagnetic term (see more details in chapter 31 of Ref. [4]).

2.3.1 Diamagnetism

Diamagnetism is present in all materials but is often obscured by paramagnetism or ferromagnetism when unpaired electrons are in presence. Based on the Hamiltonian that we have derived, the diamagnetic susceptibility is further calculated to be (known as the Larmor diamagnetic susceptibility)

$$\chi_{\text{dia}} = -\frac{n_V e^2 \mu_0}{6 m_e} \sum_{j=1}^{Z} \langle r_j^2 \rangle, \tag{2.29}$$

where n_V is the density of ions/atoms. This expression indicates that the susceptibility is negative and the magnitude will be proportional to both the number of electrons (atomic number Z) and the size/radius of the ion/atom ($\langle r_j^2 \rangle$). Note that the Larmor diamagnetic susceptibility is largely temperature independent and only considers the contribution from the completely filled electron shells.

The free electrons in conductors (metals) will also contribute to a diamagnetic response and was predicted by Landau to be

$$\chi_{\text{dia}} = -\frac{1}{3} \mu_0 \mu_B^2 g(E_F) = -\frac{1}{3} \mu_0 \mu_B^2 \left(\frac{3 n_V}{2 E_F} \right), \tag{2.30}$$

where $g(E_F)$ is the density of states of the metal of interest at its Fermi energy E_F. In a three-dimensional free electron gas system, $g(E_F) = 3n_V/2E_F$. Landau diamagnetic susceptibility is again largely temperature independent.

2.3.2 Paramagnetism

To discuss paramagnetism, one typically starts from a statistical model for those isolated atomic or ionic magnetic moments. At a finite temperature, these magnetic moments in a paramagnetic matter will be randomly oriented due to thermal

---

1 It is possible that the excited states will give rise to second-order perturbation terms, and the phenomenon is called the Van Vleck paramagnetism.

fluctuations. However, if an external magnetic field is applied, such that the Zeeman energy term (the paramagnetic term of Eq. (2.28)) is presence, then we can expect a Boltzmann distribution, as follows:

$$P \propto e^{-\left[\frac{\mu_B}{\hbar}(\mathbf{L} + g_s\mathbf{S}) \cdot \mathbf{B}\right]/k_BT} = e^{-\boldsymbol{\mu} \cdot \mathbf{B}/k_BT} = e^{-\mu_z B/k_BT}, \tag{2.31}$$

assuming that the magnetic field is applied along the z-direction. However, we have already mentioned that $\mu_z = g_J\mu_B m_J$; therefore, this probability P of observing the magnetic moment in a state with certain quantum number m_J (with $2J + 1$ possible values) will be

$$P \propto e^{-g_J m_J \mu_B B/k_BT}. \tag{2.32}$$

The expectation value of overall magnetic moment then can be calculated by

$$\langle \mu_z \rangle = g_J \langle m_J \rangle \mu_B = g_J\mu_B \cdot \frac{\displaystyle\sum_{m_J = -J}^{J} m_J \cdot \exp\left(-g_J m_J \mu_B B/k_BT\right)}{\displaystyle\sum_{m_J = -J}^{J} \exp\left(-g_J m_J \mu_B B/k_BT\right)} = g_J\mu_B J\widetilde{B}_J(x), \tag{2.33}$$

where $\widetilde{B}_J(x)$ is the Brillouin function[2] with $x = g_J J\mu_B B/k_BT$. As shown in Figure 2.5, at a finite temperature, as the magnetic field increases toward infinity ($B/k_BT \rightarrow \infty$), the Brillouin function approaches unity. This corresponds to the saturation of all magnetic moments along the magnetic field direction, where $\langle \mu_z \rangle \rightarrow g_J\mu_B J$. By defining saturation magnetization $M_s = n_V g_J\mu_B J$ (n_V is the density of the atomic/ionic magnetic moments), the overall magnetization of a (quantum) paramagnetic system under a magnetic field then can be expressed as

$$M = n_v g_J\mu_B J\widetilde{B}_J(x) = M_s\widetilde{B}_J\left(\frac{g_J J\mu_B B}{k_BT}\right). \tag{2.34}$$

When the magnetic field is low (or temperature is high), the Brillouin function can be approximated as $\widetilde{B}_J(x) \approx x(J + 1)/3J$. Therefore, the magnetization is approximated as

$$M \approx n_v g_J\mu_B J \cdot \frac{(J + 1)x}{3J} = n_v \frac{J(J + 1)(g_J\mu_B)^2 B}{3k_BT} = \frac{n_v\mu_0\mu_{\text{eff}}^2}{3k_BT} H, \tag{2.35}$$

which suggests the paramagnetic susceptibility is

$$\chi_{\text{para}} = \frac{n_v\mu_0\mu_{\text{eff}}^2}{3k_BT}. \tag{2.36}$$

---

2 $\widetilde{B}_J(x) = \dfrac{2J + 1}{2J}\coth\left(\dfrac{2J + 1}{2J}x\right) - \dfrac{1}{2J}\coth\left(\dfrac{x}{2J}\right).$

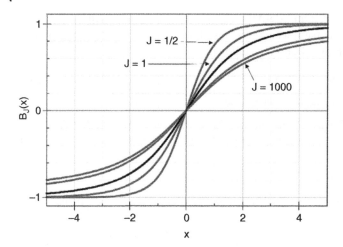

Figure 2.5 Brillouin function(s) $\widetilde{B}_J(x)$ with different total angular momentum quantum number J. Note that $J = 1/2$ corresponds to $L = 0$, $S = 1/2$, whereas J approaches infinity corresponds to a classical paramagnetic system.

This result is consistent with the so-called Curie's law of paramagnetism obtained from experiments,

$$\chi_{\text{para}} = \frac{C}{T} \propto \frac{1}{T}, \tag{2.37}$$

where C is the Curie constant, and from Eq. (2.36) we know that $C = n_v \mu_0 \mu_{\text{eff}}^2 / 3k_B$. Also note that from the previous derivations, the saturation magnetization is represented by $M_s = n_V g_J \mu_B J \propto \mu_z^{\text{max}}$, whereas the magnetic susceptibility is represented by $\chi_{\text{para}} \propto \mu_{\text{eff}}^2$.

Now we turn our focus to the paramagnetism in systems with conducting electrons. To describe such systems, we need to employ the concept of band structure and density of states. Let us consider a simple three-dimensional free electron gas system with equal numbers of spin-up and spin-down electrons, such that the net magnetization is zero in the absence of magnetic field. The density of states for spin-up $g_\uparrow(E)$ and spin-down $g_\downarrow(E)$ electrons are illustrated in Figure 2.6a. Before the application of magnetic field ($H = 0$), the densities of spin-up and spin-down electrons are the same ($n_\uparrow = n_\downarrow = n/2$).[3] When a magnetic field is applied along the

---

[3] $n_\uparrow = \int_0^{E_F} g_\uparrow(E)\, dE = \frac{1}{2}\int_0^{E_F} g(E)\, dE = \frac{n}{2}.$

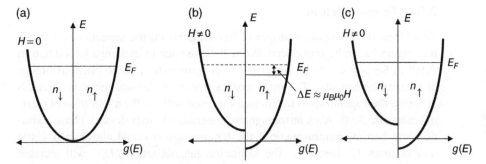

Figure 2.6 Illustration of Pauli paramagnetism.

"up" direction ($H = H_\uparrow \neq 0$), the two branches of density of states will be shifted along the energy axis due to the different Zeeman energies experienced by spin-up and spin-down electrons, as shown in Figure 2.6b. If we assume only the spin part ($S = 1/2$) is contributing to the effect, then the magnitude of this energy shift can be calculated by

$$\Delta E = g_s \frac{\mu_B}{\hbar} \mathbf{S} \cdot \mathbf{B} = \pm 2 \frac{\mu_B}{\hbar} \frac{\hbar}{2} \mu_0 H = \pm \mu_B \mu_0 H. \tag{2.38}$$

Since the Fermi energy E_F should be the same for both spin-up and spin-down electron, we should now expect more spin-up electrons than the spin-down electrons, as indicated in Figure 2.6c. The densities of electrons with opposite spins will be modified from their original magnitudes to

$$n_\uparrow = \frac{n}{2} + g_\uparrow(E_F)\Delta E = \frac{n}{2} + \frac{1}{2}g(E_F)\mu_B\mu_0 H \tag{2.39}$$

$$n_\downarrow = \frac{n}{2} + g_\downarrow(E_F)\Delta E = \frac{n}{2} - \frac{1}{2}g(E_F)\mu_B\mu_0 H. \tag{2.40}$$

The induced magnetization due to the application of magnetic field H then can be expressed as

$$M = \left(n_\uparrow - n_\downarrow\right)\mu_B = g(E_F)\mu_B^2\mu_0 H = \mu_B^2\mu_0\left(\frac{3n_V}{2E_F}\right)H, \tag{2.41}$$

where $g(E_F) = 3n_V/2E_F$ for a three-dimensional free electron gas system is used. It is also straightforward to see that, in general, the magnetic susceptibility for metallic materials, which is called the Pauli paramagnetic susceptibility, will be

$$\chi_{\text{para}} = g(E_F)\mu_B^2\mu_0. \tag{2.42}$$

This relation indicates that typically the larger the density of states at the Fermi energy $g(E_F)$, the greater the magnetic susceptibility will be for a metallic material. Also note that for metallic systems, $\chi_{\text{dia}}^{\text{Landau}} = (-1/3)\chi_{\text{para}}^{\text{Pauli}}$.

2.3.3 Ferromagnetism

To address the microscopic origin of ferromagnetism, the concept of exchange interaction has to be introduced. We will come back to exchange interaction in detail in Section 2.4. Here, we give a brief introduction of the phenomenology of ferromagnetism first. The most striking feature of ferromagnetic materials is of course their *spontaneous* magnetizations even without the application of external magnetic fields. Also, ferromagnetic materials will experience a phase transition and become paramagnetic if the temperature is raised above their Curie temperatures T_C. Below T_C, the saturation magnetization $M_s(T)$ will increase while decreasing the temperature, as shown in Figure 2.7. Magnetic properties as well as Curie temperatures for some ferromagnetic materials are listed in Table 2.4.

It is also important to note that above T_C, ferromagnetic materials behave as paramagnetic materials and their magnetic susceptibilities can be described by the Curie–Weiss law,

$$\chi = \frac{C}{T - T_C}. \tag{2.43}$$

To explain the spontaneous alignment of magnetic moments in ferromagnets (ferromagnetic materials), Pierre Weiss proposed the concept of *molecular field* (also known as Weiss exchange field since it is related to exchange interaction), which is an internal field and is proportional to the magnetization,

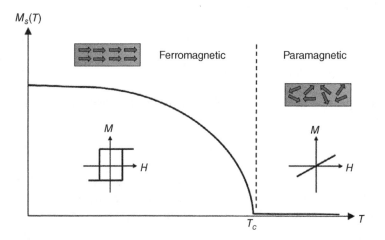

Figure 2.7 Saturation magnetization as a function of temperature for typical ferromagnetic materials. As T is above its Curie temperature T_C, a ferromagnetic-to-paramagnetic phase transition will occur, and no spontaneous magnetization can be detected. Representative M-H behaviors of the two regimes are also shown.

Table 2.4 Magnetic properties of common ferromagnetic materials.

Material	M_s (300 K) (emu/cm³)	M_s (0 K) (emu/cm³)	Moment per formula unit at 0 K (μ_B)	Curie temperature (K)
Fe	1707	1740	2.22	1043
Co	1400	1446	1.72	1388
Ni	485	510	0.606	627
Gd	–	2060	7.63	292
$Y_3Fe_5O_{12}$ (YIG)^a	130	200	5.0	560

^{a} Yttrium iron garnet (YIG) should be considered as ferrimagnetic material, but it is listed here for comparison and for future reference.

$$H_E = \lambda M, \tag{2.44}$$

where λ is the Weiss molecular field constant (material dependent and unitless). If we use this relation together with Eq. (2.34) for paramagnetism, then

$$M = M_s \widetilde{B}_J \left(\frac{g_J J \mu_B \mu_0 \left(H_{\text{applied}} + \lambda M \right)}{k_B T} \right), \tag{2.45}$$

where the original magnetic field B is replaced by the sum of applied field and Weiss molecular field. If we further set $x = g_J J \mu_B \mu_0 (H_{\text{applied}} + \lambda M)/k_B T$, then the magnetization should satisfy both

$$M = M_s \widetilde{B}_J(x) \tag{2.46}$$

and

$$M = \frac{1}{\lambda} \left(\frac{k_B T}{g_J J \mu_B \mu_0} x - H_{\text{applied}} \right). \tag{2.47}$$

The solution of M therefore can be determined by a graphical approach, as shown in Figure 2.8a. For example, when no external field is applied ($H_{\text{applied}} = 0$), Eq. (2.47) simply describes a straight line crossing the origin with a slope of $k_B T/\lambda g_J J \mu_B \mu_0$. This slope is proportional to T. If T is smaller than a certain value, then it is possible to have a nontrivial solution (crossing point with $M = M_s \widetilde{B}_J(x)$) other than $M = 0$. The lower the temperature, the larger the value of M, even with no applied magnetic field. This corresponds to the spontaneous magnetization found in ferromagnets. Using this graphical approach, one can also determine $M(T)$ for different values of J, as shown in Figure 2.8b. In fact, the $J = 1/2$ trend line derived from this approach is indeed close to experimental results from common ferromagnetic materials such as Fe, Co, and Ni.

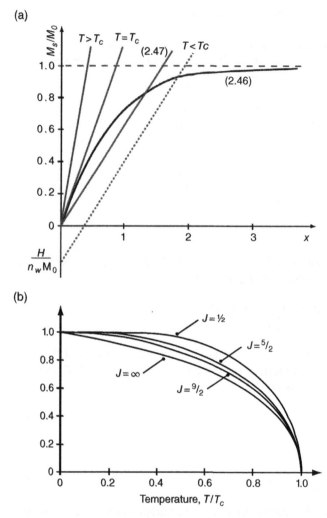

Figure 2.8 (a) Reduced magnetization M/M_s as a function of x. The crossing point of $M(x)$ described by Eqs. (2.46) and (2.47) represent the solution. (b) Solution of reduced magnetization as a function of temperature. *Source:* Figures adopted from Ref. [5].

Furthermore, to determine the temperature (slope) at which ferromagnetism disappears (two lines have only one trivial solution $M = 0$), recall that when x is small, $\widetilde{B}_J(x) \approx x(J + 1)/3J$, and therefore the slope of Eq. (2.46) will be $\approx M_s$ $(J + 1)/3J$. This suggests that the phase transition will occur when

$$\frac{k_B T}{\lambda g_J J \mu_B \mu_0} \geq \frac{M_s(J + 1)}{3J} = \frac{n_v g_J \mu_B J(J + 1)}{3J}, \tag{2.48}$$

and therefore the Curie temperature is

$$T_C = \frac{\lambda n_v \mu_0 \mu_{eff}^2}{3 k_B} = \lambda C. \tag{2.49}$$

It is also important to observe that in Table 2.4, ferromagnetic metals have non-integer Bohr magnetons per atom. This is in contrast to the isolated atom/ion case, where the localized moments per atom or per formula unit (which corresponds to μ_z^{max} that can be determined from Hund's rules) should be integral multiples of Bohr magneton, for example, as shown in Q2.5. To discuss ferromagnetism in systems with conducting electrons, one can start from the Pauli paramagnetism picture. If there exists a spontaneous spin sub-band splitting due to the internal Weiss molecular field $H_E = \lambda M$, then a spontaneous magnetization will emerge $M = (n_\uparrow - n_\downarrow)\mu_B$ due to $n_\uparrow \neq n_\downarrow$. For this to happen, the total energy change from the paramagnetic state to ferromagnetic state should be less than zero:

$$\Delta E = \Delta E_K + \Delta E_P = \frac{1}{2}g(E_F)(\delta E)^2 - \frac{1}{2}\mu_0\lambda[\mu_B g(E_F)\delta E]^2 < 0, \tag{2.50}$$

where ΔE_K and ΔE_P represent the variation of kinetic and (magnetic) potential energy, respectively. This relation can be further simplified to

$$U g(E_F) \geq 1, \tag{2.51}$$

where $U = \mu_0 \mu_B^2 \lambda$ is related the Weiss molecular field constant and exchange interaction. The previous relation is known as the Stoner criterion. For a system to be ferromagnetic, the product between the exchange interaction (related to U) and the density of state at the Fermi energy ($g(E_F)$) have to be greater than a certain value to produce spontaneous sub-band splitting and therefore spontaneous magnetization. It is also interesting to note that some paramagnetic metals, such as Pt ($U g(E_F) \approx 0.6$) and Pd ($U g(E_F) \approx 0.8$), are almost ferromagnetic [6]. Induced magnetization can exist in these metals through magnetic proximity effect when they are in contact with ferromagnetic materials.

2.3.4 Antiferromagnetism

The spontaneous magnetic ordering in materials is not limited to parallel alignment of magnetic moments. Anti-parallel alignment of magnetic moments is also possible. Such kind of magnetic ordering is known as antiferromagnetism. Therefore, in an antiferromagnetic material, the magnetic moments with the same magnitude are aligned anti-parallel with respect to each other in an ordered way and results in zero magnetization when no magnetic field is applied. Similar to ferromagnetism, there is also a phase transition temperature for antiferromagnets called the Néel temperature T_N, above which the spontaneous magnetic ordering disappears and the system becomes paramagnetic. The simplest model (Néel model) to describe an antiferromagnet is by considering a system with two sublattices (sublattice A and B), in which the magnetizations $|\mathbf{M}_A| = |\mathbf{M}_B|$ and $\mathbf{M}_A = -\mathbf{M}_B$. If we again adopt the theory of Weiss molecular field, the internal fields experienced by the two sublattices will be

$$\begin{cases} H_E^A = \lambda_{AA} M_A - \lambda_{AB} M_B \\ H_E^B = \lambda_{BB} M_B - \lambda_{AB} M_A \end{cases}, \tag{2.52}$$

where λ_{AA}, λ_{BB}, and λ_{AB} are all positive and correspond to the interactions between A–A moments, B–B moments, and A–B moments, respectively. If we further assume $\lambda = \lambda_{AB} \gg \lambda_{AA}$ (since A–B are closer than A–A or B–B), then the magnetizations in two sublattices can both be described by

$$M = M_s \widetilde{B}_J \left(\frac{-g_J J \mu_B \mu_0 \lambda M}{k_B T} \right) \tag{2.53}$$

with $M_A = M_B = M$, which is similar to the case of ferromagnetism. One can expect that the phase transition temperature will also have a similar form:

$$T_N = \frac{\lambda n_v \mu_0 \mu_{\text{eff}}^2}{3k_B}. \tag{2.54}$$

For antiferromagnets in the paramagnetic state $(T > T_N)$, by using the same approach for ferromagnets (Q2.6), the temperature dependence of the magnetization of two sublattices can be expressed as

$$\begin{cases} M_A = \dfrac{C_A}{T} \left(H_{\text{applied}} - \lambda M_B \right) \\ M_B = \dfrac{C_B}{T} \left(H_{\text{applied}} - \lambda M_A \right) \end{cases}. \tag{2.55}$$

From the definition of susceptibility,

$$\chi = \frac{M_A + M_B}{H_{\text{applied}}} = \frac{(C_A + C_B)T - 2\lambda C_A C_B}{T^2 - T_N^2} = \frac{2C}{T + T_N}, \tag{2.56}$$

where $T_N = \lambda C$ and the assumption of $C = C_A = C_B$ is used. This relation suggests that magnetic susceptibility is inversely proportional to temperature when an antiferromagnet is in the paramagnetic state. We summarize the temperature dependence of $1/\chi$ for paramagnetism, ferromagnetism, and antiferromagnetism in Figure 2.9(a). Although it seems that antiferromagnets cannot gain spontaneous magnetic ordering at positive temperatures (since the intercept from the theory is $-T_N$), in reality this is not the case. As shown in Figure 2.9b, the magnetic ordering will still emerge at positive temperatures, and only the extrapolation of the

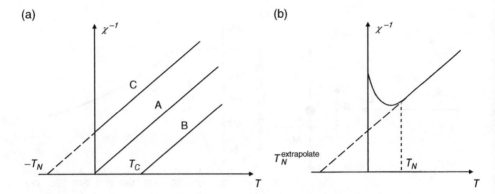

Figure 2.9 (a) Ideal temperature dependence of the inverse of magnetic susceptibility for paramagnetism (A), ferromagnetism (B), and antiferromagnetism (C). (b) Realistic temperature dependence of the inverse of magnetic susceptibility for an antiferromagnet.

Table 2.5 Properties of some representative antiferromagnetic materials.

Material	$T_N^{transition}$ (K)	$T_N^{extrapolate}$ (K)
MnO	116	−610
FeO	198	−570
CoO	291	−330
NiO	524	−1310
IrMn$_3$	690	–
Cr	311	–

trend lines will go to negative temperatures. Experimentally determined negative intercepts (denoted as $T_N^{extrapolate}$) typically deviate from the real Néel temperatures (denoted as $T_N^{transition}$), as listed in Table 2.5.

It is also important to discuss the magnetic susceptibility χ of an antiferromagnet below the Néel temperature T_N. In the antiferromagnetic state, χ is dependent on the direction of the field applied with respect to the spontaneous magnetic ordering direction. As shown in Figure 2.10a, if the magnetic field is applied parallel to the moments in the two sublattices, then the magnetization will only have small changes at low fields. In contrast, if the field is applied perpendicular to the aligned-moments direction, then the moments from both sublattices can gradually rotate toward the applied field direction, causing a larger change in terms of

(a)   (b)

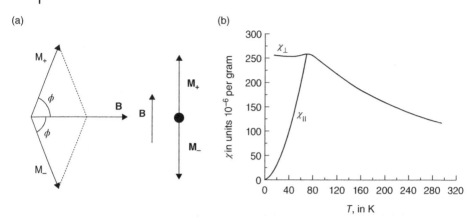

Figure 2.10 (a) Magnetization response to magnetic field applied perpendicular and parallel to the sublattice magnetic moment orientation in an antiferromagnet. (b) Corresponding perpendicular and parallel magnetic susceptibilities as functions of temperature. *Source:* Adapted from Ref. [7].

field-induced magnetization. Therefore, typically one observes $\chi_\parallel < \chi_\perp$, as shown in Figure 2.10b for their temperature dependencies.

2.3.5 Ferrimagnetism

Ferrimagnetism is similar to antiferromagnetism, except that the anti-parallel aligned magnetic moments are not having the same magnitude, i.e. $|\mathbf{M}_A| \neq |\mathbf{M}_B|$. An illustrative way to describe paramagnetism, ferromagnetism, antiferromagnetism, and ferrimagnetism is shown in Figure 2.11. Unlike antiferromagnets, ferrimagnets will have net moments even in the absence of magnetic field. Therefore, in terms of M-H response, ferrimagnetic materials behave similarly to ferromagnetic materials. The theoretical framework to discuss ferrimagnetism is similar to what we have introduced for antiferromagnetism (starting from Eq. (2.52)), except that the assumption $|\mathbf{M}_A| = |\mathbf{M}_B|$ is not used.

Although ferrimagnetic materials might behave similarly to ferromagnets, their magnetizations typically have a more complicated temperature dependence. As shown in Figure 2.12, for example, if the magnetic moments from two sublattices have different temperature dependence below the critical temperature, then it is possible that at a temperature below T_c and above 0 K, the net magnetization becomes zero. This temperature is known as the compensation temperature T_{comp}. Also note that not all ferrimagnets are necessary to have T_{comp}, as indicated in Table 2.6.

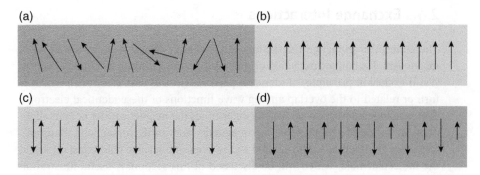

Figure 2.11 Illustrations of (a) paramagnetism, (b) ferromagnetism, (c) antiferromagnetism, and (d) ferrimagnetism.

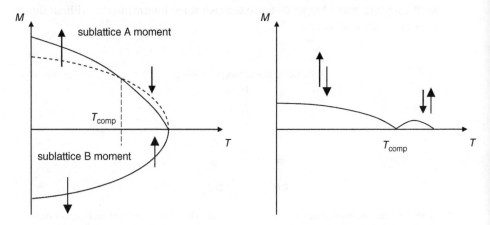

Figure 2.12 Different temperature dependence of sublattice A moment M_A and sublattice B moment M_B, which leads to the existence of the compensation temperature T_{comp}.

Table 2.6 Properties of some representative ferrimagnetic materials.

Material	T_C (K)	T_{comp} (K)	Moment per formula unit at 0 K (μ_B)
$Y_3Fe_5O_{12}$	560	–	5.0
$Gd_3Fe_5O_{12}$	564	290	16.0
$Ho_3Fe_5O_{12}$	567	137	15.2
$Dy_3Fe_5O_{12}$	563	220	18.2
Fe_3O_4	860	–	4.1

2.4 Exchange Interactions

The concept of exchange interaction and exchange coupling lie at the heart of magnetism. Exchange interaction originates from the quantum mechanical nature (Pauli exclusion principle) of identical particles (in this case electrons), which is further related to the overlap among wave functions of these identical electrons. The exchange interaction essentially determines the magnetic order or spin alignment in various ferromagnetic, ferrimagnetic, and antiferromagnetic materials. Exchange interaction is also related to the Weiss molecular field constant λ and the molecular field, which will be discussed in this section. A useful illustration to understand the hierarchy of different types of exchange interaction/coupling is shown in Figure 2.13, from which we can see that there exists two types of exchange interaction: direct and indirect, where the former is responsible for magnetic ordering of moments having wave functions overlapped, while the latter is such coupling over a longer distance through some intermediaries without direct overlap of wave functions.

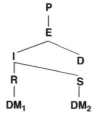

Hierarchy of exchange coupling.

P: The Pauli exclusion principle is the basis of all exchange forces.

E: An exchange interaction is a metaphorical description of the effects of the Pauli principle on the Coulomb repulsion between fermions.

I: Indirect exchange is a coupling between quantum systems so far apart that some intermediary must be involved.

D: Direct exchange is a coupling between quantum systems close enough to have overlapping wave funtions.

R: RKKY is an indirect exchange where itinerant electrons are the intermediaries.

S: Superexchange is an indirect exchange where the intermediary is a ligand.

DM_1: The Dzyaloshinsky-Moriya coupling occurs when the spin information between the indirectly coupled systems is upset asymmetrically by spin-orbit effects. In this version, itinerant electrons are the intermediaries.

DM_2: As in DM_1 except in this version the spin-orbit coupling occurs at an intermediate ligand.

Figure 2.13 Hierarchy of exchange interaction. *Source:* Figure adopted from Ref. [8].

2.4.1 Direct Exchange

To discuss direct exchange coupling, let us first consider a simple two-electron (electron a and electron b) system. The overall wave function consists of two parts: the spatial part and the spin part. By taking the Pauli exclusion principle into account, the resulting wave function can be representing the anti-bonding state:

$$\Psi_A(\mathbf{r}_1, \mathbf{r}_2) = \frac{1}{\sqrt{2}}[\psi_a(\mathbf{r}_1)\psi_b(\mathbf{r}_2) - \psi_a(\mathbf{r}_2)\psi_b(\mathbf{r}_1)] \cdot \begin{cases} \uparrow\uparrow \\ \frac{1}{\sqrt{2}}(\uparrow\downarrow + \downarrow\uparrow) \\ \downarrow\downarrow \end{cases} \quad (2.57)$$

or the bonding state:

$$\Psi_S(\mathbf{r}_1, \mathbf{r}_2) = \frac{1}{\sqrt{2}}[\psi_a(\mathbf{r}_1)\psi_b(\mathbf{r}_2) + \psi_a(\mathbf{r}_2)\psi_b(\mathbf{r}_1)] \cdot \frac{1}{\sqrt{2}}(\uparrow\downarrow - \downarrow\uparrow), \quad (2.58)$$

where ψ_a and ψ_b represent the individual wave function of electron a and electron b, respectively. The arrows $\uparrow$ and $\downarrow$ represent the spin component of the wave function. It is important to note that $\Psi_A(\mathbf{r}_1, \mathbf{r}_2) = -\Psi_A(\mathbf{r}_2, \mathbf{r}_1)$ (antisymmetric) and $\Psi_S(\mathbf{r}_1, \mathbf{r}_2) = \Psi_S(\mathbf{r}_2, \mathbf{r}_1)$ (symmetric). Also note that the spin component of $\Psi_A(\mathbf{r}_1, \mathbf{r}_2)$ has three possible states (triplet), which correspond to $S_{total} = 1$ ($m_{S_{total}} = -1, 0, +1$), while $\Psi_S(\mathbf{r}_1, \mathbf{r}_2)$ has only one possible state (singlet) with $S_{total} = 0$ and $m_{S_{total}} = 0$. Since the energies correspond to the anti-bonding state and bonding state are $E_A = \langle \Psi_A | \hat{H} | \Psi_A \rangle = \int \Psi_A^* \hat{H} \Psi_A \, d\mathbf{r}_1 d\mathbf{r}_2$ and $E_S = \langle \Psi_S | \hat{H} | \Psi_S \rangle = \int \Psi_S^* \hat{H} \Psi_S \, d\mathbf{r}_1 d\mathbf{r}_2$, we can define an effective Hamiltonian that is related to the spin information

$$\hat{H}_{eff} = \frac{1}{4}(E_S + 3E_A) - \frac{1}{\hbar^2}(E_S - E_A) \mathbf{S}_1 \cdot \mathbf{S}_2, \quad (2.59)$$

where $\mathbf{S}_1$ and $\mathbf{S}_2$ correspond to the spin angular momentum of electron a and electron b, respectively. Note that the expectation value of $\mathbf{S}_1 \cdot \mathbf{S}_2$ can be calculated by realizing that

$$|\mathbf{S}_{total}|^2 = |\mathbf{S}_1 + \mathbf{S}_2|^2 = |\mathbf{S}_1|^2 + |\mathbf{S}_2|^2 + 2\mathbf{S}_1 \cdot \mathbf{S}_2 \quad (2.60)$$

and therefore

$$S_{total}(S_{total} + 1)\hbar^2 = S_1(S_1 + 1)\hbar^2 + S_2(S_2 + 1)\hbar^2 + 2\mathbf{S}_1 \cdot \mathbf{S}_2. \quad (2.61)$$

Since $S_{total} = 1$ (triplet) or $S_{total} = 0$ (singlet) and both S_1 and $S_2 = 1/2$, the expectation value of $\mathbf{S}_1 \cdot \mathbf{S}_2$ is therefore $\hbar^2/4$ for the triplet state and $-3\hbar^2/4$ for the singlet state. Using this relation, we verify that the effective Hamiltonian in Eq. (2.59) can indeed give us $\langle \hat{H}_{eff} \rangle = E_A$ for the anti-bonding (triplet) state and $\langle \hat{H}_{eff} \rangle = E_S$ for the bonding state.

One can further simplify the effective Hamiltonian to be

$$\hat{H}_{\text{eff}} = -2J\,\mathbf{S}_1 \cdot \mathbf{S}_2, \tag{2.62}$$

where $J \equiv (E_S - E_A)/2\hbar^2$ is known as the exchange coupling constant. The constant part of the original effective Hamiltonian has been neglected. Note that when $J > 0$ $(J < 0)$, $E_S > E_A$ $(E_S < E_A)$, then the system prefers the triplet (singlet) state with parallel (anti-parallel) spins. To be more generalized, if the system consists of more than two spins, the Hamiltonian can be expressed as

$$\hat{H} = -\sum_{i,j} J_{ij}\,\mathbf{S}_i \cdot \mathbf{S}_j, \tag{2.63}$$

which is known as the famous Heisenberg Hamiltonian. The exchange constant J or J_{ij} can be either positive or negative. One can imagine that when the inter-atomic distance is small, the overlap between the two electronic wave functions is more significant; therefore, the spins prefer anti-parallel alignment. On the other hand, if the inter-atomic distance is large, where the wave function overlap is less significant, the spins then prefer parallel configuration. The Bethe-Slater curve shown in Figure 2.14 represents the magnitude of direct exchange coupling as a function of inter-atomic distance. It can be seen that $3d$ transition metal Co is located near the peak of this curve ($J > 0$) and therefore being ferromagnetic, while Mn and Cr are on the side of negative exchange ($J < 0$) and therefore being anti-ferromagnetic. The sign of J for Fe depends on the crystal structure, with $J > 0$ for α-Fe (ferrite, body center cubic structure) and $J < 0$ for γ-Fe (austenite, face center cubic structure).

The Heisenberg model is also connected to the Weiss molecular field theory. If we simply focus on one spin in the system at a time (S_i), the rest of the spins in the

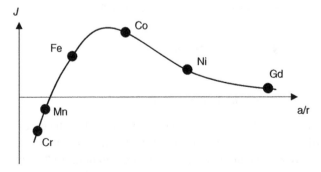

Figure 2.14 Bethe-Slater curve, where *a* represents the inter-atomic distance and *r* represents the radius of the *d* shell.

system will then create a *mean field* (H_{mf}) acting upon the spin of interest. The corresponding exchange interaction is

$$E_{ex}^i = \left\langle \hat{H}^i \right\rangle = -2JS_i \cdot \sum_j S_j = -\mu_i \cdot B_{mf} = -g\mu_B S_i \cdot \mu_0 H_{mf}. \tag{2.64}$$

Hence, this mean field can be expressed as

$$H_{mf} = \frac{2J}{\mu_0 g\mu_B} \sum_j S_j \approx \frac{2zJ}{\mu_0 g\mu_B} \langle S_j \rangle, \tag{2.65}$$

where z represents the number of nearest neighbors (assuming the nearest neighbors contribute most of the coupling). If we further use $M = n_v g\mu_B \langle S_j \rangle$, then the mean field is

$$H_{mf} = \left[\frac{2zJ}{\mu_0 n_v (g\mu_B)^2} \right] M = \lambda M. \tag{2.66}$$

Now we can see that the Weiss molecular field constant λ is directly related to the exchange coupling constant J. The Curie temperature (assuming all moments have only spin contribution) is then expressed as

$$T_C = \frac{2zJ\mu_{eff}^2}{(g\mu_B)^2 3k_B} = \frac{2zJ \cdot S(S+1)}{3k_B}. \tag{2.67}$$

2.4.2 Indirect Exchange: Superexchange

In many occasions, we do observe magnetic order even when the inter-atomic distance is too large to sustain wave function overlap and direct exchange coupling. Even without itinerant electrons, such long-range magnetic ordering can still exist. For example, in magnetic oxides such as iron garnets and spinel ferrites, the magnetic ordering actually originates from an indirect exchange mechanism called superexchange or Kramers-Anderson superexchange [9]. Superexchange describes the interaction between magnetic moments or ions too far apart to be connected by direct exchange but can be coupled over a relatively long distance through a nonmagnetic ion. We can take antiferromagnetic MnO as an example. The coupling between the moments on a pair of metallic ions (Mn^{2+}, with electronic configuration of $3d^5$ and moment of 5 μ_B) separated by an oxygen ion (O^{2-}) can be mediated by the p-orbital of O^{2-} as illustrated in Figure 2.15. Although the electrons in the d orbitals of these two Mn^{2+} ions are far away from each other, the electronic spin configuration therein should be opposite due to the overlap between their d orbitals and the p orbital of O^{2-}. The two spins in p orbital of O^{2-} can only be opposite due to the Pauli exclusion principle. This results in a

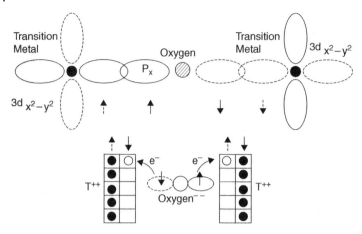

Figure 2.15 Superexchange coupling of Mn^{2+}-Mn^{2+} moments through O^{2-} in the antiferromagnetic MnO. *Source:* Adapted from Ref. [10].

long-range antiferromagnetic coupling between the magnetic moments from these two Mn^{2+} ions.

2.4.3 Indirect Exchange: RKKY Interaction

Indirect exchange couples moments over relatively large distances. It is also the dominant exchange interaction in many metals, where there is only little or almost no direct overlap of wave functions among neighboring electrons. It therefore acts through an intermediary, which in metallic materials are the conduction electrons or itinerant electrons. A simplified picture is is as follows: A magnetic ion induces spin polarization in the conduction electrons in its neighborhood. This spin polarization in the itinerant electrons is experienced by the moments of other magnetic ions within the range leading to an indirect coupling. This type of exchange is better known as the RKKY interaction, which was named after Ruderman, Kittel, Kasuya, and Yoshida.

The RKKY exchange coupling constant J oscillates from positive to negative as the separation of the magnetic moments changes and has an oscillatory nature shown in Figure 2.16. The oscillatory behavior of J can be expressed as

$$J_{\mathrm{RKKY}}(r) \propto F(2k_F r), \tag{2.68}$$

where k_F is the Fermi wave vector, and

$$F(x) = \frac{\sin x - x \cos x}{x^4}. \tag{2.69}$$

Therefore, depending on the separation between a pair of ions, their magnetic coupling can be ferromagnetic or antiferromagnetic.

Figure 2.16 Exchange coupling constant *J* from RKKY interaction as a function of inter-atomic distance *r*.

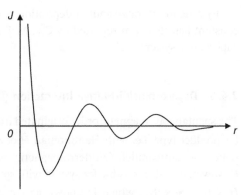

Note that for rare-earth metals, the electrons (responsible for magnetic moments) in the 4*f* shell are shielded by the 5*s* and 5*p* electrons; therefore, direct exchange is rather small, and indirect exchange through the itinerant electrons is responsible for the magnetic order in these materials. Also note that due to the oscillatory behavior of exchange coupling constant *J* as a function of moment separation, one can artificially engineer exchange coupling to be either ferromagnetic ($J > 0$) or antiferromagnetic ($J < 0$) by tuning the interatomic distance through thin film layer design. This can be achieved by growing multilayer structures with a ferromagnetic/normal metal/ferromagnetic (FM/NM/FM) sandwich structure, as shown in Figure 2.17. Parkin and Mauri demonstrated that Ru is a suitable material for realizing RKKY interaction with either ferromagnetic or antiferromagnetic coupling when Ru thickness is varied in the range of 0.5–2 nm [11]. This property is widely employed in modern-day magnetoresistive random-access memory (MRAM) devices to develop the so-called synthetic antiferromagnetic

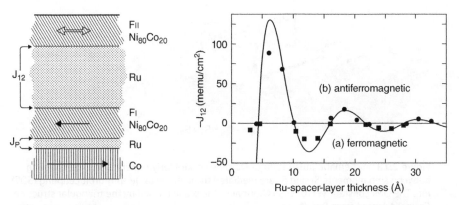

Figure 2.17 RKKY interaction leads to an oscillatory exchange coupling constant *J* as a function of Ru thickness in a $Ni_{80}Co_{20}/Ru/Ni_{80}Co_{20}$ sandwich structure. *Source:* From Ref. [11].

(SAF) structure through sputter deposition. Similar oscillatory exchange coupling constant has also been reported by C.F. Majkrzak et al. in a single-crystal Gd—Y superlattice system [12].

2.4.4 Dzyaloshinskii-Moriya Interaction (DMI)

The spontaneous magnetic order originated from exchange interactions of the scalar product type, i.e. with Hamiltonian $\hat{H} \propto \mathbf{S}_i \cdot \mathbf{S}_j$, can have parallel (ferromagnetic) or antiparallel (antiferromagnetic and ferrimagnetic) configurations. However, it is also possible for system with spins to have cross-product form, with $\hat{H} \propto \mathbf{D}_{ij} \cdot (\mathbf{S}_i \times \mathbf{S}_j)$, which is known as antisymmetric or anisotropic exchange interaction that is caused by the spin-orbit interaction. The theory of such antisymmetric exchange was first proposed by Igor Dzyaloshinskii (Dzyaloshinsky) [13] and Toru Moriya [14], and therefore such exchange interaction is also often called the Dzyaloshinskii-Moriya interaction (DMI), with

$$\hat{H}_{DMI} = -\mathbf{D}_{ij} \cdot (\mathbf{S}_i \times \mathbf{S}_j), \tag{2.70}$$

where the DMI constant vector $\mathbf{D}_{ij}$ depends on the crystal structure and broken inversion symmetry, as shown in Figure 2.18. The DMI can also be a long-range, indirect type of exchange interaction, in which the coupling of $\mathbf{S}_i$ and $\mathbf{S}_j$ can be mediated by itinerant electrons (metallic case) or ligands (insulating case) through spin-orbit interactions. In recent years, it has been found that DMI is related to the formation of exotic spin textures such as magnetic skyrmion [15] and chiral Néel domain wall (DW) [16, 17].

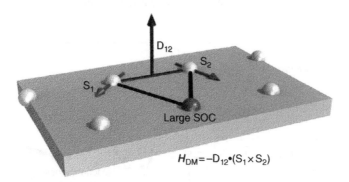

$$H_{DM} = -D_{12} \cdot (S_1 \times S_2)$$

Figure 2.18 Schematic illustration of Dzyaloshinskii-Moriya interaction. The coupling between spin moments S_1 and S_2 are mediated through the large spin-orbit coupling (SOC) ion. The DMI constant vector $\mathbf{D}_{12}$ is normal to the plane formed by the triangular structure. *Source:* Adapted from Ref. [15].

2.5 Magnetization in Magnetic Metals and Oxides

Although we have already covered the possible origins of magnetic order in different material systems, which involve different types of exchange interactions, we still have not discussed the magnetic properties of those common ferromagnetic metals, namely, Fe, Co, and Ni. In fact, we briefly mentioned in a previous section that for systems with conducting electrons, Stoner criterion determines a metallic system being ferromagnetic or not (due to spontaneous sub-band splitting). We will elaborate more on that from a band structure point of view for those common ferromagnetic metals in this section. The magnetic properties of some common magnetic insulators, such as iron garnets and spinel ferrites, will also be covered.

2.5.1 Slater-Pauling Curve

Upon observing the magnetic moments in various types of ferromagnetic alloys formed by 3*d* transition metals, one can find a general trend as shown in Figure 2.19. The observed averaged magnetic moment per atom will linearly increase from 0 to 2.5 μ_B as the number of electrons per atom (atomic number) of the alloy varies from 24 to slightly above 26. As the number of electrons per atom increases further, the moment per atom starts to decrease linearly. For example,

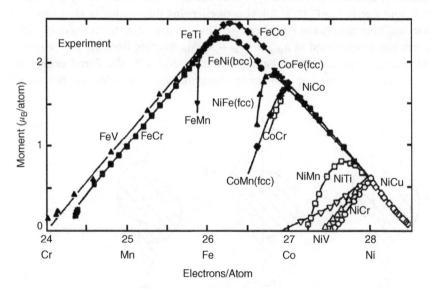

Figure 2.19 Slater-Pauling curve. The magnetic moment per atom for Fe, Co, Ni, and their alloys as a function of number of electrons per atom. *Source:* Adapted from [10].

pure Fe (bcc) has moment per atom of 2.2 μ_B. Doping Ni into Fe (forming $Fe_{1-x}Ni_x$ alloy) will first enhance the average moment to ~2.4 μ_B, then it will linearly decrease toward pure Ni value ~0.6 μ_B as Ni composition further increases. Note that there are some data from the fcc $Fe_{1-x}Ni_x$ that deviate from the trend line for bcc $Fe_{1-x}Ni_x$ alloy due to bcc-to-fcc phase transformation. The Slater-Pauling curve indicates that as long as there is no significant change in terms of crystal structure, the average magnetic moment seems to have a direct and linear dependence on the number of electron, and therefore the Fermi energy, of the ferromagnetic alloy.

2.5.2 Rigid Band Model

Besides the experimental observation of Slater-Pauling curve, let us recall that the Stoner criterion states that $U g(E_F) \geq 1$ will lead to spontaneous sub-band splitting of the opposite spins and therefore a spontaneous net magnetic moment. For transition metals, it turns out that the density of states $g(E_F)$ will be large if the Fermi energy comes across the d band (for example, Fe, Co, and Ni). In contrast, if the Fermi energy is crossing the s band instead of the d band, $g(E_F)$ will be small, and the Stoner criterion will not be satisfied (for example, Cu). Based on these knowledge, a simple "rigid" band model can be employed to describe the Slater-Pauling curve result, as shown in Figure 2.20 for the $Fe_{1-x}Ni_x$ alloy case. We assume that the density of states for both s band and d band are not affected by the variation of number of electrons (Fermi energy), with the two d sub-bands being spontaneously split due to $U g(E_F) \geq 1$. Alloying or changing the number of electrons per atom will only modify the Fermi energy. The average magnetic moment of such system can be expressed as $\mu_{alloy} \approx (N_{d\uparrow} - N_{d\downarrow})\mu_B$. Starting from the pure Ni side and gradually increasing the Fe concentration, $N_{d\uparrow} = 5$ (the Fermi energy is already above the spin-up sub-band); therefore, $\mu_{alloy} \approx (5 - N_{d\downarrow})\mu_B$. But since

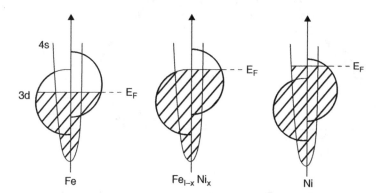

Figure 2.20 Rigid band model for $Fe_{1-x}Ni_x$ alloy. *Source:* Adapted from [10].

$N_d = N_{d\uparrow} + N_{d\downarrow} = 5 + N_{d\downarrow}$, the average moment per atom is then $\mu_{\text{alloy}} \approx (10 - N_d)\mu_B$. This reflects the linear decreasing trend of moment per atom while increasing the number of d electrons (number of electrons per atom). Note that although the rigid band model can roughly explain the experimental trend of the Slater-Pauling curve, the real band structures and densities of states for Fe, Co, and Ni are very different from the simplified picture, as illustrated in Figure 2.20.

2.5.3 Iron Oxides and Iron Garnets

Unlike ferromagnetic metals with conducting electrons, the magnetic properties of magnetic insulators typically stem from the magnetic moments of isolated ferric and rare earth ions. In the case of iron oxides, they typically have negative exchange couplings ($J < 0$) among the consisting ferric ions (Fe^{3+} or Fe^{2+}), which makes them either ferrimagnetic or antiferromagnetic. For example, the ferric ions Fe^{2+} ($4\mu_B$, Q2.5) in a FeO with rocksalt structure are negatively exchange coupled through the superexchange mechanism, which is mediated by the O^{2-} ions. Since only one type of ferric ion exists, the antiparallel alignment of those Fe^{2+} ions leads to antiferromagnetism. On the other hand, there exists two types of ferric ions (Fe^{2+} with 4 μ_B and Fe^{3+} with 5 μ_B) in a magnetite, Fe_3O_4 ($FeO\text{-}Fe_2O_3$) with inverse spinel structure; therefore, a net moment exists. Another common family of magnetic insulator is known as the iron garnet or rare-earth iron garnet, which has the form of $R_3Fe_5O_{12}$, where R stands for the rare-earth element (R = Y, Gd, Tm, etc. See Table 2.6). These ferrimagnetic insulators are complicated in terms of structure but are of great importance in some applications. For example, Yttrium iron garnet ($Y_3Fe_5O_{12}$, also known as YIG) has a relatively much lower magnetic damping constant ($\sim 10^{-4}$) while compared to those from ferromagnetic metals ($\sim 10^{-3}$), which is advantageous for various microwave, optical, and acoustic applications.

2.6 Phenomenology of Magnetic Anisotropy

When a physical property of a material is a function of direction, we say that property has anisotropy or is anisotropic. In contrast, the property is isotropic if there is no directional dependence. Magnetic anisotropy is the key to many important features of MRAM as well as other applications, such as permanent magnets. In previous sections, we only covered the origin of magnetization and magnetic moments, but we have not mentioned the origin of magnetic anisotropy. In this section, we first briefly discuss the phenomenology of magnetic anisotropy and its possible origins. Then we discuss the concept of domain and DWs, which is related to the competition between magnetic anisotropy energy and exchange interaction energy.

2.6.1 Uniaxial Anisotropy

Without addressing its origin, the simplest model to describe a magnetic system with anisotropy is the uniaxial model, which indicates that the magnetization prefers to point along a certain axis called the *easy axis* (with two preferred directions). For example, if we set the easy axis to be along z-direction, then a phenomenological uniaxial magnetic anisotropy energy density can be expressed as

$$u_a = \frac{U}{V} = \sum_n K_{u,n} sin^{2n}\theta = K_{u0} + K_{u1} sin^2\theta + K_{u2} sin^4\theta + ..., \qquad (2.71)$$

where V is the volume of the magnetic system and angle θ is the angle between the magnetization and the z-axis. $K_{u,n}$ represents the nth order anisotropy energy density. If we drop the constant term (since it is not angle dependent) and only keep the first-order term, then the expression can be further simplified to

$$u_a \approx K_{u1} \sin^2\theta. \qquad (2.72)$$

The uniaxial magnetic anisotropy energy profile then can be illustrated as shown in Figure 2.21, where the sign of K_{u1} will determine the shape of this energy landscape. It can be seen that for a positive K_{u1}, magnetization M (which is considered as a constant magnitude vector) tends to align along the positive z-direction or negative z-direction, therefore having the easy axis behavior as we expected. On the

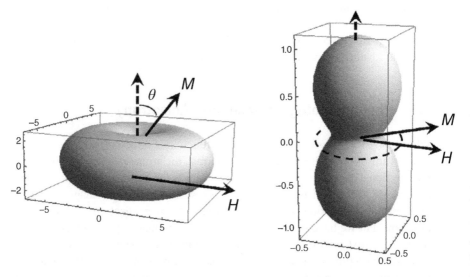

Figure 2.21 Visualization of uniaxial anisotropy energy density profile as a function of θ with $K_{u1} > 0$ (easy axis) and $K_{u1} < 0$ (easy plane).

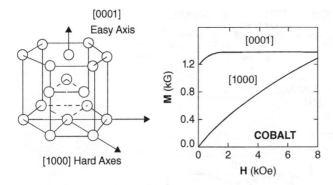

Figure 2.22 hcp Co has uniaxial magnetic anisotropy with easy axis along the [0001] axis. *Source:* Adapted from [10].

other hand, when K_{u1} is negative, M prefers to lie in the x-y plane, therefore having the *easy plane* behavior. We will use this model to discuss the switching behavior of M when a magnetic field H is applied in Chapter 5.

In the case of crystalline ferromagnetic materials, Co with hcp crystal structure shows uniaxial magnetic anisotropy with its easy axis along the c-axis ([0001] axis) of its unit cell, as shown in Figure 2.22. The first- and the second-order anisotropy energy densities are $K_{u1} = 4.1 \times 10^5$ J/m^3 and $K_{u2} = 1.5 \times 10^5$ J/m^3(SI unit), respectively. Also note that the axes on the easy plane (x-y plane) can be considered as the hard axes, which suggests that a stronger magnetic field is required to saturate magnetization along those directions (for example, [1000] direction for Co).

2.6.2 Cubic Anisotropy

Crystalline Fe and Ni have bcc and fcc crystal structures, respectively. They possess cubic anisotropy instead of uniaxial anisotropy, as shown in Figure 2.23. For bcc Fe, the magnetization is relatively easier to get saturated by applying a magnetic field along [100] directions (along x, y, and z axes), whereas for Ni the [111] axes are the easy axes.

The anisotropy energy density for systems with cubic anisotropy can be expressed as

$$u_a = K_0 + K_1\left(\alpha_1^2\alpha_2^2 + \alpha_2^2\alpha_3^2 + \alpha_3^2\alpha_1^2\right) + K_2\left(\alpha_1^2\alpha_2^2\alpha_3^2\right) + ..., \qquad (2.73)$$

where $\alpha_i = m_i = M_i/M_s$ is the projection of magnetization onto the ith-axis. For the Fe case, $K_1 = 4.8 \times 10^4$ J/m^3 and $K_2 = -1.0 \times 10^4$ J/m^3. For the Ni case, $K_1 = -4.5 \times 10^3$ J/m^3 and $K_2 = -2.3 \times 10^3$ J/m^3. Their anisotropy energy density profiles based on these numbers are illustrated in Figure 2.24.

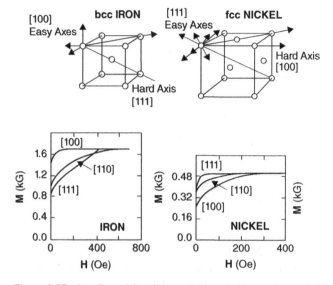

Figure 2.23 bcc Fe and fcc Ni have cubic anisotropy. *Source:* Adapted from [10].

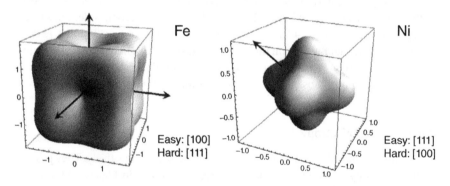

Figure 2.24 Visualization of cubic anisotropy energy densities for Fe (bcc) and Ni (fcc).

2.7 Origins of Magnetic Anisotropy

Although the phenomenological picture of magnetic anisotropy is useful in describing different anisotropic behaviors of various materials systems, it does not give us any insights on the microscopic origin(s) of such direction or angle-dependent effect. For the following sections, we will address some commonly seen origins of magnetic anisotropy.

2.7.1 Shape Anisotropy

In previous sections, we see that the magnetic anisotropy can be related to the crystal structure for single crystal systems. However, it is also common for materials scientists and physicists to deal with materials with polycrystalline or amorphous structure. In those cases, since macroscopically there are no preferred orientations of magnetic moment, ideally those systems can be considered as magnetically isotropic. Nevertheless, some other factors will come into play and serve as the source of magnetic anisotropy, even without crystalline structures. Shape anisotropy is one commonly seen source.

Recall that in Chapter 1 we discussed the existence of demagnetization field (demag field), which purely originates from the magnetostatic property of the magnetic object and therefore its geometry. Let us consider a ferromagnetic rod as shown in Figure 2.25, where the magnetization can be measured by applying magnetic field either along the long axis or along the short axis of the rod. We find that the magnetization will be easier (harder) to saturate if the magnetic field is applied along the long (short) axis of the rod. Therefore, the long axis is the easy axis of this magnetic rod system. It can be readily understood by what we have learned in Chapter 1: The surface magnetic charges with opposite signs are further away from each other in the **H**-parallel-to-long-axis case, while compared to the **H**-perpendicular-to-long-axis case. Therefore, the demag field will be greater if **H** is applied perpendicular to the long axis. We will discuss this in more detail for the thin film case and its importance for MRAM applications in Chapter 5.

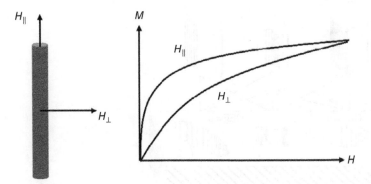

Figure 2.25 The easy (long) axis and the hard (short) axis of a ferromagnetic rod with elongated geometry. The magnetization will be harder to be saturated along the short axis due to a greater demagnetization field to overcome.

2.7.2 Magnetocrystalline Anisotropy (MCA)

The previously mentioned magnetic anisotropy for crystalline Fe, Co, and Ni is called the MCA. MCA originates from the interplay between spin-orbit interaction (atomic factor) and crystal electric field effect (environment factor). As illustrated in Figure 2.26, the atomic magnetic moments originated from the spin and orbital angular momentum will be affected by the crystal field from the neighboring orbitals. For example, if the orbital quantum number of the moment of interest is nonzero, which results in an atomic orbital with twofold or fourfold symmetry, then placing such orbital in a crystal electric field with twofold or fourfold symmetry will make the orbital hard to rotate freely; hence, MCA will be found. In contrast, if such orbital is placed in an isotropic environment, then no anisotropy will be induced. Another example is looking at different d-orbitals located at an octahedral site of a crystal. The electrons in different d-orbitals will experience different magnitudes of Coulomb interactions, which lifts the degeneracy of the five d-orbitals shown. In this case, d_{xy}, d_{yz}, and d_{xz} are more energetically favorable than d_z^2 and $d_{x^2-y^2}$. The preferred orientations of magnetic moments (as determined by the d-orbital) will therefore be affected by the environment (the crystal field from the octahedral site). In sharp contrast, the s-orbital will not experience the crystal field effect, since it is isotropic.

It is also important to note that for 3d transition metal ions, the crystal electric field effect dominates the spin-orbit interaction; therefore, the angular orbital momentum contribution is typically "quenched" by the environment, which leads

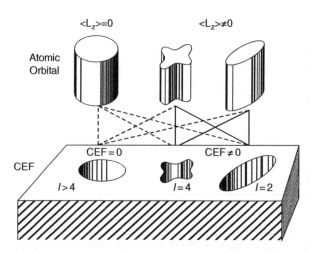

Figure 2.26 Schematic illustration of MCA: the atomic orbital and crystal electric field (CEF) effects. *Source:* Adapted from Ref. [10].

to $L = 0$, and the resulting effective moment is expressed as $\mu_{\text{eff}} = g_s \mu_B \sqrt{S(S+1)}$; i.e. only the spin part contributes to the moment. For $4f$ rare-earth ions, the spin-orbit interaction term is relatively larger than that from the crystal field effect. Therefore, total angular momentum has to be considered, and the Hund's rules prediction is valid; i.e. $\mu_{\text{eff}} = g_J \mu_B \sqrt{J(J+1)}$.

2.7.3 Perpendicular Magnetic Anisotropy (PMA)

In magnetic thin film engineering, hard-disk drive development, as well as MRAM development, perpendicular magnetic anisotropy (PMA) of the magnetic record-ing media is crucial. It is one of the key approaches to enhance the data storage density. For those magnetic recording media, magnetic thin films of less than 1 μm of thickness are typically employed. However, we learned in a previous section that the demagnetization field from the magnetostatic effect will tend to make the magnetic moments in such films possess in-plane anisotropy (shape ani-sotropy) rather than out-of-plane anisotropy (PMA). Therefore, materials selection and engineering are necessary to introduce PMA to the thin films.

Two types of PMA can be induced in magnetic thin films: bulk PMA and inter-facial PMA. Bulk PMA can be found in rare-earth-transition metal (RE-TM) thin films, such as CoGd, CoTb, and FeCoTb. These films can be prepared by co-sputter deposition. The moment of each sublattice in these materials are antiferromagne-tically coupled, resulting in ferrimagnetism. As for interfacial PMA, buffer layer/CoFeB/MgO, Pt/Co (Pd/Co, Ni/Co) multilayer systems are some of the common options. In buffer/CoFeB/MgO heterostructures, the PMA originates from the hybridization of Fe(Co) d-orbital and O p-orbital [18]. An annealing process around 300 °C is usually necessary to allow the boron to diffuse out of the CoFeB layer and let the CoFeB layer crystallize into bcc structure with (100) texture (with respect to MgO) [19]. To assist the boron diffusion, it is also important that the buffer layer acts as a boron sink to induce PMA. In contrast, Pt/Co (or Ni/Co, Pd/Co) multilayer system does not require a thermal annealing process. The origin of PMA in Pt/Co multilayer is still unclear. Some groups attribute the origin to the interfacial MCA between Pt and Co since the anisotropy energy alters when the Pt crystallographic orientation is different [20], while some groups suggest that the PMA originates from the inter-diffusion between Pt and Co or the formation of Pt-Co alloy [21].

2.8 Magnetic Domain and Domain Walls

In reality, the moments in magnetic materials can have different orientations even if they are coupled ferromagnetically. Weiss postulated that this circumstance can originate from the multiple tiny magnetized blocks inside these materials, which is

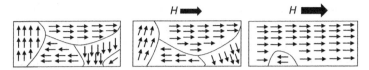

Figure 2.27 Illustration of magnetic domains in a ferromagnetic material.

the result of competition among different magnetic energies. These blocks with particular magnetization are called *magnetic domains*, and the region/boundaries that separate the domains are called *domain walls (DWs)*, as shown in Figure 2.27. Under the application of an external field, magnetization in domains and DWs will be affected. To configure the alignment of magnetic moments in a domain and the corresponding motion of a DW, equilibrium among different magnetic energies plays an important role. The relevant energy terms are exchange energy, magnetostatic energy, magnetic anisotropy energy, magnetoelastic energy, and Zeeman energy (external field).

2.8.1 Domain Wall

A typical DW can be treated as the region with a continuous variation of the magnetic moment alignment between two magnetic domains, as shown in Figure 2.28 for a one-dimensional case. The key parameter of describing a DW is its width (or thickness). DW width can be further determined by the equilibrium between exchange energy and anisotropy energy. The size of a DW can range from ~10 nm to ~1000 nm, depending on the values of the exchange and anisotropy energies of the materials.

It is also important to note that there are two major types of DWs, namely, Bloch wall and Néel wall (as shown in Figure 2.29). In normal circumstances, the DW moments will tend to point perpendicular to the DW plane normal due to lower

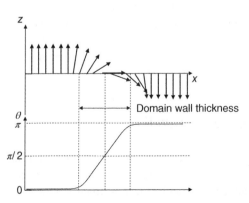

Figure Figure 2.28 Illustration of a DW in between two magnetic domains.

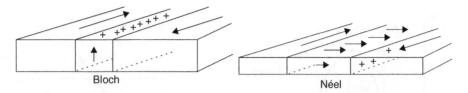

Figure 2.29 Bloch wall and Néel wall in magnetic thin films. *Source:* Adapted from Ref. [10].

total energy, which is the Bloch type. However, Néel noticed that the result might be different when the magnetic film becomes thinner than a specific thickness, at which the magnetic moments tend to lie in-plane to reduce the magnetostatic energy. Néel walls with the magnetic moments therein pointing along the normal of the wall plane will form consequently. In recent studies researchers also found that the interfacial DMI in some PMA systems (such as Pt/Co/oxide, Ta/CoFeB/MgO, etc.) will favor the formation of Néel wall as well, which will further affect the dynamics of such DW motion.

2.8.2 Single Domain and Superparamagnetism

For MRAM applications, we typically want the magnetic layer used for memory storage to have a single domain rather than multiple domains. This is because the multidomain nature will result in intermediate states of the memory element, which is undesirable. Achieving single domain, for example, for a magnetic memory element, is typically done by shrinking the size of it. By taking the competition among exchange interaction, anisotropy energy, and magnetostatic energy into account, the critical size of gaining a single-domain nature of a magnetic particle is typically ~100 nm. That is one of the reasons why state-of-the-art MRAM devices are commonly smaller than 100 nm in terms of lateral dimensions.

However, it is also important to address that one cannot indefinitely shrink the size of such magnetic particles and still maintain its ferromagnetic properties. This is because in previous discussions, we have not considered the factor of thermal activation; i.e. the ambient thermal energy may become greater than the energy barrier separating the two distinct magnetic states (two different magnetic moment orientations of the single-domain particle). If the size of the magnetic particle is too small (~ the order of 10 nm) such that the thermal energy from the surrounding is greater than its energy barrier (which is proportional

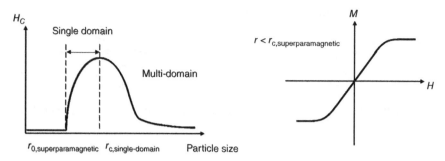

Figure 2.30 Coercive field of a magnetic particle as a function of its size. The corresponding *M-H* loop for a particle size smaller than the superparamagnetic limit is also shown.

to the size of the magnet), then the particle will behave like a paramagnetic material, since thermal agitation will cause the jump between the two magnetic states. This is known as superparamagnetism. Therefore, a reasonable design of the size r of the magnetic particle or magnetic memory element is that $r_{0,\,\text{superparamagnetic}} < r < r_{c,\,\text{single-domain}}$, as indicated in Figure 2.30.

Homework

Q2.1 The ferromagnetic resonance (FMR) corresponds to the excitation of magnetic precession in real ferromagnetic materials. FMR is actually related to both the applied field B and the demagnetization field $\mu_0 H_d$ inside the material. For a ferromagnetic thin film, $\mu_0 H_d = \mu_0 M$. In 1948, Charles Kittel provided a formula to describe the resonance (precession) condition for magnetic thin films [22]:

$$\omega = \gamma\sqrt{B(B + \mu_0 M)}. \tag{2.74}$$

Consider that you have a sample of Co thin film with $M = 1000$ emu/cm^3. If you have an electromagnet that can provide magnetic field from 0 up to 0.2 Tesla ($0 < B < 0.2$ T), then what's the frequency range (in units of Hz) of seeing FMR?

A2.1: $M = 1000$ emu/cm^3 = 10^6 A/m. $\omega = 2\pi f = \gamma\sqrt{B(B + \mu_0 M)} \Rightarrow$
$f = (\gamma/2\pi)\sqrt{B(B + \mu_0 M)}$. $B = 0$–2 Tesla corresponds to $f = 0$–7.55 GHz.

Q2.2 Estimate the Bohr magneton, which is the magnitude (SI units) of magnetic moment from an electron orbiting around the hydrogen nucleus, through a classical approach. (Hint: Some quantities might be useful during the calculations, for instance, electron charge $e = 1.6 \times 10^{-19}$ C, electron mass $m_e = 9.1 \times 10^{-31}$ kg, Bohr radius $r_0 = 0.53$ Å, and the binding energy of a hydrogen atom E = 13.6 eV.)

A2.2 $\mu = IA = efA = e(\omega/2\pi)\pi r_0^2 = evr_0/2 = e\left(\sqrt{2E/m_e}\right)r_0/2$

$= 9.274 \times 10^{-24}$ A $\cdot$ m^2

Q2.3 Followed by previous problem, consider a material made up of such kind of atomic magnetons, what will be the magnetization value in terms of A/m and emu/cm^3, respectively? Please give a rough estimation. (Hint: How many atoms are there in a cubic meter space?) Does the number close to the saturation magnetization values that you can find for typical ferromagnetic materials?

A2.3 Typically the density of atoms in solids is $n_V \sim 10^{28}$ to 10^{29} m^{-3}. Magnetization can be calculated by $M = n_V \cdot \mu \approx 10^{28(29)} \cdot 9.274 \times 10^{-24}$ A/m $\approx 10^{5(6)}$ A/m $\approx 10^{2(3)}$ emu/cm^3. This is close to the measured values of saturation magnetizations in Fe (1707 emu/cm^3), Co (1400 emu/cm^3), and Ni (485 emu/cm^3).

Q2.4 Consider a typical potential energy for the electron in a hydrogen atom: U $(r) = -e^2/4\pi\varepsilon_0 r$. Please give a rough estimation of the magnitude of ΔE_{SO} for a p electron (such that orbital quantum number is not zero) in a hydrogen atom. Please write down your estimation in units of both eV and J (Hint: Do not worry about the exact form or number of $\langle 1/r^n \rangle$, $\langle S \rangle$, and $\langle L \rangle$; just consider their orders of magnitude for this problem. $\langle r \rangle \sim$ Bohr radius, $\langle S \rangle \sim \langle L \rangle \sim \hbar$.)

A2.4 $\Delta E_{SO} \approx \frac{e^2 \hbar^2}{8\pi\varepsilon_0 m_e^2 c^2 a_0^3} \approx 1.2 \times 10^{-22}$ J $\approx 7.5 \times 10^{-4}$eV.

Q2.5 Determine the values of S, L, and J for the ground states of isolated Fe^{3+} and Fe^{2+} ions using Hund's rules. Further determine their effective magnetic moments per ion, μ_{eff}, and the maximum allowed μ_z. Now, consider the fact that orbital quenching exists and $L = 0$ for both ions, determine their μ_{eff} and maximum μ_z.

A2.5

Scenario 1 (without orbital quenching):

	Fe^{3+}	Fe^{2+}
3d shell	3d^5	3d^6
S, L, J	5/2, 0, 5/2	2, 2, 4
μ_{eff}	5.92 μ_{B}	6.71 μ_{B}
μ_z^{max}	5 μ_{B}	6 μ_{B}

Scenario 2 (with orbital quenching):

	Fe^{3+}	Fe^{2+}
3d shell	3d^5	3d^6
S, L, J	5/2, 0, 5/2	2, 0, 2
μ_{eff}	5.92 μ_{B}	4.90 μ_{B}
μ_z^{max}	5 μ_{B}	4 μ_{B}

Q2.6 Show that at $T > T_C$, when a small magnetic field is applied, the magnetic susceptibility follows the Curie–Weiss law.

A2.6 When the applied field is small,

$$M = M_s \widetilde{B}_J(x) \approx M_s x(J+1)/3J = M_s g_J(J+1)\mu_B\mu_0\left(H_{\text{applied}} + \lambda M\right)/3k_B T$$

$$\Rightarrow M = n_V g_J \mu_B J \cdot g_J(J+1)\mu_B\mu_0\left(H_{\text{applied}} + \lambda M\right)/3k_B T = \frac{C}{T}\left(H_{\text{applied}} + \lambda M\right)$$

$$\Rightarrow \chi \equiv \frac{M}{H_{\text{applied}}} = \frac{C}{T} \cdot \frac{T}{T - \lambda C} = \frac{C}{T - T_C}.$$

Q2.7 Let us assume $S = 1$ for BCC Fe (α-Fe). Based on the observed $T_C = 1043$ K, what is the magnitude of J (in units of J and eV)?

A2.7 For a BCC structure, the coordination number $z = 8$. Using the provided $S = 1$ and $T_C = 1043$ K, J is calculated to be 1.35×10^{-21} J (8.44×10^{-3} eV). In fact, this magnitude is much greater than the dipolar interaction energy ($\sim 10^{-6}$ eV) of two spin moments separated by a regular inter-atomic distance, which indicates that exchange interaction is dominating over dipolar interaction in determining the magnetic order in materials.

Q2.8 When ferromagnetic layer CoFeB thin film is sandwiched between Ta and MgO, the CoFeB layer could possess PMA if the thickness t_{CoFeB} is less than

2 nm. As shown in Figure 2.31a and b, the presence of a Ta insertion layer will have a dramatic influence on anisotropy: Without Ta insertion, the CoFeB layer has in-plane magnetic anisotropy, while with Ta the anisotropy becomes out of plane. Assume that these two representative films both have uniaxial anisotropy with the measured $M_s = 1280\,\text{emu/cm}^3$ for CoFeB; please estimate anisotropy energy density constant K (in units of erg/cm^3) for both films using the M-H curves provided.

A2.8 K (without Ta) $\sim -2.24 \times 10^6\,\text{erg/cm}^3$ and K (with Ta) $\sim 1.28 \times 10^6$ erg/cm^3. Note that negative K corresponds to in-plane anisotropy, while positive K corresponds to PMA.

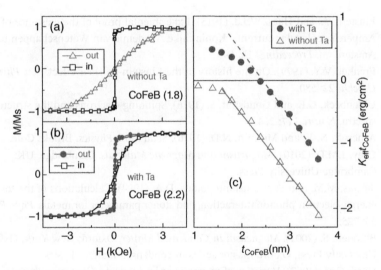

Figure 2.Q8 (a) $MgO(5)/Co_{40}Fe_{40}B_{20}(1.8)/MgO(5)/Ta(5)$ and (b) $MgO(5)/Co_{40}Fe_{40}B_{20}$ $(1.1)/Ta(0.2)/Co_{40}Fe_{40}B_{20}(1.1)/MgO(5)/Ta(5)$. (c) Dependences of K_{eff} t_{CoFeB} on t_{CoFeB} for the samples $MgO(5)/Co_{40}Fe_{40}B_{20}(t_{CoFeB})/MgO(5)/Ta(5)$ and with and without the 0.2 nm Ta insertion layer in the middle of the CoFeB layers. *Source:* Adapted from [23].

Q2.9 The K that you calculated in previous problem is typically called effective perpendicular anisotropy energy density and can be expressed as (in cgs units)

$$K_{\text{eff}} = \left(K_V - 2\pi M_s^2\right) + \frac{2K_s}{t_{\text{CoFeB}}}, \tag{2.75}$$

where K_V is the CoFeB volume anisotropy energy density, and in this case $K_V \sim 0$. K_s is the interfacial anisotropy contribution. Please explain why we typically plot $K_{eff}t_{CoFeB}$ as a function of t_{CoFeB} (Figure 2.31c) rather than K_{eff} versus t_{CoFeB}. What information can we extract from this type of graph?

A2.9 By plotting $K_{eff}t_{CoFeB}$ as a function of t_{CoFeB}, one can extract demagnetization field of the film from the slope and the interfacial anisotropy energy density K_s from the y-intercept.

References

1 Einstein, A. and Haas, W.J.d. (1915). Experimental proof of the existence of Ampère's molecular currents, Koninklijke Akademie van Wetenschappen te Amsterdam. *Proceedings* 18: 696.
2 Frenkel, V.Y. (1979). On the history of the Einstein-de Haas effect. *Sov. Phys. Uspekhi* 22: 580.
3 Uhlenbeck, G.E. and Goudsmit, S. (1926). Spinning electrons and the structure of spectra. *Nature* 117: 264.
4 Ashcroft, N.W. and Mermin, N.D. (1976). *Solid State Physics*. Brooks Cole.
5 Coey, J.M.D. (2010). *Magnetism and Magnetic Materials*. Cambridge, UK: Cambridge University Press.
6 Sigalas, M.M. and Papaconstantopoulos, D.A. (1994). Calculations of the total-energy, electron-phonon interaction, and stoner parameter for metals. *Phys. Rev. B.* 50: 7255.
7 Blundell, S. (2001). *Magnetism in Condensed Matter*. Oxford; New York: Oxford University Press, Oxford master series in condensed matter physics.
8 Hurd, C.M. (1982). Varieties of magnetic order in solids. *Contemp. Phys.* 23: 469.
9 Anderson, P.W. (1950). Antiferromagnetism. Theory of Superexchange interaction. *Phys. Rev.* 79: 350.
10 O'Handley, R.C. (2000). *Modern Magnetic Materials: Principles and Applications*. New York: Wiley.
11 Parkin, S.S.P. and Mauri, D. (1991). Spin engineering: direct determination of the Ruderman-Kittel-Kasuya-Yosida far-field range function in ruthenium. *Phys. Rev. B.* 44: 7131.
12 Majkrzak, C.F., Cable, J.W., Kwo, J. et al. (1986). Observation of a magnetic Antiphase domain structure with long-range order in a synthetic Gd-Y superlattice. *Phys. Rev. Lett.* 56: 2700.
13 Dzyaloshinsky, I. (1958). A thermodynamic theory of "weak" ferromagnetism of antiferromagnetics. *J. Phys.Chem. Solids* 4: 241.

14 Moriya, T. (1960). Anisotropic Superexchange interaction and weak ferromagnetism. *Phys. Rev.* 120: 91.

15 Fert, A., Cros, V., and Sampaio, J. (2013). Skyrmions on the track. *Nat. Nanotechnol.* 8: 152.

16 Emori, S., Bauer, U., Ahn, S.M. et al. (2013). Current-driven dynamics of chiral ferromagnetic domain walls. *Nat. Mater.* 12: 611.

17 Ryu, K.S., Thomas, L., Yang, S.H., and Parkin, S. (2013). Chiral spin torque at magnetic domain walls. *Nat. Nanotechnol.* 8: 527.

18 Yang, H.X., Chshiev, M., Dieny, B. et al. (2011). First-principles investigation of the very large perpendicular magnetic anisotropy at Fe|MgO and Co|MgO interfaces. *Phys. Rev. B.* 84: 054401.

19 Ikeda, S., Miura, K., Yamamoto, H. et al. (2010). A perpendicular-anisotropy CoFeB-MgO magnetic tunnel junction. *Nat. Mater.* 9: 721.

20 Lin, C.J., Gorman, G.L., Lee, C.H. et al. (1991). Magnetic and structural properties of Co/Pt multilayers. *J. Magn. Magn. Mater.* 93: 194.

21 Carcia, P.F. (1988). Perpendicular magnetic anisotropy in Pd/Co and Pt/Co thin-film layered structures. *J. Appl. Phys.* 63: 5066.

22 Kittel, C. (1948). On the theory of ferromagnetic resonance absorption. *Phys. Rev.* 73: 155.

23 Liu, T., Cai, J.W., and Sun, L. (2012). Large enhanced perpendicular magnetic anisotropy in CoFeB/MgO system with the typical Ta buffer replaced by an Hf layer. *AIP Adv.* 2: 032151.

3

Magnetic Thin Films

3.1 Introduction

In the first two chapters we covered basic electromagnetism, magnetic properties, and the origins of magnetism in different types of material systems. However, for newcomers and the engineers-would-be in the field of magnetoresistive random-access memory (MRAM), knowledge from the experimental side of magnetism is also indispensable. For MRAM applications, magnetic materials in the thin film form is omnipresent. To develop thin film stacks with desirable magnetic and electrical properties, researchers typically have to start from thin film deposition steps (materials growth) and then choose suitable measurement protocols to characterize the magnetic properties of the deposited films (materials characterization). If the measured results are not desirable, then one can adjust the film growth parameters and make sure satisfactory results can be produced. In this chapter, we will briefly introduce some common magnetic materials growth and characterization approaches, especially for thin film forms.

3.2 Magnetic Thin Film Growth

Many common thin film growth methods can be adopted to deposit magnetic thin films or their heterostructures. Most of them involve the usage of high vacuum or ultra-high vacuum chambers to achieve well-controlled growth of layer stacks. For example, thermal evaporation, sputter deposition, molecular beam epitaxy, etc., all can be employed to grow high-quality magnetic films for various applications. We will introduce briefly two important approaches for modern magnetic thin film research.

Magnetic Memory Technology: Spin-Transfer-Torque MRAM and Beyond,
First Edition. Denny D. Tang and Chi-Feng Pai.
© 2021 The Institute of Electrical and Electronics Engineers, Inc.
Published 2021 by John Wiley & Sons, Inc.

3.2.1 Sputter Deposition

Sputtering is a type of physical vapor deposition (PVD) technique to deposit thin film. It has been widely used in industrial-level manufacturing, especially for preparing high-quality films with ~ nm scale of thickness. As shown in Figure 3.1, in a high vacuum chamber with typical base pressure $< 10^{-7}$ Torr, a strong direct current (DC) or alternating current (AC) (radio frequency, RF) voltage is applied between the target and the substrate to ionize inert gases (usually Ar) being flowed into the chamber. After ionization, positive Ar ions are attracted by cathode (which usually locates at the target position) and accelerate toward targets. The bombardment between energetic Ar ions and target atoms allows the target atoms to be ejected from the target. After traveling some distances, the escaped target atoms reach the substrate and start to condense into film. As more and more atoms reach the substrate, they start to bind with each other and form a tightly bound atomic layer. By controlling the sputtering time, atomic-layer precise thin film can be obtained. Typical growth (deposition) rate can be as low as several nanometers per minute. RF voltage is usually used when the target has a poor conductivity (some semiconductors and insulators). By using a high-frequency oscillation, charges accumulating at the target can be neutralized and thus can avoid charging on target. However, using RF sputtering will lead to a lower sputtering rate.

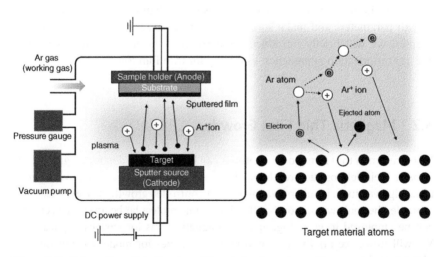

Figure 3.1 Schematics of sputter deposition. A simple system with one vacuum chamber and one sputter source is shown. The vacuum system depends on the vacuum requirement, which can be a turbo molecular pump or a cryogenic pump with a rotary roughing pump.

(a)   (b)

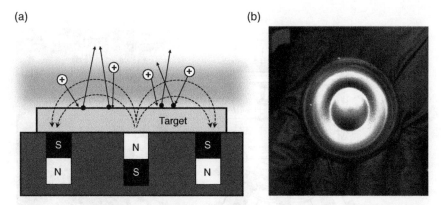

Figure 3.2 (a) Schematics of magnetron sputtering. (b) An example of using a 2-in. target. It can be seen that a ring-like region has been consumed more than the other part.

Compared to conventional sputtering, magnetron sputtering places magnets under the target to improve the sputtering efficiency. For example, the strong magnetic field from the magnet confines the electron into a spiral motion around the target, as shown in Figure 3.2a. The confinement of electron around the target not only leads to a higher density of plasma, and thus a higher deposition rate, but also prevents damage from electrons impacting the substrate or the growing film. Note that to sputter ferromagnetic materials and their alloys, stronger permanent magnets have to be used to enhance the magnetic flux; otherwise, the magnetic field will not be strong enough to penetrate through the target, and the plasma will therefore be hard to ignite. Note that due to the configuration of the magnetic field lines, only a portion of the target material will be sputtered, as shown in Figure 3.2b.

A modern-day sputtering system consists of many ingenious designs to facilitate the sputtering deposition process. For example, a commercial system has a control panel to let users set up their own deposition parameters such as working gas flow, deposition power, substrate temperature, etc. The combination of all these materials growth parameters is called a "recipe" for a specific growth. Automatic sample transferring systems are also getting more common these days, in which the sample can be loaded in a chamber with a high vacuum (called load lock) and then be transferred to the ultra-high vacuum's main chamber for materials deposition. Figure 3.3 shows an example. Depending on the chamber size and the magnetron sputter source size, the sputter deposition can be uniform across a broken piece of wafer or up to 12 in. wafers for industrial-level tools (Figure 3.4). A nice review on the development of sputtering deposition technology is given by J.E. Greene [1].

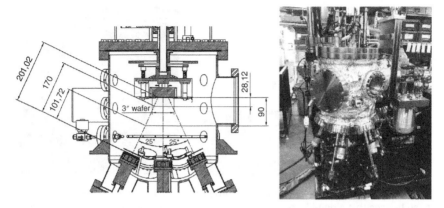

Figure 3.3 A typical multisource (gun) magnetron sputter system in modern labs. The system shown consists of a main chamber and a load lock design to facilitate sample loading.

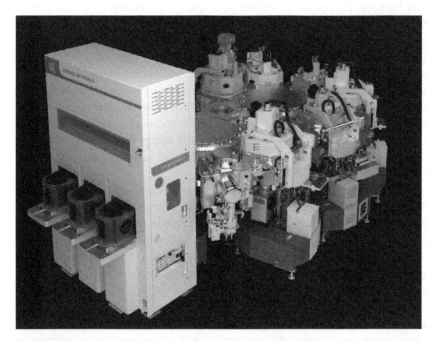

Figure 3.4 Industrial-level sputter deposition system (ENDURA® CLOVER™ MRAM PVD from Applied Materials) for making high-quality magnetic tunnel junction (MTJ) thin films on 12-in. wafers. *Source:* Figure reproduced from www.appliedmaterials.com/products/MRAM.

3.2.2 Molecular Beam Epitaxy (MBE)

Although modern sputter deposition systems are already feasible to provide high-quality thin films with the precision of ~0.1 nm, most of them are not yet "epitaxial," which means that they are not single crystalline. To further produce single-crystalline, well-textured, epitaxial arrangements of atomic layers, a common alternative is the so-called molecular beam epitaxy (MBE), which was invented by Alfred Y. Cho and John R. Arthur, Jr. around 1968 (see [2]). In this type of deposition approach, an even higher vacuum than the sputter case is required, typically with 10^{-9} to 10^{-12} Torr of base pressure (ultra-high vacuum). This can result in high-purity epitaxial films. Figure 3.5 shows a schematic illustration of the working principles of MBE. Effusion cells (also known as Knudsen cells) are used to heat up the materials until they sublime, which form the molecular beams. The temperature requirement to achieve stable molecular flow therefore depends on the materials of interest. The gaseous materials then condensate onto substrate and form an atomic-layer thin film. During deposition, the epitaxial growth of materials is typically monitored by reflection high-energy electron diffraction (RHEED). The growth rate of such a deposition approach can be as low as a monolayer (ML) per second. Note that unlike sputter deposition, there is no working gas being introduced into the system.

While most of the recent magnetic thin film studies, including industrial-level MRAM layer stack growth, are achieved via sputter deposition, MBE provides opportunities to study materials and their heterostructures with even better interfacial quality. For example, one of the first RKKY studies based on a GdY superlattice system [3] (see Chapter 2) and high tunneling magnetoresistance tunnel junction [4]

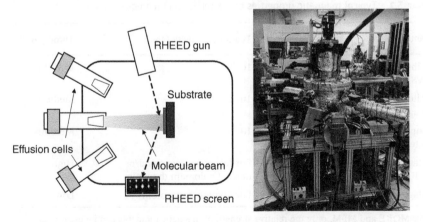

Figure 3.5 Schematics of molecular beam epitaxy and a representative system. *Source:* Picture courtesy of Prof. Jung-Chun Huang of National Cheng-Kung University, Tainan, Taiwan.

(see Chapter 4) were achieved by MBE growth. For nonconventional materials systems such as topological insulators [5–7], a single crystalline of materials growth is required; therefore, MBE is the standard tool for such sample preparation.

3.3 Magnetic Thin Film Characterization

It is of great importance to develop suitable experimental methods to measure magnetization with reasonable precision, especially for thin films. Note that for MRAM applications, the ferromagnetic layers in the multilayer stacks typically have a thickness of 1–10 nm, which means that the overall magnetization is extremely small ($\sim 10^{-7}$ to 10^{-5} emu or less, depending on the size of the tested sample). To detect such minute magnetization, special types of measurement techniques are required. In this section, we will introduce one of the most commonly employed magnetometry (the term means that one can directly measure the value of magnetization) approach, namely, vibrating sample magnetometer (VSM). Readers who are interested in other types of magnetometry measurements can refer to the comprehensive book by Coey (see Chapter 10 therein) [8]. A magneto-optical approach for magnetic thin film characterization will also be discussed, though typically only the direction of the magnetization can be detected. Nevertheless, while compared to magnetometry, the optical characterization of magnetic thin films allows for a fast-track, high-throughput inspection of several key magnetic properties of the layer stack of interest. It is therefore favorable for industrial in-line testing, especially when large area wafers are the main subjects of study. Table 3.1 summarizes some common characterization approaches.

Table 3.1 Typical magnetic properties characterization techniques.

Method	Typical sensitivity (emu)[a]	Throughput[b]
Vibrating sample magnetometer (VSM)	10^{-6}	Medium
Alternating gradient force magnetometer (AGM)	10^{-7}	Medium
Superconducting quantum interference device (SQUID)	10^{-8}	Medium
Magneto-optical Kerr effect (MOKE)	For hysteresis loops and/or magnetic domain dynamics	High
Magnetic force microscopy (MFM)	For domain pattern only	Low

[a] For MOKE and MFM, only the relative strength of magnetic moments can be quantified.
[b] The throughput depends on some user-determined parameters such as the data acquisition speed.

3.3.1 Vibrating-Sample Magnetometer (VSM)

VSM is an instrument to measure magnetic properties of magnetic materials in different forms (particles, bulks, or thin films), for example, saturation magnetization, hysteresis loop, anisotropy field, etc. Figure 3.6 shows a simple sketch of VSM. When a sample is placed in a uniform magnetic field H, a magnetization M will be induced in the sample. The vibration of the sample will cause a time-dependent magnetic flux change, which will further induce electric signals in the pickup coil proportional to the magnetization of sample based on Faraday's law of induction. The first VSM was designed and demonstrated by Foner [9], in which a loudspeaker transducer was used to generate the vibration of the tested samples. Although the concept of VSM seems to be simple, extra care such as choosing suitable vibration frequency and designs of pickup coils is needed to improve the sensitivity, which can reach as fine as $\sim 10^{-6}$ emu.

Also note that other types of magnetometers (less common and typically more expensive) can provide higher sensitivities than VSM, such as alternating gradient force magnetometer (AGM) and superconducting quantum interference device (SQUID). AGM utilizes an alternating magnetic field gradient to exert an alternating force on the sample. The force will be proportional to the magnitude of the gradient field as well as the magnetic moment of the sample. This resulting force is further detected through a piezoelectric sensing component for further analysis. The sensitivity of an AGM system can reach as fine as $\sim 10^{-7}$ emu. SQUID, on the other hand, adopts the Josephson junctions to detect small magnetic flux

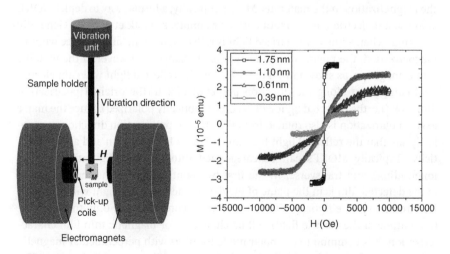

Figure 3.6 Schematics of a VSM and a representative dataset of VSM-obtained M versus H results from a series of W/CoFeB(t)/MgO samples as prepared by magnetron sputtering.

variations. Since the Josephson effect relies on the existence superconductivity, such measurement unit typically needs to be soaked in liquid Helium to maintain its function. Typical commercial SQUID products can be used to determine magnetization of samples with sensitivity down to 10^{-8} emu.

An example of VSM-obtained M versus H plot is shown in Figure 3.6, which includes the results from a series of W/CoFeB/MgO heterostructures with different CoFeB thicknesses. One can see that as the CoFeB thickness increases from 0.39 nm to 1.75 nm, the total magnetic moment increases from 0.5×10^{-5} emu to 3.2×10^{-5} emu accordingly. The hard-axis behavior observed from these films indicates that samples in this series have perpendicular magnetic anisotropy; for example, the film with 1.1 nm of CoFeB needs around 4000 Oe to saturate its magnetization along in-plane direction.

3.3.2 Magneto-Optical Kerr Effect (MOKE)

The magneto-optical Kerr effect (MOKE) describes the phenomenon of polarization change or intensity change of an electromagnetic wave (light) upon reflection from a material surface. For metallic thin films, since the incident light will interact with only the top surface (~10 nm), this effect is also known as surface MOKE or SMOKE. Therefore, unlike magnetometry, which can measure the bulk magnetic properties and the magnetic moments of the whole sample, MOKE is sensitive only to the top surface of the detected films, depending on the penetration depth of the particular light in the materials in use. Microscopically, the origin of MOKE is the spin-orbit interaction between the electric field of the light and the magnetization in the materials. Macroscopically, a formal way to depict MOKE is to use a dielectric tensor treatment; for example, see Zak et al. [10]. Generally speaking, when a linearly polarized light is reflected off a metallic surface without magnetization, the polarization direction is unchanged. However, if the metallic film is magnetic, as shown in Figure 3.7, then the reflected light will gain electric field component along the direction perpendicular to the original polarization direction (i.e. the reflected light will become elliptically polarized). Since the major axis of polarization is also rotated from the original polarization direction, we typically say that the reflected light has gained both a Kerr rotation and a Kerr ellipticity. Typically MOKEs can be categorized into three configurations: polar, longitudinal, and transverse. In the first two configurations, the magnetization of the detected film is in the plane of incidence; therefore, the light will gain both Kerr rotation and ellipticity. For the transverse configuration, only the intensity (magnitude of the electric field) will be changed. For magnetic thin film characterization, it is common to use polar mode for films with perpendicular magnetic anisotropy and to use longitudinal mode for those with in-plane anisotropy. The polarization change of the reflected light can be analyzed by introducing an extra polarizer in the optical path.

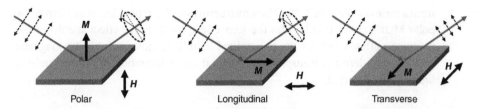

Figure 3.7 Schematic illustrations of different types of MOKE measurement. *M* represents the thin film magnetization direction, and *H* represents the applied field direction. The dashed lines represent polarization direction of the electric field of propagating electromagnetic waves.

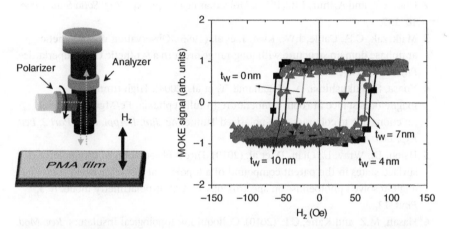

Figure 3.8 An example of MOKE setup (with an LED light source and a CCD detector) and the measured MOKE signals from W(t_W)/CoFeB(1 nm)/MgO(2 nm) films with perpendicular magnetic anisotropy (PMA).

The MOKE signals can be measured by using either a laser with a photodetector (single spot) or an LED light source with a microscope (wide-field, direct imaging). The former one can be used to obtain magnetic properties from a relatively small region on the film or patterned devices (for example, see Emori et al. [11]), depending on the spot size of the laser, whereas the latter is more suitable to study magnetic domain pattern evolution or domain wall motion (see Je et al. [12]). Since typically no mechanical parts are involved in the detection scheme and the films are not necessary to be cut into small pieces, data acquisition by MOKE can have a higher throughput than by using conventional magnetometers. However, to increase signal-to-noise ratio, it is also not uncommon that extra optical components such as optical chopper and modulator are used in MOKE setups. Figure 3.8 shows representative results of out-of-plane hysteresis loops obtained from a series of W/CoFeB/MgO heterostructures with perpendicular anisotropy by polar MOKE. Note that besides the Kerr effect, the Faraday effect stands for the magneto-optical effect for the transmitted light. It is most useful if the magnetic material of interest is transparent, such that the polarization of the transmitted light can be detected.

References

1 Greene, J.E. (2017). Review article: tracing the recorded history of thin-film sputter deposition: from the 1800s to 2017. *J. Vac. Sci. Technol. A* 35: 05C204.

2 Cho, A.Y. and Arthur, J.R. (1975). Molecular beam epitaxy. *Prog. Solid State Chem.* 10: 157.

3 Majkrzak, C.F., Cable, J.W., Kwo, J. et al. (1986). Observation of a magnetic antiphase domain structure with long-range order in a synthetic Gd-Y superlattice. *Phys. Rev. Lett.* 56: 2700.

4 Yuasa, S., Fukushima, A., Nagahama, T. et al. (2004). High tunnel magnetoresistance at room temperature in fully epitaxial Fe/MgO/Fe tunnel junctions due to coherent spin-polarized Tunneling. *Jpn. J. Appl. Phys. Part 2: Lett. Exp. Lett.* 43: L588.

5 Hsieh, D., Wray, L., Qian, D. et al. (2010). Direct observation of spin-polarized surface states in the parent compound of a topological insulator using spin- and angle-resolved photoemission spectroscopy in a Mott-polarimetry mode. *N. J. Phys.* 12.

6 Hasan, M.Z. and Kane, C.L. (2010). Colloquium: topological insulators. *Rev. Mod. Phys.* 82: 3045.

7 Chen, Y.L., Analytis, J.G., Chu, J.H. et al. (2009). Experimental realization of a three-dimensional topological insulator, Bi2Te3. *Science* 325: 178.

8 Coey, J.M.D. (2010). *Magnetism and Magnetic Materials*. Cambridge, UK: Cambridge University Press.

9 Foner, S. (1959). Versatile and sensitive vibrating-sample magnetometer. *Rev. Sci. Instrum.* 30: 548.

10 Zak, J., Moog, E.R., Liu, C., and Bader, S.D. (1990). Universal approach to magneto-optics. *J. Magn. Magn. Mater.* 107.

11 Emori, S., Bauer, U., Ahn, S.M. et al. (2013). Current-driven dynamics of chiral ferromagnetic domain walls. *Nat. Mater.* 12: 611.

12 Je, S.-G., Kim, D.-H., Yoo, S.-C. et al. (2013). Asymmetric magnetic domain-wall motion by the Dzyaloshinskii-Moriya interaction. *Phys. Rev. B* 88: 214401.

4

Magnetoresistance Effects

4.1   Introduction

In Chapter 3 we addressed the physical properties of magnetic materials, especially for them in the thin film forms, prepared either by sputter deposition or by epitaxial growth. In fact, even richer physics can be discovered from these materials systems when we pass electron flows through them. These extra physical properties or the so-called transport properties (since the electrons or charge carriers are transporting in the materials) that are related to the electrical resistance of these materials in the presence of magnetic field or magnetization are typically coined as different types of magnetoresistance (MR). MR behavior and the corresponding MR ratio, which is defined as the percentage of change of MR value with respect to applied magnetic field, can be very different among different materials systems. For example, as shown in Table 4.1, the earliest found anisotropic MR (AMR) in ferromagnetic (FM) materials system can reach a typical MR ratio of ~1%. However, by suitable film stack engineering, the MR ratio can reach greater than 100% for the tunneling MR (TMR) case.

Finding materials systems or thin film layer stack structures with a large MR ratio is the key to develop workable magnetoresistive random-access memory (MRAM), since the large MR effect can improve the memory readout. For example, for state-of-the-art industrial-level spin-transfer torque magnetic random-access memory (STT-MRAM), the TMR ratio in such devices can reach as high as 200%. In this chapter, we will go through the basics of three types of MR effects that can be found in ferromagnetic materials and their heterostructures, namely, AMR, giant magnetoresistance (GMR), and TMR. These three MR effects are the most relevant and fundamental to MRAM as well as the related spintronics applications that we see these days.

Magnetic Memory Technology: Spin-Transfer-Torque MRAM and Beyond,
First Edition. Denny D. Tang and Chi-Feng Pai.
© 2021 The Institute of Electrical and Electronics Engineers, Inc.
Published 2021 by John Wiley & Sons, Inc.

Table 4.1 Comparison of different magnetoresistances.

Magnetoresistance	Materials system	Typical MR ratio	Origin	Reference(s)
Ordinary MR	NM	Field-dependent	Lorentz force	[1, 2]
Anisotropic MR (AMR)	FM	~1%	Spin-orbit interaction	[3]
Giant MR (GMR)	FM/NM/FM	~ 50%	Spin-dependent transmission	[4, 5]
Tunneling MR (TMR)	FM/I/FM	≧100%	Spin-dependent tunneling	[6, 7]
Colossal MR (CMR)	Mn-based perovskite oxide	~100 000%	Double-exchange and hopping (metal-to-insulator transition)	[8]
Spin Hall MR (SMR)	NM/FMI	~ 0.001% to 1%	Spin Hall effect	[9–11]
Unidirectional SMR	NM/FM	~ 0.001%	Spin Hall effect	[12]

FM, FMI, NM, and I stand for ferromagnetic metal, ferromagnetic insulator, normal metal (nonmagnetic), and insulator, respectively.

4.2 Anisotropic Magnetoresistance (AMR)

In a ferromagnetic material, the measured longitudinal resistance or, resistivity ρ, will vary depending on the angle φ between the applied current direction and the magnetization direction. This AMR effect was first discovered by William Thomson, later known as Lord Kelvin. The origin of the AMR is spin-orbit interaction, which connects the transport properties of conducting electrons (itinerant electrons) in the material to the magnetization (localized electrons). In general cases, ρ depends on φ as

$$\rho(\varphi) = \rho_\perp + \left(\rho_\parallel - \rho_\perp\right)\cos^2\varphi, \tag{4.1}$$

where $\rho_\perp$ and $\rho_\parallel$ represent the resistivities of the material when the magnetization is perpendicular and parallel to the current, respectively. Typical AMR ratio of common ferromagnetic materials, which can be expressed as

$$\frac{\Delta R}{R} = \frac{R_\parallel - R_\perp}{R_\parallel} = \frac{\rho_\parallel - \rho_\perp}{\rho_\parallel}, \tag{4.2}$$

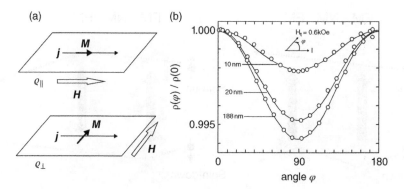

Figure 4.1 (a) Schematic illustration of AMR measurement. (b) AMR of Co with three different thicknesses. *Source:* Figure reproduced from Gil et al. [3].

which is about a few percent or less. Figure 4.1 shows an example. Note that due to the resistance or resistivity variation with respect to the external magnetic field, AMR can be employed in field sensors and hard-disk drive (HDD) read head.

4.3 Giant Magnetoresistance (GMR)

A ferromagnetic metal/normal metal/ferromagnetic metal (FM/NM/FM) sandwich structure, with the magnetization direction in both FM layers, can be controlled by external magnetic field. The electrical resistance across this trilayer structure is a variable of the relative configuration of the magnetization in the two FM layers. When the magnetization directions in the two FM layers are parallel to each other (parallel, or P, configuration), the resistance will be lower than that for the magnetization direction antiparallel to each other (antiparallel, or AP, configuration). This is the simplest phenomenological picture of the famous GMR effect. The magnitude of the GMR of a particular FM/NM/FM heterostructure can be characterized by the GMR ratio $(R_{AP} - R_P)/R_{AP}$, where R_{AP} and R_P stand for resistances of AP state and P state, respectively.

A pedagogic way to explain GMR effect is to treat the first FM layer that electrons encounter as a spin polarizer; i.e. the conduction electrons become (partially) spin-polarized after passing through the first FM layer. The spin-polarized electrons then go through the NM spacer without significant scattering. While impinging onto the second FM layer, the electrons with spin orientation aligned parallel to the local magnetic moment will have higher probability to pass through the FM layer. In contrast, the electrons with spin

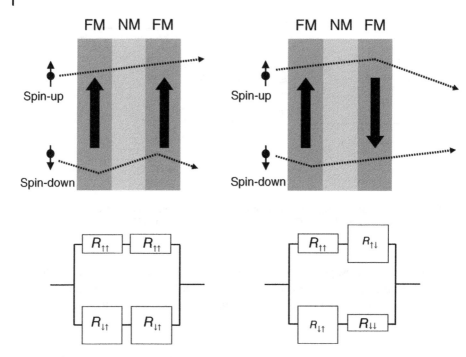

Figure 4.2 Illustration of GMR based on the two-channel model.

orientation opposite to the local moment will tend to encounter more scattering. Figure 4.2 illustrates this.

A two-channel model, which considers the spin-up and spin-down electrons as charge carriers in two parallel circuits (also shown in Figure 4.2), provides a semi-quantitative perspective of understanding GMR. When spin-up (down) electrons transport through the FM layer with a magnetization up (down) state, the channel resistance is $R_{\uparrow\uparrow}(=R_{\downarrow\downarrow})$. On the other hand, when spin-up (down) electrons transport through the FM layer with a magnetization down (up) state, the channel resistance is $R_{\uparrow\downarrow}(=R_{\downarrow\uparrow})$, and typically $R_{\uparrow\downarrow} > R_{\uparrow\uparrow}$. A little math will give us the GMR ratio as

$$\frac{\Delta R}{R} = \frac{R_{AP} - R_P}{R_{AP}} = \frac{\left(R_{\uparrow\downarrow} - R_{\uparrow\uparrow}\right)^2}{\left(R_{\uparrow\downarrow} + R_{\uparrow\uparrow}\right)^2}. \tag{4.3}$$

A more detailed theory of GMR in a multilayer system regarding the transport of conduction electrons in diffusive regime (Boltzmann equation approach) was proposed by Valet and Fert [13].

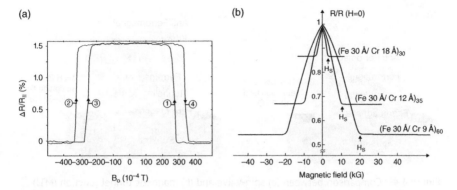

Figure 4.3 Experimental evidence of GMR in (a) Fe/Cr/Fe sandwich structure and in (b) Fe/Cr superlattices. *Source:* Figures reproduced from Binasch et al. [5] and Baibich et al. [4].

The experimental evidence of GMR was first given by Binasch et al. in the Fe/Cr/Fe system [5] and by Baibich et al. in the Fe/Cr superlattice structure [4], as shown in Figure 4.3. Depending on the materials system being measured and the configuration of measurements, namely, current-in-plane (CIP) configuration or current-perpendicular-to-plane (CPP) configuration, the GMR ratio can vary from a few percent to a few tens of percent [14]. The FM/NM/FM structures that utilize CPP configuration are also known as "spin-valves" since the two-terminal resistance of the device can be switched between high-resistance (AP configuration) and low-resistance states (P configuration) with the application of external magnetic field. Later, GMR proved itself to be a powerful means of sensing magnetic field, which led to its revolution in HDD read head and the earliest MRAM technologies.

4.4 Tunneling Magnetoresistance (TMR)

Similar to GMR in FM/NM/FM sandwich structures (spin-valves), FM/insulator/FM magnetic tunnel junctions (MTJs, see Figure 4.4) could also possess an MR effect via quantum mechanical tunneling between two FM electrodes with a suitable insulator as a tunnel barrier. The resulting effect is the so-called TMR, which typically has a TMR ratio an order of magnitude larger than GMR with suitable materials combinations and film stack engineering.

TMR of around 14% was first experimentally discovered by Julliere in 1975 in a Fe/(Oxidized-)Ge/Co system at cryogenic temperature [16], even earlier than the

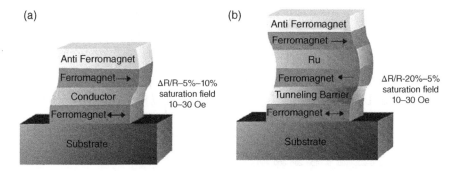

Figure 4.4 Comparison between (a) spin-valve and (b) magnetic tunnel junction (MTJ) devices. The anitferromagnetic layer in both structures serves as a pinning layer. The FM/Ru/ FM layer in (b) is called the synthetic antiferromagnetic (SAF) layer, which employs the RKKY interaction. *Source:* Figures reproduced from Wolf et al. [15].

discovery of GMR. Julliere's theory further suggests that the TMR ratio can be expressed as

$$\frac{\Delta R}{R} = \frac{R_{AP} - R_P}{R_P} = \frac{2P_1 P_2}{1 - P_1 P_2}, \tag{4.4}$$

where P_1 and P_2 represent the spin polarization values of the two ferromagnetic electrodes. The spin polarization P is related to the density of states N_s for spin-up and spin-down electrons at the Fermi level as

$$P = \frac{N_\uparrow - N_\downarrow}{N_\uparrow + N_\downarrow}. \tag{4.5}$$

From Julliere's elegant formula, it can be readily seen that with both *P1* and *P2* close to unity, the TMR ratio can reach values much greater than 1. However, the importance of the role of a tunnel barrier and of its interplay between the band structures of FM electrodes is neglected in this simple expression. Later experiments and improved theories (for instance, see Slonczewski [17]) showed that the quality of the interfaces as well as the type of barrier should be taken into account to achieve a high TMR.

The giant TMR at room temperature was first found independently by Moodera et al. [18] and Miyazaki et al. [19] in MTJs with Al_2O_3 as a tunnel barrier in 1995, 20 years after Julliere's work. A TMR of 11.8% in $CoFe/Al_2O_3/Co$ and 18% in $Fe/Al_2O_3/Fe$ junctions were reported, as shown in Figure 4.5. In 2004, another paradigm shift arrived when the room temperature TMR of almost 200% was discovered in MgO-based magnetic tunnel junctions, by Parkin et al. using sputtering technique [7] and Yuasa et al. using epitaxial growth method [6] (see Figure 4.6). Although

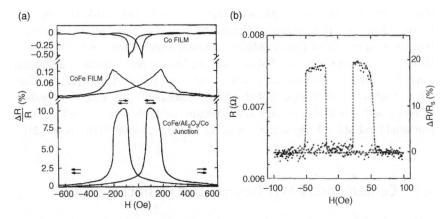

Figure 4.5 Giant TMR at room temperature in (a) CoFe/Al$_2$O$_3$/Co and (b) Fe/Al$_2$O$_3$/Fe MTJs. *Source:* Figures reproduced from Moodera et al. [18] and Miyazaki et al. [19].

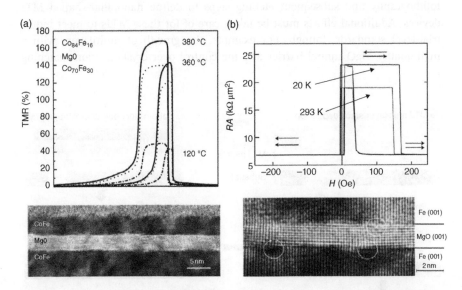

Figure 4.6 Giant TMR at room temperature in (a) CoFe/MgO/CoFe and (b) Fe/MgO/Fe MTJs. The TEM images indicate that the orientation and lattice match between MgO and FM layers play crucial roles in achieving high TMR. The temperature in (a) is the annealing temperature, in (b) is the sample temperature under test. *Source:* Figures reproduced from Parkin et al. [7] and Yuasa et al. [6].

several theoretical works that predicted high TMR in MgO-based MTJ were published in 2001 [20, 21], it turned out that the realization of high TMR in this system requires good control of Fe(001)/MgO(001) lattice match during material growth. Using amorphous CoFeB as the FM layer and performing post-fabrication annealing can further enhance the room temperature TMR to greater than 200% [22, 23]. Currently, sputtered CoFeB/MgO/CoFeB trilayer treated with the suitable annealing process is still the most common MTJ structure for both research and tentative industrial-level MRAM applications, with the highest reported TMR ~ 600% [24].

4.5 Contemporary MTJ Designs and Characterization

For real memory device applications, the fabrication of MTJ (into a CPP configuration) is based on standard back end of line (BEOL) complementary metal–oxide–semiconductor (CMOS) technology, as illustrated in Figure 4.7. This typically starts from the MTJ layer stack deposition by a sputtering system, followed by photolithography and subsequent etching steps to define nanopillar-shaped MTJ devices. Additional efforts must be taken care of for these MTJs to meet industrial-level standards, though. For example, the growth of multilayers with a high-quality MgO tunnel barrier and the control over crystallization matching

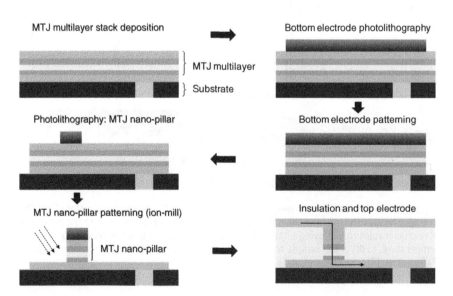

Figure 4.7 Simplified process flow of making MTJ devices.

of ferromagnetic layers are critical to obtain high TMR ratios and strong perpendicular magnetic anisotropy (PMA) of the magnetic layer. To achieve these goals, an advanced sputtering machine needs to be developed. The lithography and the subsequent ion-milling (etching/patterning) processes are also critical to achieve high-yield, high-TMR MTJs at nano-scale.

Set aside these complications, at the end of this chapter, we would like to introduce some basic examples of MTJ layer stack design and characterization for modern MRAM applications. The first one is the first demonstration of MgO-based MTJ with FM layer being perpendicular magnetic tunnel junction (p-MTJ). The second one is a full p-MTJ structure with synthetic antiferromagnetic (SAF) layer, showing how complicated the MTJ layer design can be for modern MRAM. Finally we will briefly discuss a technique to characterize TMR from MTJ layer stack structures without patterning those films into micron-scale or nano-scale devices, which can facilitate the film optimization process.

4.5.1 Perpendicular MTJ (p-MTJ)

Contemporary MTJs typically use MgO as their tunnel barriers. CoFeB ferromagnetic alloy layers are typically used for the FM electrodes. With suitable buffer layer, such as Ta, in 2010 Ikeda et al. showed that the Ta/CoFeB/MgO heterostructure can gain decent PMA after an annealing process, provided that the CoFeB layer is thin and around 1 nm-thick [25]. Ta/CoFeB/MgO/CoFeB/Ta layer design with thin CoFeB layers therefore can be made into a perpendicular MTJ (p-MTJ), which can be scaled down to ~40 nm in diameter, as shown in Figure 4.8. Unlike the conventional in-plane magnetized MTJ case, which relies on shape anisotropy to define the easy-axis (EA) direction, this p-MTJ design allows for gaining EA along the film normal, thereby increasing the potential memory density and feasibility of realistic MRAM application. It is surprising that the MgO-based materials system can result in high TMR (~100%) and PMA of the adjacent FM layer at the same time, which makes it the most fundamental layer structure in various modern STT-MRAM designs.

4.5.2 Fully Functional p-MTJ

If the readers look closely, then they will find in Figure 4.8 that the p-MTJ mentioned in the previous section has only one stable state (low resistance state or the P state) at zero magnetic field (similar to the MR loop shown in Figure 4.5). This is because in the absence of magnetic field, the magnetizations in two CoFeB layers tend to have ferromagnetic coupling and therefore parallel state is more energetically favorable. To make the above-mentioned p-MTJ into a fully functional memory element, one has to introduce some extra layers to ensure that both P state and AP state can be stable when the field is absent. This is typically achieved

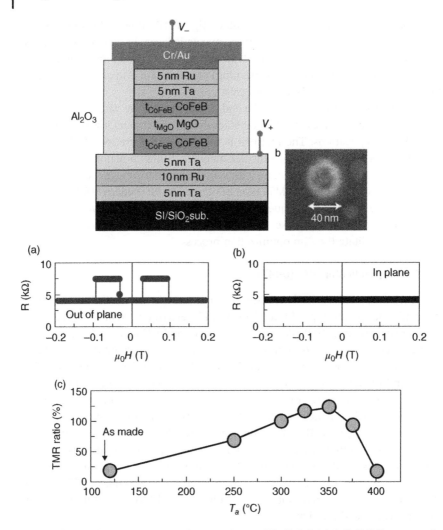

Figure 4.8 p-MTJ with diameter ~ 40 nm made out of Ta/CoFeB/MgO/CoFeB/Ta symmetric layer structure. The anisotropy is obviously along out-of-plane direction, which is evidenced by the TMR measurement results. The TMR can reach ~100% by tuning the annealing recipe. *Source:* Figure reproduced and modified from Ikeda et al. [25].

by effectively "pinned" one of the CoFeB layers using a SAF layer. In this case, the SAF layer is the $[Co/Pd]_6/Co/Ru/Co/[Co/Pd]_2$ multilayer structure shown in Figure 4.9, which utilizes the RKKY interaction mentioned in Chapter 2. An extra Ta/CoFeB/MgO layer can also be added to the structure to enhance the PMA of the recording layer (double interface). The resulting MTJ can possess high TMR,

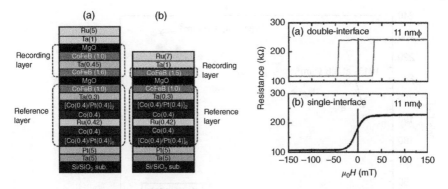

Figure 4.9 An example of fully functional p-MTJ design with (a) double CoFeB/MgO interface and (b) single CoFeB/MgO interface. The double interface case has better PMA and can achieve bi-stable states of TMR even after patterning down to an 11-nm in diameter pillar. *Source:* Figure reproduced from Ikeda et al. [26].

bi-stable resistance states at zero field, and scalability down to 11 nm in terms of diameter.

In later chapters we will introduce the writing mechanism of these p-MTJ memory devices, which is based on the principle of current-induced spin transfer torque [27] to switch the CoFeB layer magnetization. To reduce the switching current of manipulating CoFeB magnetization, the resistance area (RA, essentially corresponding to the thickness of the tunnel barrier) should be reduced. However, reducing RA will also degrade the TMR. Recent industrial efforts have been pushing the trade-off between these two factors to a new limit, i.e. TMR ratio ~ 200% at RA ~ 5 $\Omega\mu m^2$, as shown in Figure 4.10. The quest for even higher TMR ratio and moderately low RA is the key to develop next-generation efficient MRAM.

4.5.3 CIPT Approach for TMR Characterization

To make fully functional MTJ devices, lengthy fabrication processes are required, including the deposition of a multilayer film stack (blanket), lithography process to define pillar-shaped MTJ device, etching process to pattern the pillar, passivation to separate the bottom and top electrodes, etc. It might take a team of well-experienced engineers several months or years to develop suitable recipes for each step of the fabrication flow (Figure 4.7). To bypass these time-consuming processes and to characterize the TMR properties of the deposited film stacks in the blanket (unpatterned) form, a new probing method needs to be developed. The current-in-plane tunneling (CIPT) approach turns out to be an efficient way of measuring the properties of a MTJ by placing a set of probes on an unpatterned film sample. For CIPT measurements, the sheet resistance of an unpatterned film is measured using

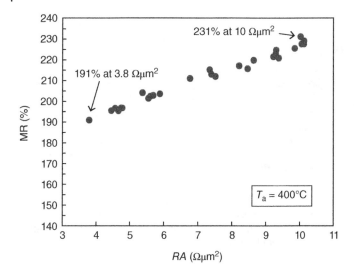

Figure 4.10 TMR ratio versus resistance area (RA) of the state-of-the-art patterned p-MTJ devices. With the refinement of the fabrication process and the advancement of deposition/etching tools, the industrial-level MTJs can reach TMR as high as ~200% in modern times. *Source:* Figure courtesy of Dr. Chang-Man Park from Tokyo Electron Ltd.

a standard four-point-probe method. A current is sent through a sample by two current probes (denoted as I+ and I− in Figure 4.11), and the voltage difference will be picked up by two other voltage probes (denoted V+ and V−). The resistance of the sample can then be calculated.

It is important to note that, however, the distances among these current and voltage probes are critical in determining the current distribution as the current flows from one current probe to the other. If the two probes are too close to each

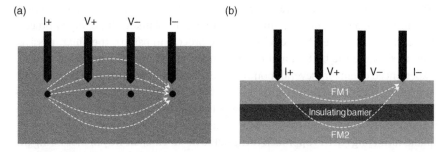

Figure 4.11 (a) Top view (schematics) and (b) side view of CIPT measurement of a MTJ layer stack without patterning the film into a two-terminal device for CPP measurement.

Figure 4.12 CIPT versus standard CPP TMR measurements from AlO_x-based MTJs with different AlO_x thicknesses. *Source:* Figure reproduced from Worledge et al. [28].

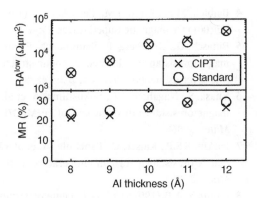

other, then no current will be tunneled through the barrier, causing no TMR effect. In contrast, if the two probes are too far away from each other, then the current will simply flow through the two FM layer parallel with a negligible contribution from the tunneling part. Typical probe distance of getting observable TMR effects with CIPT approach is ~10–100 µm. Figure 4.12 shows an example of a TMR measurement result from both CIPT and standard CPP (patterned device) approaches.

Homework

Q4.1 Assuming that Julliere's formula of TMR ratio is correct, what kind of ferromagnetic electrode should you choose to maximize the TMR ratio of an MTJ?

A4.1 One should look for materials with a spin polarization value of $P \to 1$, such as half-metallic Heusler alloy, which has only one type of spin at the Fermi level. However, in reality, using materials with $P = 1$ does not ensure high TMR, since the effect from the barrier and the temperature dependence of the P value have been ignored (for example, see [29]).

References

1 Ashcroft, N.W. and Mermin, N.D. (1976). *Solid State Physics*. Brooks Cole.
2 O'Handley, R.C. (2000). *Modern Magnetic Materials: Principles and Applications*. New York: Wiley.
3 Gil, W., Görlitz, D., Horisberger, M., and Kötzler, J. (2005). Magnetoresistance anisotropy of polycrystalline cobalt films: geometrical-size and domain effects. *Phys. Rev. B* 72: 134401.

4 Baibich, M.N., Broto, J.M., Fert, A. et al. (1988). Giant magnetoresistance of (001) Fe/(001) Cr magnetic Superlattices. *Phys. Rev. Lett.* 61: 2472.

5 Binasch, G., Grunberg, P., Saurenbach, F., and Zinn, W. (1989). Enhanced magnetoresistance in layered magnetic-structures with antiferromagnetic interlayer exchange. *Phys. Rev. B* 39: 4828.

6 Yuasa, S., Nagahama, T., Fukushima, A. et al. (2004). Giant room-temperature magnetoresistance in single-crystal Fe/MgO/Fe magnetic tunnel junctions. *Nat. Mater.* 3: 868.

7 Parkin, S.S.P., Kaiser, C., Panchula, A. et al. (2004). Giant tunnelling magnetoresistance at room temperature with MgO (100) tunnel barriers. *Nat. Mater.* 3: 862.

8 Ramirez, A.P. (1997). Colossal magnetoresistance. *J. Phys. Condens. Matter* 9: 8171.

9 Chen, Y.T., Takahashi, S., Nakayama, H. et al. (2013). Theory of spin Hall magnetoresistance. *Phys. Rev. B* 87: 144411.

10 Cho, S., Baek, S.H.C., Lee, K.D. et al. (2015). Large spin Hall magnetoresistance and its correlation to the spin-orbit torque in W/CoFeB/MgO structures. *Sci. Rep.* 5: 14668.

11 Kim, J., Sheng, P., Takahashi, S. et al. (2016). Spin hall magnetoresistance in metallic bilayers. *Phys. Rev. Lett.* 116: 097201.

12 Avci, C.O., Garello, K., Ghosh, A. et al. (2015). Unidirectional spin Hall magnetoresistance in ferromagnet/normal metal bilayers. *Nat. Phys.* 11: 570.

13 Valet, T. and Fert, A. (1993). Theory of the perpendicular magnetoresistance in magnetic multilayers. *Phys. Rev. B* 48: 7099.

14 Pratt, W.P., Lee, S.F., Slaughter, J.M. et al. (1991). Perpendicular giant magnetoresistances of Ag/Co multilayers. *Phys. Rev. Lett.* 66: 3060.

15 Wolf, S.A., Awschalom, D.D., Buhrman, R.A. et al. (2001). Spintronics: a spin-based electronics vision for the future. *Science* 294: 1488.

16 Julliere, M. (1975). Tunneling between ferromagnetic films. *Phys. Lett. A* 54: 225.

17 Slonczewski, J.C. (1989). Conductance and exchange coupling of two ferromagnets separated by a tunneling barrier. *Phys. Rev. B* 39: 6995.

18 Moodera, J.S., Kinder, L.R., Wong, T.M., and Meservey, R. (1995). Large magnetoresistance at room temperature in ferromagnetic thin film tunnel junctions. *Phys. Rev. Lett.* 74: 3273.

19 Miyazaki, T. and Tezuka, N. (1995). Giant magnetic tunneling effect in Fe/Al2O3/Fe junction. *J. Magn. Magn. Mater* 139: L231.

20 Butler, W.H., Zhang, X.G., Schulthess, T.C., and MacLaren, J.M. (2001). Spin-dependent tunneling conductance of Fe | MgO | Fe sandwiches. *Phys. Rev. B* 63: 054416.

21 Mathon, J. and Umerski, A. (2001). Theory of tunneling magnetoresistance of an epitaxial Fe/MgO/Fe(001) junction. *Phys. Rev. B* 63: 220403.

22 Djayaprawira, D.D., Tsunekawa, K., Nagai, M. et al. (2005). 230% room-temperature magnetoresistance in CoFeB/MgO/CoFeB magnetic tunnel junctions. *Appl. Phys. Lett.* 86: 092502.

23 Hayakawa, J., Ikeda, S., Lee, Y.M. et al. (2006). Effect of high annealing temperature on giant tunnel magnetoresistance ratio of CoFeB/MgO/CoFeB magnetic tunnel junctions. *Appl. Phys. Lett.* 89: 232510.

24 Ikeda, S., Hayakawa, J., Ashizawa, Y. et al. (2008). Tunnel magnetoresistance of 604% at 300K by suppression of Ta diffusion in CoFeB/MgO/CoFeB pseudo-spin-valves annealed at high temperature. *Appl. Phys. Lett.* 93: 082508.

25 Ikeda, S., Miura, K., Yamamoto, H. et al. (2010). A perpendicular-anisotropy CoFeB-MgO magnetic tunnel junction. *Nat. Mater.* 9: 721.

26 S. Ikeda, H. Sato, H. Honjo, E. C. I. Enobio, S. Ishikawa, M. Yamanouchi, S. Fukami, S. Kanai, F. Matsukura, T. Endoh, and H. Ohno (2014). *2014 IEEE International Electron Devices Meeting*, p. 33.2.1.

27 Slonczewski, J.C. (1996). Current-driven excitation of magnetic multilayers. *J. Magn. Magn. Mater.* 159: L1.

28 Worledge, D.C. and Trouilloud, P.L. (2003). Magnetoresistance measurement of unpatterned magnetic tunnel junction wafers by current-in-plane tunneling. *Appl. Phys. Lett.* 83: 84.

29 Shan, R., Sukegawa, H., Wang, W.H. et al. (2009). Demonstration of half-metallicity in fermi-level-tuned Heusler alloy $Co_2FeAl_{0.5}Si_{0.5}$ at room temperature. *Phys. Rev. Lett.* 102: 246601.

5

Magnetization Switching and Field MRAMs

5.1 Introduction

This chapter describes magnetization behavior under a static magnetic field. Section 5.2 covers both reversible magnetization rotation and irreversible magnetization switching. Under an applied field, magnetization rotates from its easy axis position to a new angle position. The rotation angle and switching are analyzed based on Stone-Walfarth's *Astroid* (sometimes called the *Asteroid*).

This analysis is entirely static; i.e. transient behavior is not discussed. Nonetheless, the static analysis is sufficient for the purpose of understanding the key design issues of field magnetoresistive random-access memory (MRAM). We will come to discuss the transient behavior in Chapter 6.

Field MRAM is named after its magnetic tunnel junction (MTJ) free-layer magnetization switching induced by a magnetic field that is generated by the write current on the word line and bit line of the MRAM array. We will describe the operation issues of Astroid mode switching and the solutions in Section 5.3.

5.2 Magnetization Reversible Rotation and Irreversible Switching Under External Field

In the following few sections, we will study the magnetization under an applied external field. The angle of the magnetization in a film may rotate or switch with a single domain model. The magnetization M of the entire film or device is assumed to be uniform and the same, a single vector. Thus, the mathematical treatment is greatly simplified. In other words, edge curling and domains are not considered. In addition, the demagnetizing field H_D, is also uniform. Under such assumptions, a simple closed-form film energy equation on an angle of M

Magnetic Memory Technology: Spin-Transfer-Torque MRAM and Beyond,
First Edition. Denny D. Tang and Chi-Feng Pai.
© 2021 The Institute of Electrical and Electronics Engineers, Inc.
Published 2021 by John Wiley & Sons, Inc.

of the entire film can easily be formulated. The rotation angle is the lowest energy point of the energy equation. This model is called a *uniform rotation model* or *coherent rotation model*. It is a good approximation of the magnetization behavior under external field, and we will proceed with this model in this chapter for tutorial purposes. For more accurate calculation, one can use micromagnetic simulations, which take into account the edge curling, etc.

When a field rotates the magnetization less than 90°, the magnetization reverses to its initial position once the external field is removed. The situation is called reversible rotation. On the other hand, if the rotation angle is greater than 90°, the magnetization will be pulled to the 180° by the anisotropy, or switches irreversibly.

5.2.1 Magnetization Rotation Under an External Field in the Hard Axis Direction

When an external field, H_y, is applied in the hard axis direction of a full film, M responds to the external field by rotation from the initial position (along the easy axis) to lower the energy state. Figure 5.1a depicts the easy and hard axes; the magnetization rotates away from the axis under an external hard-axis field. In this case, the energies involved are the crystalline anisotropy energy $\frac{1}{2}H_K \cdot M_s \cos \theta^2$ and the magnetostatic energy $H_y \cdot M_s \cos (\pi/2\text{-}\theta)$, where θ is the angle between the easy axis and M, or

$$
\begin{aligned}
\varepsilon &= -\frac{1}{2}H_K M_s \cos \theta^2 - H_y M_s \cos (\pi/2 - \theta) \\
&= -\frac{1}{2}H_K M_s \cos \theta^2 - H_y M_s \sin \theta,
\end{aligned}
\tag{5.1}
$$

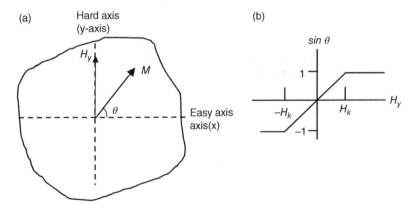

Figure 5.1 (a) Easy, hard axes, magnetization with angle θ from easy axis, (b) Magnetization rotates angle θ under the hard axis field H_y. The easy axis is along the x-axis.

where ε is the density of energy in a unit volume of the ferromagnetic material. The torque on M is $d\varepsilon/d\theta$, and at zero torque

$$d\varepsilon/d\theta = H_K \cdot M_s \cos\theta \cdot \sin\theta - H_y \cdot M_s \cos\theta = 0, \tag{5.2}$$

or

$$\sin\theta = \frac{H_y}{H_K}. \tag{5.3}$$

The angle θ of M is

$$\theta = \sin^{-1}\left(\frac{H_y}{H_K}\right) = \sin^{-1}\left(\frac{M_y}{M_s}\right). \tag{5.4}$$

Figure 5.1b shows the magnetization angle dependence on the external field. Notice that a zero-torque point is essentially the same as the lowest energy point. One may obtain the angle position of M using either approach.

When a field is applied to a patterned film, the rotation angle of the magnetization will be different. Nonetheless, the angle can be derived in the same manner, with additional consideration of the shape anisotropy. Because the shape anisotropy cancels a portion of the applied field inside the ferromagnet. We will leave this derivation exercise to readers. Please refer to Q5.1 at the end of this chapter.

5.2.2 Magnetization Rotation and Switching Under an external Field in the Easy Axis Direction

Let's consider a case that magnetization M_s of a full film initially rests along the easy axis (x-direction of Figure 5.1a). Apply a field H_x in the –x direction. Then, M_s rotates to angle θ.

The energy of the film is

$$\begin{aligned}
\varepsilon &= -\frac{1}{2} \cdot H_K \cdot M_s \cos^2\theta - H_x \cdot M_s \cos(\pi - \theta) \\
&= -\frac{1}{2}H_K M_s \cos^2\theta + H_x \cdot M_s \cos\theta.
\end{aligned} \tag{5.5}$$

Thus, the minimal energy point can be derived as

$$d\varepsilon/d\theta = H_K \cdot M_s \sin\theta \cos\theta - H_x \cdot M_s \sin\theta. \tag{5.6}$$

There are two θ angles that satisfy $d\varepsilon/d\theta = 0$: $\theta = 0$ or $\theta = \cos^{-1}\left(\frac{H_x}{H_K}\right)$. But, only one is a stable position at which the torque is zero, or $d^2\varepsilon/d\theta^2 < 0$.

$$d^2\varepsilon/d\theta^2 = H_K \cdot M_s \cos 2\theta - H_x \cdot M_s \cos\theta < 0. \tag{5.7}$$

For example, when $H = \frac{1}{2}H_K$, the stable solution is $\theta \sim 1.3$ rad.

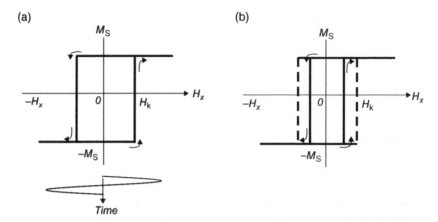

Figure 5.2 (a) The *M-H* hysteresis loop when sweeping an easy-axis field (H_x) to a full film, when (a) $H_y = 0$, (b) $H_y > 0$. The switching field is reduced from dashed line to solid line.

This angle increases with an increasing applied field. When the magnitude of H is equal or exceeds H_K, θ reaches $\pi/2$, the maximum energy barrier point. Then $\boldsymbol{M}$ switches to $\theta = \pi$ as the film lowers its energy state. This process is irreversible; namely, the $\boldsymbol{M}$ does not return to its initial angle after the applied field is removed.

Sweeping an easy-axis field with a larger than H_k value will produce an M-H loop as shown in Figure 5.2a, while sweeping a hard-axis field will produce Figure 5.1b. Strictly speaking, this analysis is value at zero Kelvin temperature, where there was no thermal energy. At room temperature, the magnetization fluctuates around its initial position. The switching field, called Coercive field, H_C, is smaller than H_K.

5.2.3 Magnetization Rotation and Switching Under Two Orthogonal External Fields

When both $-H_x$ and H_y are applied to the film, the switching threshold can be calculated with the energy equation:

$$\varepsilon = -\frac{1}{2}H_K \cdot M_s \cos^2\theta + H_x \cdot M_s \cos\theta - H_y \cdot M_s \sin\theta = 0 \qquad (5.8)$$

The stable position is at $d\varepsilon/d\theta < 0$ and $d^2\varepsilon/d\theta^2 < 0$. These two equations can be written as

$$d\varepsilon/d\theta = \frac{1}{2}H_K \cdot M_s \sin 2\theta - H_x \cdot M_s \sin\theta - H_y \cdot M_s \cos\theta = 0 \qquad (5.9)$$

$$d^2\varepsilon/d\theta^2 = H_K \cdot M_s \cos 2\theta - H_x \cdot M_s \cos\theta + H_y \cdot M_s \sin\theta = 0. \qquad (5.10)$$

The condition that Eq. (5.9) = 0 is the threshold between a reversible rotation Eq. (5.10) < 0 and irreversible switching Eq. (5.10) > 0. In the reversible condition, once the applied field is removed, M returns to its initial position. In the case of irreversible switching, M switches to a new low-energy position and stay, even after the field is removed. Solving these two equations, one finds the switching threshold of M to be

$$H_x^{2/3} + H_y^{2/3} = (H_K)^{2/3}. \qquad (5.11)$$

This is the so-called Stoner-Walfarth switching Astroid. It is illustrated in Fig. 5.3. Outside of the Astroid, the field strength is strong enough to switch the M to the opposite direction. Inside the Astroid, the strength of field is not sufficient to switch the magnetization. Once the field is removed, the magnetization returns to its initial position. The Astroid provides a simple graphic tool for predicting reversible rotation and irreversible switching of a magnetization under a field with arbitrary direction and magnitude.

The switching Astroid also tells us that when the hard axis field H_y is nonzero, the easy-axis switching threshold H_x is reduced, since $H_x = (H_K^{2/3} - H_y^{2/3})^{3/2}$. This point becomes more obvious when readers complete Q5.2. This is the basis of x-y bit selection in the field MRAM array.

5.2.4 Magnetization Behavior of a Synthetic Anti-ferromagnetic Film Stack

Figure 5.3a illustrates a synthetic anti-ferromagnetic film stack: two ferromagnetic films spaced by a nonmagnetic film, such as a very thin Ruthenium layer. The magnetizations of the two ferromagnetic layers are M_1 and M_2, and they couple

Figure 5.3 Stoner-Walfarth switching Astroid. $H_x^{2/3} + H_y^{2/3} = H_K^{2/3}$.

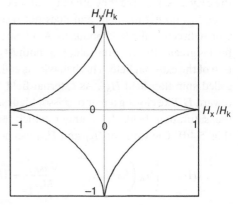

through RKKY inter-layer exchange coupling. For this case, M_1 and M_2 are anti-parallel.

In the absence of an external field, M_1 and M_2 both lie on the easy axis, and in opposite directions. When an external field H, at angle φ from the easy axis, is applied to the film stack, the energy per unit area of the film stack is the sum of each film's anisotropy energy, coupling energy and magnetostatic energy,

$$\varepsilon = -H_K \cdot \left(M_1 \cdot \cos^2\theta_1 + M_2 \cdot \cos^2\theta_2\right) + J_{RKKY} M_1 \cdot M_2 \cdot \cos(\theta_1 - \theta_2)$$
$$-H \cdot (M_1 \cdot \cos(\theta_1 - \varphi) + M_2 \cdot (t_2/t_1) \cdot \cos(\theta_2 - \varphi)\,), \tag{5.12}$$

where t is the ferromagnetic film thickness, θ is the angle of the magnetization, and the subscripts 1 and 2 stand for the top and bottom ferromagnetic layers of the SAF, respectively. For the case of anti-ferromagnetic coupling, J_{RKKY} is positive. The term H_K keeps M_1 and M_2 aligned to the easy axis, while J_{RKKY} keeps M_1 and M_2 in anti-parallel. The magnetostatic energy term pushes both M_1 and M_2 toward same direction as H.

The torque on M_1 is

$$\Gamma_1 = \frac{\partial \varepsilon}{\partial \theta_1}, \tag{5.13}$$

and on M_2 is

$$\Gamma_2 = \frac{\partial \varepsilon}{\partial \theta_2}. \tag{5.14}$$

At equilibrium, both torques Γ_1 and Γ_2 are zero. Solving Eq. (5.13) = 0 and Eq. (5.14) = 0, one finds the equilibrium angular positions of θ_1 and θ_2. These equations can only be solved numerically.

For the case that $M_1 t_1 \sim M_2 t_2 = Mt$, at low external field, M_1 and M_2 remain on the easy axis, just like they are in the absence of external field (Fig. 5.4b). The film stack has no net moment and does not respond to the external field, no matter which direction the field points to. As illustrated in Figure 5.4c, when the external field is greater than a threshold, H_{SF}, both M_1 and M_2 tilt slightly toward the direction of the external field. This behavior is called *spin flop*, and the threshold field is called spin-flop field H_{SF}. As external field further increases, M_1 and M_2 rotate toward H, like closing a pair of scissor blades. Eventually, H reaches a value called the saturation field, H_{sat}, under which M_1 and M_2 align themselves to H (Fig. 5.4d). The value of H_{SF} and H_{sat} has been derived as [1]

$$H_{SF} = \left[H_K\left(8\pi M_s N_y \frac{t}{b} - \frac{2J_{RKKY}}{M_s t} + H_K\right)\right]^{1/2}, \tag{5.15}$$

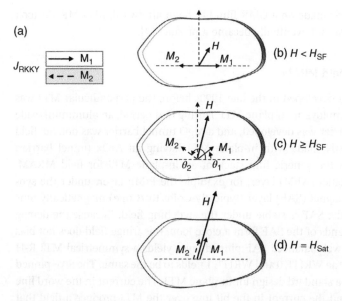

Figure 5.4 (a) Synthetic anti-ferromagnetic film stack, (b–d) magnetization under external field.

and

$$H_{ssat} = 8\pi M_s N_x \frac{t}{b} - \frac{2J_{RKKY}}{M_s t} - H_K, \tag{5.16}$$

where H_K is the crystalline anisotropy, a and b are length and width, and N_x and N_y are unit-less demagnetizing factor of an elliptic-shape film.

When the external field is greater than H_{SF}, M_1 and M_2 scissors, their moments no longer cancel each other and there is a net moment. As a result, the net moment of the pair responds to the direction of the external field. When the applied field rotates, the net moment of the pair follows. This is the principle of write operation of toggle-mode magnetic RAM.

5.3 Field MRAMs

The concept of Field MRAM evolved from Stoner-Walfarth switching of a ferromagnetic device under two orthogonal fields (x, y), as described in Section 5.2.3. This x-y access scheme fits into traditional semiconductor memory random access scheme. The first demonstration of this random-access memory device concept was made on a GMR film [4]. Soon afterward, MTJ MRAM took over GMR MRAM and eventually became a product [3].

5.3.1 MTJ of Field MRAM

Field MRAM was developed in the late 1990s before the perpendicular MTJ was developed, so it employs an in-plane MTJ. During that period, an aluminum oxide (AlOx) tunnel barrier was developed, and a MgO tunnel barrier was not. So, field MRAMs are mostly made with in-plane MTJ having an AlOx tunnel barrier. Figure 5.5 shows the generic film stack of an in-plane MTJ for field MRAM. The anti-ferromagnet (AFM) layer, for example the PtMn layer, under the synthetic anti-ferromagnet (SAF) layer (typically CoFe/Ru/CoFe) magnetically pins the SAF so that the SAF is stable under the switching field. Because the demag fields at the two ends of the SAF form a close loop, this fringe field does not bias the free layer. A well-balanced SAF pinned layer yields a symmetrical MTJ R-H loop. This allows the WRITE 0 and WRITE 1 fields to be the same. The SAF pinned layer is currently a standard design for in-plane MTJ. The current in the word line under the MTJ and the current in the bit line over the MTJ produce a field that switches the magnetization of the free layer. (See Figure 5.6).

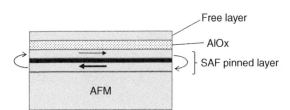

Figure 5.5 Film stack of a generic in-plane MTJ for field MRAM cell.

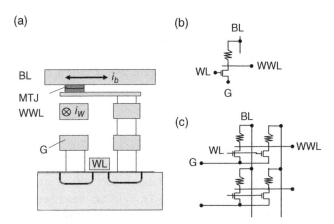

Figure 5.6 (a) Structure of 1M-1T cell of field MRAM, (b) cell schematic, (c) 2 × 2 array. Resistor in schematics (b) and (c) represents MTJ.

5.3.2 Half-Select Bit Disturbance Issue

When writing an array with Astroid mode to a selected bit in an x-y array with current in word line (x) and bit line (y), every bit on the selected word line and the selected bit line experiences a field, including all half-selected bits. They see only one field (from word line or from bit line, not both), which is insufficient to switch the bit. But, as mentioned in Section 5.2.3, they still may switch under the subthreshold field with low, but nonzero, probability. This half-selected bit disturbance issue has plagued the field of MRAM development for a long time.

Two elaborated efforts have been launched to address this issue. One was based on asymmetric-shape MTJ [2] and segmented-write architecture. The effort failed. The other succeeded to deliver a product in 2006. The product is called Toggle MRAM [3]. To gain a better understanding of field MRAM issues and solutions, readers are referred to [5].

The Toggle MRAM product is implemented with in-plane MTJ. The free layer is the SAF stack (shown in Figure 5.7). The toggle cell relies on spin flop energy for its thermal stability. The threshold dependence of spin-flop is very abrupt, much sharper than the Stoner-Walfarth switching of conventional single-free-layer field MRAM. As a result, Toggle MRAM solves the half-selected cell write disturbance issue of the conventional field MRAM and becomes a highly reliable MRAM.

Toggle MRAM's thermal stability is good, easily offers 10-year data retention, and enjoys a device endurance of 10^{12} write cycles. Unfortunately, the write current is large, in the order of 10 mA each. The write energy is orders larger than that of the spin-transfer torque magnetic random-access memory (STT-MRAM). In addition, the product requires special magnetic shield, inside the chip package, to improve the product immunity to the magnetic field disturbance in the user environment. No further scaling effort has been exercised after a 16 Mb product at 130-nm node.

The strength and weakness and production of field MRAM will be described in Section 9.4.

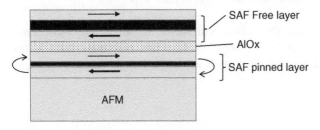

Figure 5.7 Film stack of Toggle-MRAM.

Homework

Q5.1 Formulate the magnetization rotation angle of a long and narrow film stripe. Include a demag field. The external field is along the narrow direction of the film stripe.

A5.1 Section 1.5 describes the demag field. Although the demag field is not uniform across the film stripe, for convenience, we choose to use the average demag field $\overline{H_{demag}}$ and abbreviate the notation as H_D. The demag field $H_D = NM_s$, where N is the demag factor. For a long and narrow stripe along the x-direction, the demag factor $N_x \ll N_y$. The external field H is in the film plane (x-y), and along the narrow direction, say y-direction, the dominant demag field is $H_D \approx N_y M_{s,y}\hat{y} = N_y(M_s \sin \theta)\hat{y}$, where θ is the angle between M_s and the easy-axis (x-direction). The demag field is along y-direction. Thus, the energy equation of Eq. (5.1) can be rewritten as

$$\varepsilon = -\frac{1}{2} H_K M_s \cos \theta^2 - \left(H_y - N_y(M_s \sin \theta)\right) \cdot \cos (\pi/2 - \theta).$$

From the energy minimum, one finds that

$$\theta = \sin^{-1}\left[\frac{H_y}{H_K + N_y M_s}\right]. \tag{5.17}$$

Equations (5.4) and (5.17) are identical, if we treat $(H_K + N_y M_s)$ as H_K', the anisotropy of a patterned film. The magnetic poles at the edge reduce the applied field in the film. Thus, the equivalent anisotropy of a patterned film is usually the combination of H_K (crystalline anisotropy) and H_D (shape anisotropy).

Since H_D is linearly proportional to the film thickness and material M_s and decreases with film size, the anisotropy of small-size patterned film is dominated by the shape anisotropy. For an elliptic- or oval-shape film, the anisotropy, or the "easy-axis," is along the long axis. For a circular-shape film, the shape anisotropy in every direction is the same due to symmetry and, thus, is the same as the crystalline anisotropy H_K.

Q5.2 Refer to Figure 5.3. The initial position of M lies in the x-direction, the easy-axis. What is the angle of M when the apply field is (a) $H = -0.25 H_K \hat{x} + 0.25 H_K \hat{y}$, (b) $H = -H_K \hat{x} + H_K \hat{y}$ to a Ferromagnetic film?

(a)   (b)

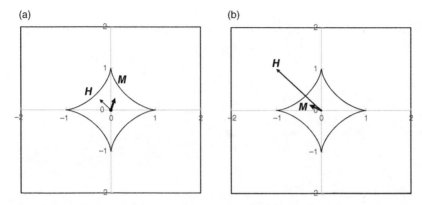

Figure 5.Q2 A normalized applied field vector and magnetization in Astroid, corresponding to problem (a) *H* inside Astroid and (b) *H* outside Astroid.

A5.2 Solve Eq. (5.9); one gets (a) $\theta = 1.25$ (radian) field is inside the Astroid, no switch. (b) $\theta = 2.8$ (radian), greater than $\pi/2$. *M* switches. These two solutions are shown in Fig. Q5.2.

Astroid analysis provides a good first-order estimate for samples that exhibit coherent rotation behavior and at $0°K$ temperature. At room temperature, thermal agitation energy contributes to the switching of magnetization in ferromagnetic film. Thus, there is a finite probability that MTJ switches when the write field is inside the Astroid. Similarly, there is a finite probability that switching may fail when the field is outside the Astroid.

References

1 Osborne, J.A. (1942). *Phys. Rev.* 67: 361.
2 Ounadjada, K. (2003). US patent 6798691.
3 Freescale (2006). Freescale begins selling the world's first MRAM chip. Press release (July 10).
4 Tang, D.D., Wang, P.K., Speriosu, V.S. et al. (1995). *IEDM Tech. Digest*: 997.
5 Tang, D. and Lee, Y.J. (2010). *Magnetic Memory*. Cambridge Press.

6

Spin Current and Spin Dynamics

6.1 Introduction to Hall Effects

The family of Hall effects plays a crucial role not only in magnetic-related studies but also in the development of modern condensed matter physics. This chapter begins with a review of the ordinary Hall effect (OHE) and the anomalous Hall effect (AHE). The latter can be further viewed as the spin Hall effect (SHE) in ferromagnetic materials. From such effects, we introduce the concept of spin-polarized current and pure spin current, namely, the flows of spin angular momentum carried by flows of spin-polarized conduction electrons. Through the spin-transfer torque mechanism, the spin angular momentum, carried by a stream of either spin-polarized current or pure spin current, exerts torques onto the magnetic moment on their path. Both spin-polarized current and pure spin current offer new ways to control the magnetization by simple current injection, without the help of an applied external field. More important, spin-transfer torque can be employed to induce magnetization reversals and oscillations, or to induce magnetic domain wall motion, in various types of micron-sized and nano-sized devices.

6.1.1 Ordinary Hall Effect

In 1879, Edwin Hall [1] found that when electric current flows longitudinally in a nonmagnetic film stripe, such as normal metal or semiconductor, in the presence of a magnetic field perpendicular to the plane of the stripe, a transverse voltage can be measured across the two sides of the stripe (Figure 6.1). This voltage is now called the Hall voltage. The Hall voltage is the result of the Lorentz force skewing the trajectory of the positively charged holes sideways in a direction transverse to the current flow direction and skewing the negatively charged electron sideways in the opposite direction. The spatial separation of electrons from holes builds up an electrostatic potential (electric field) transverse to the direction of the current flow

Magnetic Memory Technology: Spin-Transfer-Torque MRAM and Beyond,
First Edition. Denny D. Tang and Chi-Feng Pai.
© 2021 The Institute of Electrical and Electronics Engineers, Inc.
Published 2021 by John Wiley & Sons, Inc.

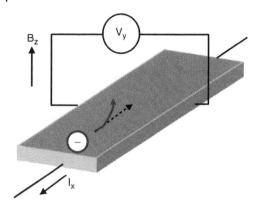

and the magnetic field. This potential is proportional to the applied magnetic field. The Hall resistivity is further defined as

$$\rho_H = V_H/I = R_H B, \tag{6.1}$$

where B is the applied magnetic field, and V_H and R_H are the Hall voltage and Hall coefficient, respectively. Notice that the Hall voltage is transverse to the current direction. By measuring the sign of the Hall voltage, one can determine the dominant type of charge carrier in normal metals as well as semiconductors, either being electron ($R_H < 0$) or hole ($R_H > 0$). For instance, it is a common practice to determine whether a semiconductor is n-type (electron transport) or p-type (hole transport) by measuring its Hall coefficient.

6.1.2 Anomalous Hall Effect and Spin Hall Effect

In 1881, Hall found that, for ferromagnetic materials, a similar transversal voltage can be detected even without the application of an external magnetic field [2]. The Hall resistivity was found as

$$\rho_H = V_H/I = R_H B + 4\pi R_A M_S, \tag{6.2}$$

where M_S is the saturation magnetization of the ferromagnetic material. Hall voltage consists of two terms. The first term is B-field dependent, which corresponds to the OHE. The second term depends on the magnetization in the material and is characterized by the anomalous Hall coefficient R_A. It was not understood then, and the phenomenon was called the AHE. Typically the contribution from AHE is greater than that from OHE; i.e. $R_A > R_H$. It is noted that since AHE is directly proportional to the magnetization in magnetic materials, it can be utilized to probe the magnetization direction or dynamics of magnetic materials (see Figure 6.2).

In late twentieth century, the AHE is understood as the result of spin-dependent transport [3]. In essence, electrons moving longitudinally through a ferromagnetic conductor acquire a transverse velocity. The transversal velocity of spin-up

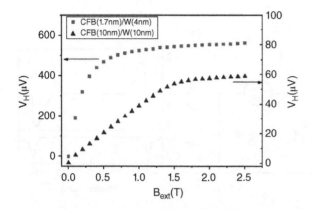

Figure 6.2 Examples of anomalous Hall voltage measurement on CoFeB/W bilayer films with different thicknesses. The CoFeB layers have in-plane anisotropy. Using AHE, one can determine the out-of-plane field required to saturate magnetization along the out-of-plane direction. From the data, one can also see that the anomalous Hall coefficient is greater than the ordinary Hall coefficient ($R_A > R_H$).

electron is opposite to that of spin-down electrons due to spin-orbit interaction. Spin-up electrons therefore accumulate on one side of the ferromagnetic material, spin-down electrons the other side. In a ferromagnet, the population of majority-spin electron (their spin is in same direction as magnetization of the ferromagnet) is greater than that of the minority-spin electron. Thus, more majority-spin electrons accumulate in one side than the minority-spin electrons on the other side, resulting in an unequal charge on two sides; thus, a measurable voltage is developed. That is the anomalous Hall voltage.

The same spin-dependent transverse velocity is also present in nonmagnetic materials. In the direction orthogonal to the current direction, spin-up electrons accumulate in one side of the conductor, spin-down electrons in the other side, as depicted in Figure 6.3a. The separation of spin-up and spin-down electrons, however, does not develop a transversal Hall voltage in nonmagnetic material, since the populations of the two spins on the two sides are equal. Experimentally, the separation can be observed with the Kerr imaging technique [4]. Figure 6.3b shows the experimental observation of spin-dependent electron accumulation on the two sides in a GaAs sample. In 1999, Hirsch called this phenomenon the *spin Hall effect* (SHE) [5]. Figure 6.4 shows a comparison among OHE, AHE, and SHE.

Both AHE and SHE originate from the spin-orbit interaction in materials. The former is observed in ferromagnets, while the latter is the case for nonmagnetic materials (normal metals or semiconductors). The SHE, as a result, is more prominent in materials with large spin-orbit interaction such as 5d transition metals: β-Ta, Pt, β-W, etc.

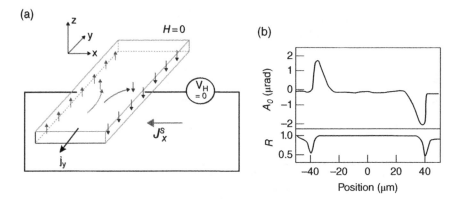

Figure 6.3 (a) The spin Hall effect takes place in nonmagnetic material. Under no applied field, electrons of up-spin gain a transversal velocity toward left and down-spin toward right. Due to equal numbers of up-electron and down-electrons, the Hall voltage is zero. (b) Kerr measurement showing electrons with one spin accumulates on one side of a 300 μm x 77 mm GaAs sample as current flows through the sample; the electrons of other spin accumulate on the other side. The top figure shows the rotation angle, and the bottom figure shows the reflectivity as a function of sample lateral position [4]. (Retrace from original.)

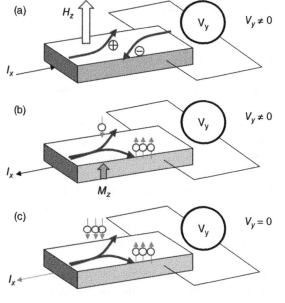

Figure 6.4 Different Hall effects. (a) ordinary Hall effect, (b) anomalous Hall effect in ferromagnet, and (c) spin Hall effect.

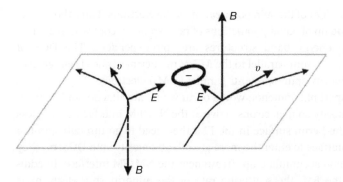

Figure 6.5 Schematics of electron scattering by a negatively charged center. The electron spin experiences a magnetic field $B^* \sim v \times E$ perpendicular to the plane of the electron trajectory. Note that this magnetic field has opposite directions for electrons scattered to the left and to the right of the charged center [12].

Different mechanisms were proposed to explain the spin-dependent transversal trajectory; they are (a) spin skew scattering [6], (b) side jump [7], and (c) intrinsic mechanism [8]. The first two involve spin-dependent scattering (extrinsic) [9], and the last one does not (which is related to the band structure and its Berry curvature) [10]. An excellent review of SHE in metals was given by Hoffmann [11].

A recent theory was proposed by Dyakonov [12, 13]. Dyakonov offers a mechanism involving relativity. When an electron is scattered by the electric field of a charged scattering center, a magnetic field B exists in the electron's moving frame. This field is perpendicular to the plane of the electron trajectory, and B has opposite signs for electrons moving to the right and to the left of the charged center. The B is seen by the electron spin. The Zeeman energy of the electron spin in this field corresponds to the spin-orbit interaction. It is illustrated in Figure 6.5.

6.2 Spin Current

As described, spin Hall effect splits the electron transport trajectory according to its spin polarity. We can view that the total electron current is made up of two components. One component is contributed by the spin-up electrons and the other spin-down electrons. Their current paths skew toward opposite directions, and they carry opposite angular momentum. One can view this phenomenon as a lateral flow of angular momentum. That is the spin current we are going to discuss in this section.

6.2.1 Electron Spin Polarization in NM/FM/NM Film Stack

In a normal or nonmagnetic metal (NM), the density of state (DOS) at the Fermi level of spin-up electrons $n_\uparrow$ is the same as that of the spin-down electrons $n_\downarrow$. The scattering is independent of the spin polarization of the electrons. Thus, those leaving an NM are made up of equal populations of both spins. In contrast, in a ferromagnet (FM), the energy band structures are spin-dependent. The DOS of majority-spin electrons (spin parallel to the M_s) at the Fermi surface is larger than the DOS of the minority-spin (spin anti-parallel to M_s) electrons.

Interesting transport phenomenon would occur when one attaches an NM to an FM. When the minority-spin electrons arrive at the NM/FM interface, seeing less available DOS at the Fermi surface in the FM, they need to go through spin-flip scattering at the interface to enter the more abundant majority-spin DOS; or, they are reflected back and accumulate up stream near the NM/FM interface. In addition, once inside the FM, the scattering rate of the minority-spin electrons is greater, and they are more likely to go through spin-flip scattering. This effect is the spin-filtering effect and is also related to the origin of giant magneto-resistance. Thus, electrons leaving the FM are predominantly majority-spin electrons. In other words, while the population of spin-up and spin-down electrons of the incoming electrons are the same, the outgoing electrons are polarized with spin parallel to the M_s of the FM.

Figure 6.6 illustrates the spin polarization process when electrons pass through the NM/FM/NM stack. The layer stack essentially acts as a spin filter. The FM is magnetized down, and the electrons of majority (minority) spin are spun down (up). Excess minority-spin (up) electrons accumulate up-stream and excess majority-spin (down) electrons down-stream within the spin diffusion length from the interface of FM.

In 1973, Tedrow et al. studied electrons tunneling from an FM into a superconducting aluminum and found the injected electrons are indeed spin polarized [14]. In 1994, Y. Lassailly observed that very thin Au–Co–Au film (Co ~ 2 nm) acts like a spin filter [15]. The transmission coefficient of minority-spin electrons is about 0.7 times that of majority-spin electrons. The figure of merit while discussing the spin filtering effect is spin polarization factor P, which is defined as

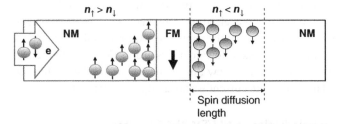

Figure 6.6 When current flows through an NM/FM/NM stack, minority-spin electrons accumulate up-stream near the NM/FM interface, and majority-pin electrons accumulate near the down-stream FM/NM interface, within the spin diffusion length region.

$$P = \frac{(n_\uparrow - n_\downarrow)}{(n_\uparrow + n_\downarrow)}, \tag{6.3}$$

where $n_\uparrow$ and $n_\downarrow$ are the populations of electrons with majority spin and minority spin, respectively. Taking Lassailly's data as an example, one obtains a polarization factor $P = (1–0.7)/(1 + 0.7) \sim 0.18$. Although this value is smaller than those in the recent literature, $P \sim 0.35$ for Co [16], the filter action is evident. Among ferromagnetic films, CoFeB is known to have the highest polarization factor, $P \sim 0.65$, followed by Fe, $P \sim 0.40$ [17]. These early works established the fact that an FM film polarizes electron current in terms of its spin degree of freedom.

In the NM/FM/NM structure shown in Figure 6.6, $n_\uparrow \neq n_\downarrow$ within the spin diffusion length near the FM/NM interface. A vector called *spin current*, which describes the transport of spin angular momentum, can be further defined as

$$\mathbf{J}^S = \hbar/2(n_\uparrow v_\uparrow - n_\downarrow v_\downarrow), \tag{6.4}$$

where $v_\uparrow$ and $v_\downarrow$ are velocity of spin-up and -down electrons, respectively. The spin current is in unit of angular momentum per unit area per unit time or (N/m). (Strictly speaking, $\mathbf{J}^S$ is spin angular momentum current *density*.) It represents a flow of angular momentum carried by conduction electrons, each with angular momentum of $\hbar/2$. The unit of charge current $J^C = e \ (n_\uparrow v_\uparrow + n_\downarrow v_\downarrow)$ is charge density in unit of charge density times velocity $(Coulomb/s^*m^2)$. The unit conversation between charge current and spin current is $\hbar/2e$, and its dimension is $(Joule^*s/C)$. Notice that $v_\uparrow$ and $v_\downarrow$ may not have to be in the same direction. For example, when $n_\uparrow = n_\downarrow$, a spin current may still flow $(n_\uparrow v_\uparrow - n_\downarrow v_\downarrow \neq 0)$ if $v_\uparrow \neq v_\downarrow$. Also note that spin current can be roughly categorized into two groups, namely, *spin-polarized current* and *pure spin current*, as shown in Figure 6.7. Spin-polarized current represents the case for spin-filtering effect, i.e. $n_\uparrow \neq n_\downarrow$, but $v_\uparrow$ and $v_\downarrow$ are along the same direction. Pure spin current, on the other hand, represents the case of pure spin separation (such as the SHE) without a net charge flow, where $n_\uparrow = n_\downarrow$ but $v_\uparrow$ and $v_\downarrow$ are pointing along different directions.

6.2.2 Spin Current Injection, Diffusion, and Inverse Spin Hall Effect

In Section 6.1.2, we discuss SHE: a longitudinal charge current flow J^C can generate a transversal spin current J^S. In this section, we discuss the experimental studies that show $J^S = \hbar/2(n_\uparrow v_\uparrow - n_\downarrow v_\downarrow) \neq 0$ indeed exists in regions where the net charge current $J^C = e(n_\uparrow v_\uparrow + n_\downarrow v_\downarrow)$ is zero. We will further show that spin current can generate charge current, which is the *inverse spin Hall effect* (ISHE). ISHE has become an important experimental approach to study spin current properties in various types of film stacks [19–21].

Figure 6.8a schematically illustrates the side view of a nonlocal spin valve (NLSV), which consists of two FM contacts on top of an NM bar. This structure

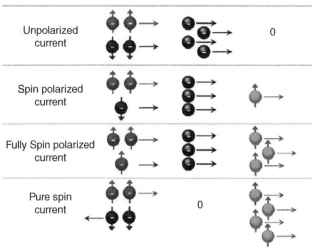

Figure 6.7 Various types of spin currents. *Source:* Adapted from Feng et al. [18].

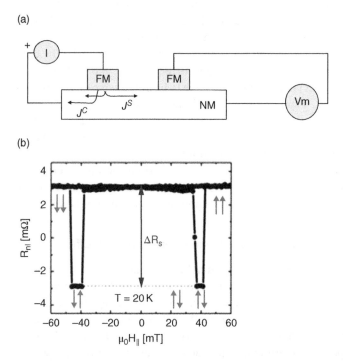

Figure 6.8 Spin current in an NLSV with ferromagnetic injector and detector. When spin-polarized electrons are injected from the left FM electrode into the NM, the electrons are drained to the left. The spin current diffuses and reaches the FM detector to the right FM electrode. A nonlocal resistance $R_{nl} = V_m/I$ is therefore detected. The polarity of R_{nl} depends on the relative magnetization alignment of the FM injector and detector [18].

and its variations have played important roles in the experimental discovery of many spin-dependent transport phenomena and was first demonstrated by Johnson and Silsbee in their seminal paper [22]. The left FM contact serves as spin injector and the right one as spin detector. The basic idea is to inject a spin polarized current I from the FM contact into an NM Hall bar, which generates a net spin accumulation near the injection contact. The concomitant charge current J^C can be drained toward the left side contact, while the spin accumulation diffuses from the contact in all directions and results in a pure spin current J^S on the right side of the injection contact. This spin current diffuses toward the right and is detected by a spin-sensitive detector (e.g. the other FM contact), which develops a voltage V_m, and a nonlocal resistance is defined as $R_{nl} = V_m/I$. When the magnetization of the two FMs are of opposite polarity, the voltage obtained at the detector is smaller (see Figure 6.8b). Therefore, similar to a regular spin valve, the measured resistance depends on the relative orientation of the two FM electrodes.

Borrowing the concept of spin injection from the NLSV structure, one can also use the generated pure spin current J^S to study other spin-dependent transport properties, for example the ISHE. The structure for detecting ISHE is illustrated in Figure 6.9, which consists of an FM spin injector on top of an NM Hall bar. While a charge current flows on the left side of the injector electrode FM, a pure spin current flows toward the right side. The pure spin current is made up of spin-up (majority spin corresponds to the magnetization of FM) electrons flowing east-bound, and the spin-down electron flows westbound. Through SHE, electrons of both spin polarizations skew toward electrode A, resulting in a finite electrical potential between A and B [24]. Thus, a stream of nonzero pure spin current induces a charge current in between terminals A and B – the *inverse spin Hall effect*.

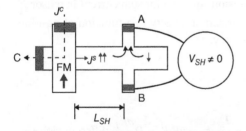

Figure 6.9 The spin injector is FM, and the Hall cross is made of NM. Charge current is injected from FM and drained through electrode C. Once injected, the majority-spin (spin-up) electrons diffuse eastbound and, through SHE, skew toward electrode A. Similarly, the spin-down electron flows westbound and skew toward electrode A. Note that in NM, $n_\uparrow v_\uparrow + n_\downarrow v_\downarrow = 0$ and $n_\uparrow v_\uparrow - n_\downarrow v_\downarrow \neq 0$ along the x-direction. The accumulation of electrons at electrode A induces an electric potential difference between electrode A and B; thus, a measurable Hall voltage is built up between A and B. The measurement was conducted at temperature 4.2 K [23].

By measuring the inverse spin Hall voltage signal V_{SH} as a function of the distance between the spin injector and the Hall cross (A, B), the spin current diffusion length can be characterized. The ISHE voltage decays as $e^{-x/\lambda}$, where λ is the spin diffusion length. Valenzuala found that spin diffusion length is over 705 nm of a 25 nm thick aluminum film at 4.2 K. At room temperature, the spin diffusion length of a transition metal is typically in the order of 10–100 nm, and Cu spin diffusion length is up to 0.3 μm due to its weak intrinsic spin-orbit interaction [25, 26].

In fact, since the ISHE voltage is directly related to the SHE, many of the early studies utilizes ISHE to determine the strength of SHE in different transition metal systems, rather than directly probing the SHE. This was the case for the discovery of large SHE in 5d transition metal Pt at room temperature [21].

6.2.3 Generalized Carrier and Spin Current Drift-Diffusion Equation

D'yakonov and Perel have generalized the drift-diffusion equation to couple both spin and charge carrier current [27, 28] into a unified formulation. The well-known drift-diffusion equation of electron charge current is

$$\frac{\boldsymbol{J}^c}{e} = \mu n \boldsymbol{E} + D \nabla n, \tag{6.5}$$

where $\boldsymbol{J}^c$ is the charge current density, $\boldsymbol{E}$ is electric field, and ∇ is *del* (the gradient of $\nabla n = \dfrac{\partial n}{\partial x}\hat{x} + \dfrac{\partial n}{\partial y}\hat{y} + \dfrac{\partial n}{\partial z}\hat{z}$ in Cartesian coordinate, and $\hat{x}$ is unit vector in x-direction, etc.). All are vectors; n is electron density as a function of position (x, y, z), D is diffusion constant, and μ is electron mobility.

Similarly, one can define a drift-diffusion equation for spin current in tensor J_{ij}^S, with the j component of the spin polarization density P, a function of position, flowing in the i-direction, and $(I, j) \in (x, y, z)$:

$$\frac{J_{ij}^S}{\hbar} = -\mu E_i n P_j + Dn\frac{\partial P_j^*}{\partial x_i}. \tag{6.6}$$

* {D'yakonov's original expression is $\dfrac{J_{ij}^S}{\hbar} = -\mu E_i P_j - D\dfrac{\partial P_j}{\partial x_i}$, different sign.}

The charge and spin currents are coupled through spin-orbit interaction; for materials with inversion symmetry, this coupling can be included by modifying Eqs. (6.5) and (6.6) as follows:

$$\frac{\boldsymbol{J}^c}{e} = \mu n \boldsymbol{E} + D \nabla n + \Theta_{SH}\mu(\boldsymbol{E} \times \boldsymbol{P}) + \Theta_{SH}D(\nabla \times \boldsymbol{P}) \tag{6.7}$$

**

$$\frac{J_{i,j}^S}{\hbar} = -\mu E_i n P_j + Dn\frac{\partial P_j}{\partial x_i} - e_{i,j,k}\left(\Theta_{SH}\mu n E_k + \Theta_{SH}D\frac{\partial n}{\partial x_k}\right), \tag{6.8}$$

where Θ_{SH} is the *spin Hall angle*. We will come back to its definition later in this section. The third term in Eq. (6.8) describes the direct SHE, where a transverse spin current $-e_{i,j,k}(\Theta_{SH}\mu n E_k)$ is generated in response to an electric field E_k. For example, let (i, j, k) be (y, z, x). The third term of Eq. (6.8) is $-(\Theta_{SH}\mu n E_x)$, which contributes to $\frac{J_{y,z}^S}{\hbar}$, a spin current in the y-direction and electron spin in the z-direction.

These two equations should be complemented by the continuity equation of spin current,

$$\frac{\partial P_j}{\partial t} + \frac{\partial J_{ij}^S}{\partial x_i} + \frac{P_j}{\tau_s} = 0, \tag{6.9}$$

where τ_s is spin relaxation time.

**[In D'Yakonov's original paper, the third term of Eq. (6.7) is $\Theta_{SH}(\boldsymbol{E} \wedge \boldsymbol{P})$, indicating any possible direction in the plane normal to the vector $\boldsymbol{E} \times \boldsymbol{P}$, but not a vector. This means that the spin current flowing toward the film surface will be polarized in-plane, in direction transversal to the charge current direction.]

Similarly, the third term in Eq. (6.7) describes the *anomalous Hall effect* in the presence of a net spin polarization, while the fourth term in Eq. (6.7) describes the *Inverse spin Hall effect*, i.e. the generation of a charge current in response to gradient in the spin accumulation.

The spin Hall angle Θ_{SH} is the ratio of spin current conductivity $\sigma_{xy}^S = \hbar\mu n\Theta_{SH}$ and charge current conductivity $\sigma_{xx}^c = e\mu n$, in the absence of an electric field. It can readily be expressed as

$$\Theta_{SH} = \frac{\sigma_{xy}^S}{\sigma_{xx}^c}\frac{e}{\hbar}. \tag{6.10}$$

This is a measure of the charge current to spin current conversion efficiency. Researchers are looking for material with large spin Hall angle with some success; see Chapter 3.

Similar to the right-hand rule, when electron current flows in the x-direction, electron spin in the y-direction forms a spin current that flows in the +z direction, and electron with spin polarization in the –y direction flows in the –z-direction. Notice that each electron carries $\hbar/2$ angular momentum; this is illustrated in Figure 6.10. Eq. (6.10) can be rewritten as

$$|J^S| = \frac{\hbar}{2e}\Theta_{SH}|J^C|. \tag{6.11}$$

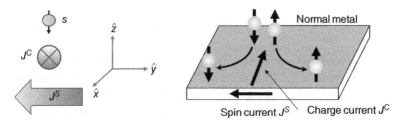

Figure 6.10 Direction of particle current, spin current, and spin polarization.

While the unit of charge current J^C is {charge in unit volume x velocity} or (Coulomb/m²S), the unit of spin current J^S is a flow of angular momentum in unit area or (N/m). The unit conversion between charge current and spin current is $\hbar/2e$, and its dimension is (Joule.S/C).

6.3 Spin Dynamics

When excited, a magnetic moment in a ferromagnet precesses around an effective field. This effective field can be the sum of various different contributions such as external field, demagnetizing field, anisotropy field, etc. During this process, the energy of the precessing moment is dissipated into the lattice or into other forms of energy, and the precession dynamics gradually cease. This is called the damping process. Without additional supply of energy, the precession eventually comes to an end, and the moment aligns to the effective field. This momentum behavior is captured by the Landau-Lifshitz-Gilbert (LLG) equation, a phenomenological equation. On the other hand, with a continuous supply of energy at the precession resonant frequency to counter the damping, the precession dynamics can be sustained. This *ferromagnetic resonance* is a powerful tool for studying the damping coefficient of different magnetic materials and film stack structures.

When the anti-damping torque generated from the spin current is large enough to overcome the damping torque, the spin current can sustain precession and/or induces magnetization reversal to the local magnetization. The spin current induced magnetization reversal was first pointed out by Butler [29] and Slonczewski [30] in their seminal papers. We study the interaction of the spin current and the local magnetization in the following sections.

6.3.1 Landau-Lifshitz and Landau-Lifshitz-Gilbert Equations of Motion

In a ferromagnet, when an external magnetic field is applied to a magnetization m, it exerts a torque $-m \times H_{\text{eff}}$ to the magnetization. The effective field H_{eff} is the negative derivative of the total energy with respect to the magnetization at position **r**:

$$H_{\text{eff}}(\boldsymbol{r}) = -\frac{\partial E_{\text{tot}}(\boldsymbol{r})}{\partial \boldsymbol{M}(\boldsymbol{r})}. \tag{6.12}$$

The total energy E_{tot} of Eq. (6.11) is a sum of the crystalline anisotropy energy, demagnetization energy, Zeeman energy, and the energy induced by the external field. Therefore, $H_{\text{eff}}(\boldsymbol{r})$ can be considered as the sum of all kinds of magnetic field contributions.

The response upon the application of an external field is magnetization precession. The magnetization precession is a dynamic process. Understanding the behavior of this dynamic process requires solving the Landau-Lifshitz dynamic equations. The dynamic equation of motion for the magnetization $\boldsymbol{M}(\boldsymbol{r})$ is a macroscopic extension of the spin Hamiltonian and can be written as

$$\frac{\partial \boldsymbol{M}(\boldsymbol{r})}{\partial t} = -\gamma \, \boldsymbol{M}(\boldsymbol{r}) \times H_{\text{eff}}(\boldsymbol{r}) - \gamma \frac{\lambda}{M_s} \, \boldsymbol{M}(\boldsymbol{r}) \times [\boldsymbol{M}(\boldsymbol{r}) \times H_{\text{eff}}(\boldsymbol{r})], \tag{6.13}$$

where γ is the gyromagnetic ratio of electron and λ is a phenomenological damping parameter. The first term on the right side of Eq. (6.13) represents the precession of the moment, and the second term is a phenomenological damping term. To be more specific, the first term describes the precession of magnetization around the effective field, which provides the magnetic torque $\boldsymbol{M} \times H_{\text{eff}}$ for the rotation of magnetization. The second term describes the energy dissipation, which damps the precessional motion and acts as a torque pulling the magnetization toward the effective field. After reaching the equilibrium state, the magnetization aligns along the direction of the effective field. Figure 6.11 shows the motion of magnetization as it is described by the Landau-Lifshitz dynamic equations.

Alternatively, if the damping effect is strong and is acting like a torque, Gilbert expressed the damping term as $\dfrac{\alpha}{M_s} \boldsymbol{M}(\boldsymbol{r}) \times \dfrac{\partial \boldsymbol{M}(\boldsymbol{r})}{\partial t}$, where $\alpha > 0$ is the dimensionless Gilbert damping constant. Therefore, Eq. (6.13) can also be written as

$$\frac{\partial \boldsymbol{M}(\boldsymbol{r})}{\partial t} = -\gamma \, \boldsymbol{M}(\boldsymbol{r}) \times H_{\text{eff}}(\boldsymbol{r}) + \frac{\alpha}{M_s} \, \boldsymbol{M}(\boldsymbol{r}) \times \frac{\partial \boldsymbol{M}(\boldsymbol{r})}{\partial t}. \tag{6.14}$$

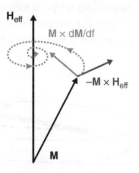

Figure 6.11 The precession and the damping of the magnetization around an effective field according to the Landau-Lifshitz-Gilbert dynamic equation.

This is the so-called Landau-Lifshitz-Gilbert (LLG) equation. Also note the sign change of the second term on the right side of Eq. (6.14). If we further normalized both sides of Eq. (6.14) by the saturation magnetization M_s, then the LLG equation can be simplified to this more commonly seen form:

$$\frac{\partial \boldsymbol{m}(\boldsymbol{r})}{\partial t} = -\gamma\, \boldsymbol{m}(\boldsymbol{r}) \times H_{\text{eff}}(\boldsymbol{r}) + \alpha\, \boldsymbol{m}(\boldsymbol{r}) \times \frac{\partial \boldsymbol{m}(\boldsymbol{r})}{\partial t}. \tag{6.15}$$

When the external magnetic field suddenly changes direction, the magnetization will precess around the new direction of the effective field and eventually align to the direction of the new effective field as the damping dissipates the precession energy.

6.3.2 Ferromagnetic Resonance

Ferromagnetic resonance (FMR) is a method to measure various magnetic properties by detecting the precessional dynamics of the magnetization in a ferromagnetic sample. Microscopically, a magnetic field H creates a Zeeman splitting of the energy levels, and a fixed frequency radio wave signal is applied to excite magnetic dipole transitions between these split levels, while the DC magnetic field H varies (Figure 6.12a). The RF absorption near the resonant frequency is measured. The resonance signal resembles a Lorentzian line shape (Figure 6.12b). The resonance field H_{res} depends on the angles of the field, anisotropy parameters, g-factor, and saturation magnetization of the sample [31].

When a ferromagnetic film is excited with a DC field $\boldsymbol{H}$ and an AC transverse field $h_0 e^{j\omega t}$, the combination of these two applied fields is referred to as $\boldsymbol{H_a}$. The magnetization precesses around the DC field. Unlike what we saw in the previous section, this precession dynamics does not die down due to damping. Instead, the precession is sustained with the energy from the AC field h_0. When the frequency

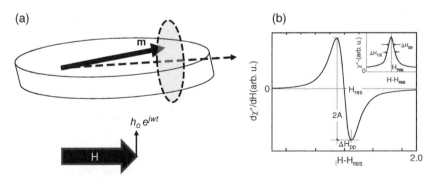

Figure 6.12 (a) Magnetization precesses under applied DC and AC fields around resonant frequency. (b) FMR spectra.

of the AC field is the same as the intrinsic precession frequency of the ferromagnetic film, the film exhibits a resonant absorption of the AC power. The resonant frequency is typically in the GHz range when the field is in a few thousands Oe. Note that in the case of FMR, the externally applied H_a and the crystalline anisotropy field in the ferromagnet H_K makes up the H_{eff}.

In a thin ferromagnetic film, the H_{eff} field contains one more term – the demagnetization field NM. Thus,

$$H_{eff} = (H_k + NM(t)) + H_a, \tag{6.16}$$

N can be decomposed as N_x, N_y, and N_z for film with the principal axes parallel to the x-, y-, z-axes of a Cartesian coordinate system. Let the applied field be $H_a = H\hat{z} + h_0 e^{j\omega t}\hat{x}$, and $H \gg h_0$. M is magnetic moment of the film $M = |M_S|V \cong M_z V$, and V is the film volume. The time varying vector components of the magnetization is M_x and M_y, since $H \gg h_0$. Ignoring the damping for the time being, Eq. (6.14) becomes

$$\frac{dM(t)}{dt} = \gamma M(t) \times H_{eff}. \tag{6.17}$$

From Eq. (6.17), the resonant frequency is

$$\omega = \omega_0 = \gamma\sqrt{AB}, \tag{6.17a}$$

$$\text{where } A = H - H_{k,y} - (N_z - N_y)M, \tag{6.17b}$$

$$\text{and } B = H - H_{k,x} - (N_z - N_x)M, \tag{6.17c}$$

where $M = |M_S|/V$. For a large film in the x-y plane, the film anisotropy is in the plane along the x-direction, i.e. $H_{k,z} = H_{k,y} = 0$, and $N_z = 4\pi$, $N_x = N_y = 0$, whence

$$\omega_0 = \gamma(H - 4\pi M) \tag{6.18}$$

For a film in the x-z plane, the DC field H_z field is in plane, then $N_y = 4\pi$, $N_x = N_z = 0$, and

$$\omega_0 = \gamma[H(H + 4\pi M)]^{\frac{1}{2}}. \tag{6.19}$$

When damping is included in Eq. (6.17), the resonant frequency shifts slightly. The Gilbert damping constant α in the Landau–Lifshitz-Gilbert equation can be further measured from the FMR line width ΔH (see the homework). The two quantities are related to each other by $\alpha = \gamma\Delta H/(2\omega)$, where ω is the frequency at which the swept field line width is measured (Figure 6.12). The absorption line width ΔH is directly connected to the relaxation processes. In thin films, Gilbert damping is commonly used to describe the relaxation. But several other possible relaxation paths are also known, e.g. spin-pumping effect (Section 6.3.3), etc., which can also contribute to the detected line width. The anisotropy constants can also be deduced from angle-dependent measurements of single crystalline samples. For magnetic multilayers, also the interlayer exchange coupling constant can be determined by FMR in absolute units.

6.3.3 Spin Pumping and Effective Damping in FM/NM Film Stack

A thin FM in contact with NM displays magnetization dynamic effects that are often nonlocal in nature. As will be mentioned in Section 6.4, a pure spin current or a spin-polarized current, entering from NM into FM, can exert a torque on the magnetization of the FM, leading to current-induced magnetic switching. The inverse of this process is the interfacial spin "pumping." The precession of the magnetization in the FM pumps spin into the adjacent NM layers [32]. Figure 6.13 illustrates the spin pumping concept. Following Section 6.3.2, a precessing magnetization in an FM can be described as $M_S \cong M_Z \, \hat{z} + m_0$ $[cos(\omega t)\hat{x} + \sin(\omega t)\hat{y}]$, where $M_Z \gg m_0$, when the precession cone angle is small. The moving component of the magnetization is in the x-y plane. The strong exchange in FM makes the entire sheet of magnetization at the interface precess in unison. This moving magnetization component exerts a torque to the electrons in the NM adjacent to the FM/NM interface through exchange interaction. The spin moment in the NM adjacent to the interface precesses. Thus, a "spin current" J^S flows into NM even though there is no net electron flow into the NM from the FM. Note that J^S is a flow of spin angular momentum generated by the precessing magnetization in the FM layer adjacent to the FM/NM interface. The "spin accumulation" takes place adjacent to the interface electrons in the NM side. The spin

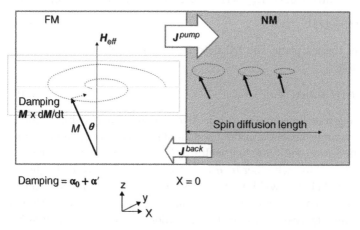

Figure 6.13 The precession of magnetic moment in an FM "pumps" spin current J^S into the NM and effectively increases the damping coefficient.

accumulation decays as $e^{-x/\lambda_{sf}}$, where x is the distance from the interface, and λ_{sf} is the spin (flip) diffusion length. If the thickness of NM is thinner than spin diffusion length, the spin current reflects back as J^{back}.

The magnitude of the spin current J^S at the FM/NM interface is given by [32, 33]:

$$J^S \hat{s} = \frac{\hbar}{8\pi} Re\left(2g^{\uparrow\downarrow}\right)\left[\boldsymbol{m} \times \frac{\partial \boldsymbol{m}}{\partial t}\right] \tag{6.20}$$

where $g^{\uparrow\downarrow}$ is the spin-mixing conductance, $\boldsymbol{m}$ is the unit vector of the magnetization in the FM, and $\hat{s}$ is the unit vector of the spin polarization in the NM conductor. Experimentally the spin-mixing conductance can be determined from the line width broadening in an FMR experiment when an NM is attached to an FM.

The spin pumping process is different from spin filtering (Section 6.4.1). Pumping involves no charge current, only a spin current (in other words, no net transfer of charge, only angular momentum transfer from FM into NM). The direction of spin current is perpendicular to the FM/NM interface plane. As long as the precession in the FM is sustained (e.g. by an AC magnetic field at resonant frequency), the spin current will continue. Through AMR and ISHE, an inverse spin Hall voltage can be detected [34, 35].

Two effects can be observed due to the spin pumping effect. As an electron travels in any metal, it scatters. Each scatter may or may not result in a spin flip, which is characterized as spin relaxation time or spin diffusion length. In some metals in which the spin relaxation is fast, the spin current transfers angular momentum from an FM into an NM, affecting the magnetization dynamics of the FM as an enhanced Gilbert damping in the Landau-Lifshitz-Gilbert equation. The enhancement of the damping constant is found to be

$$\alpha = \alpha_0 + \alpha' = \alpha_0 + \frac{a}{d_{FM}} \tag{6.21}$$

where α is the effective damping coefficient of the NM/FM/NM stack, α_0 is the damping of the FM without an adjacent NM, α' is the damping contributed by the spin current into the NM, a is a positive constant, and d_{FM} is the thickness of the FM layer [32, 36].

Figure 6.14 shows the experimentally observed enhanced damping in CoFeB film when the film contacts with heavy transition metals such as Ru and Ta. The effective damping is inversely proportional to the CoFeB film thickness, as described by Eq. (6.21). In other metals, a slow spin-relaxation rate results in a "spin accumulation" in the metal. Hence, there can be a backflow of spin current into the ferromagnet and, due to its spin-dependent conductivity, leads to a charging of the ferromagnet. This spin current has resulted in the electrical detection of

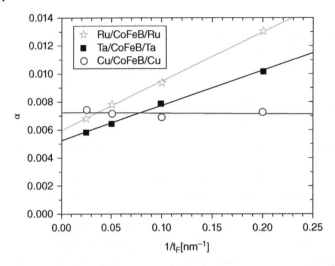

Figure 6.14 Measurement results of the damping constant of three film stacks: Ru/CoFeB/Ru, Ta/CoFeB/Ta, and Cu/CoFeB/Cu. Other than the Cu/CoFeB/Cu stack, all others show that damping constant increases with $1/t_F$, where t_F is the thickness of the CoFeB [37].

DC voltage caused by spin pumping. Cu is known to be an NM with very long spin relaxation time, $\tau_s = 10^{-12}$s in Cu [38]. With a Fermi velocity of $v_F = 1.57 \times 10^6$ m/S in Cu [39], this gives a spin-flip mean free path $\lambda_s \approx 1$ μm. (measurement: Cu ~350 nm, Ta, ~10 nm and Ru, 14 nm at 4.2K). As a result of the long spin relaxation time in Cu, there is no observable enhancement in damping when a CoFeB film is in contact with Cu [37].

6.3.4 FM/NM/FM Coupling Through Spin Current

The last section shows that spin current carries angular momentum from an FM to an NM. In an FM/NM/FM film stack, the two FM films can dynamically couple through spin current. This phenomenon is illustrated in Figure 6.15. An FM1/NM/FM2 stack is excited at a fixed AC frequency, and a DC H field is scanned (FMR measurement). Layer F1 is thinner, its resonant field is H1, and its precessing magnetic moment pumps spin current from F1 into the NM spacer. Layer F2 is thicker; it resonates at field H2 and pumps spin current from F2 to NM. Note that the two spin currents flow in opposite direction, due to the film thickness difference, H1≠H2. The angle of the applied DC *H* field determined the resonant H1 and H2 when the FMR is excited by a fixed frequency AC field (ref. to

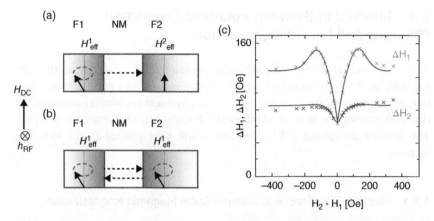

Figure 6.15 Two FM films with an NM spacer in between: F1/NM/F2 stack. H_{DC} and h_{RF} are applied to excite FMR. (a) F1 resonates, while F2 does not. F1 pumps spin current into the spacer, which is absorbed near the interface of F2 and results in enhanced damping. (b) Both films at resonance, inducing spin currents in opposite direction. (c) Effective line width of magnetic bilayers with tunable resonance fields. At H1 = H2, the resonance field of the two ferromagnetic layers coincide, and both ferromagnets emit equal and opposite spin currents, which cancel, leading to zero transfer of angular momentum and reduction in damping [40].

Section 6.3.2), due to the fact that the demagnetization field of these two films is different (Eq. [6.17a–c]). So, one may tune the fields of H1 and H2 by adjusting the DC field angle. When the resonant field H2 is detuned from H1, the DC H field that excites resonance in F1 does not excite resonance in F2 (Figure 6.15a). In this case, only one FM pumps spin current to the NM. If H1 = H2, both FM films pump spin current, as illustrated in Figure 6.15b. These two spin currents flow in opposite directions, partially cancel each other. One finds that the resonant line width ΔH is smallest when H1 = H2. When H1 is detuned from H2, the ΔH is larger; see Figure 6.15c.

In contrast to the well-known RKKY oscillatory exchange interaction in the ground state, this coupling is dynamic in nature and long ranged. Precessing magnetizations interact with each other through the spacer by exchanging nonequilibrium spin currents. When the resonance frequencies of the ferromagnetic bands differ, their motion remains asynchronous, and net spin currents persist. However, when the ferromagnets have identical resonance frequencies, the coupling quickly synchronizes their motion and equalizes the spin currents. Since these currents flow in opposite directions, the net flow across both FM_1/NM and NM/FM_2 interfaces vanishes in this case. The lifetime of the arising collective motion is limited only by the intrinsic local damping.

6.4 Interaction Between Polarized Conduction Electrons and Local Magnetization

The magnetization in a ferromagnetic film can change direction without an external field, but by a stream of high-density electrons carrying a particular polarization of magnetic (or spin) moments. This spin moment (or torque) exchange (or transfer) mechanism is what Slonczewski described in his seminal paper [30]. The spin-torque-transfer (STT) mechanism will be explained in the next few sections.

6.4.1 Electron Spin Torque Transfer to Local Magnetic Magnetization

As a stream of spin-polarized electrons is injected into an FM film, through the moment exchange, the polarization of the magnetization of an FM film may switch to the spin-polarization direction of the injected electrons. Berger first predicted the existence of this "spin transfer torque" when he explained the experimental observation of the current-induced movement of magnetic domain wall [41]. In 1996, Slonczewski [30] and Berger [29] independently predicted an electron current passing perpendicularly through a stack of an NM/FM1/NM/FM2/NM metallic multilayer (see Figure 6.16) will be able to generate a spin transfer torque to switch the magnetization in the magnetic layer stack.

As shown in Figure 6.16, a stream of right-going unpolarized electrons is first spin-polarized by the first FM layer, FM1. The magnetization direction of this FM1 layer is fixed along unit vector *s*. When the spin-polarized electrons with their spins parallel to *s* reach the NM/FM2 interface and find that the magnetization of FM2 *m* is not parallel to its spin *s*, a spin re-alignment is required for the electron to

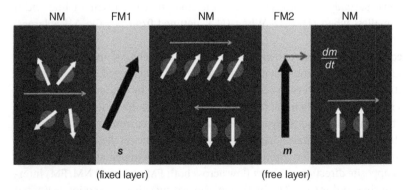

Figure 6.16 Illustration of spin-transfer torque in an NM/FM1(fixed)/NM/FM2(free)/NM layer structure.

enter/pass through FM2. If the electron is not able to re-align its spin due to scattering by a defect in the NM, its entry into FM2 is impeded, and it can be reflected at the interface. The net effect is that a torque will be transfer onto FM2 due to the conservation of (spin) angular momentum.

The magnetization switching in the FM2 occurs when the spin-polarized electron current density is greater than a threshold value, which was predicted to be in the order of mid-10^6–mid-10^7 A/cm^2 [41]. This phenomenon is called STT switching, spin-transfer-torque switching, or spin-moment-transfer switching.

The verification of the spin-transfer switching was first carried out in GMR film stacks in the 1990s. The switching-current density is in the range from mid-10^6 to mid-10^9 A/cm^2 (for example, see Refs. [42–44]). Recently, studies of STT switching on magnetic tunnel junctions show that the switching-current density is in the range of mid-10^6 A/cm^2. For a comprehensive review of the physics of spin-torque transfer, see Refs [45, 46].

6.4.2 Macrospin Model

The Macrospin model treats the magnetization $M = M_s V$ in a device as one uniform magnet, or simply considers its single domain behavior. It is based on LLG equation analysis [47].

When a flux of spin-polarized electrons is injected into a ferromagnet, the dynamic motion of the magnetization given in Eq. (6.13) is modified: Not only does magnetization M experience the torque of the usual anisotropy field and the external field in Eq. (6.13), it also experiences the torque through the exchange interaction from the injected polarized electrons passing through M. The effect of spin transfer torque on the magnetization dynamics can be taken into account as a current-density dependent term as follows:

$$\frac{\partial M}{\partial t} = -\gamma M \times H_{\text{eff}} - \gamma \frac{\alpha}{M_s} M \times (M \times H_{\text{eff}}) + \Gamma, \tag{6.22}$$

where the torque from the spin-polarized current is

$$\Gamma = \gamma \left(\frac{\alpha_J}{M_S} M \times (M \times s) + \beta_J M \times s \right), \tag{6.22a}$$

where the vector s is the moment of the polarized free electrons and

$$\alpha_J = \hbar J P / 2 M_s e t, \tag{6.23}$$

where J is the current density, P is the electron spin polarization, M_s is the saturation magnetization of the magnetic layer, and t is the magnetic layer thickness. Eq. (6.22) is also called the Landau-Lifshitz-Gilbert-Slonczewski (LLGS) equation. The third term on the right side is the torque produced by the spin-polarized electrons, i.e. spin torque or spin-transfer torque. Eq. (6.22a) describes the two components of the spin-torque term. α_j and β_j are coefficients of the in-plane and out-of-plane torque, respectively, produced by the spin-polarized current on the magnetization; both are a function of the density of the spin-polarized electrons. The first term is in-plane torque, which is in the plane formed by M and s (or $M \wedge s$ plane). The second term is out-of-plane torque, which is pointing out of the $M \wedge s$ plane. In a spin valve (FM/NM/FM), the second term (out-of-plane torque) is negligible and is usually ignored.

In the absence of injected spin-polarized electrons, $s = 0$, the third term on the right side of Eq. (6.22) vanishes and M simply precesses around H_{eff}. When spin-polarized electrons are injected into a FM film, $s > 0$ and Γ is nonzero. Consider two special cases. The first case is that M (or H_{eff}) and s are parallel. Then, $(M \times H_{eff})$ and $(M \times s)$ have the same sign, the torque is in same direction, and, thus, the second and third terms both damp the precession of the moment.

The second case is that M (or H_{eff}) and s are antiparallel, so the third term counteracts the damping. Thus, the second and third terms in Eqs. (6.22) and (6.22a) can be combined and written as follows:

$$\frac{1}{\gamma}\frac{\partial M(t)}{\partial t} = -M \times H_{eff} - \left(\alpha - \frac{\alpha_J}{|H_{eff}|}\right)\frac{M}{M_s} \times (M \times H_{eff}). \tag{6.24}$$

The effective damping is reduced by the current density of the polarized electrons. Thus, at a certain current flux density, the effective damping becomes negative. Figure 6.17 illustrates the dynamic motion of the magnetization when a flux of spin-polarized electrons is injected. At a moderate flux level, the magnetization adjusts its precession cone angle θ and eventually settles at an equilibrium angle; the precession is sustained and $\alpha - \dfrac{\alpha_J}{|H_{eff}|} = 0$. Such a DC-current-driven FMR emits a radio frequency (RF) signal. The sustained precession and its RF emission was first predicted theoretically and observed by Berger. At a higher flux level, $\alpha - \dfrac{\alpha_J}{|H_{eff}|} < 0$, the magnetization spirals away from the precession axis, the cone angle grows beyond $\pi/2$, and this results in magnetic reversal. The latter case is called the STT switching.

In a ferromagnet, the effective field H_{eff} is the sum of various components of anisotropy field and the demagnetizing field H_D, i.e. $H_{eff} = H_K + H_D$.

For an in-plane magnetized free layer, the effective field is dominated by the demagnetizing field ($H_K \ll H_D = NM_s$), where N is the demagnetization factor. While in an out-of-plane case, it is the other way around ($H_K \gg H_D$).

When $\alpha_J > \alpha|H_{eff}|$, the magnetization spirals away from the precession axis and starts to switch to the direction of the magnetization of pinned layer. Therefore, at zero temperature (Kelvin), the critical current density is given as [47]

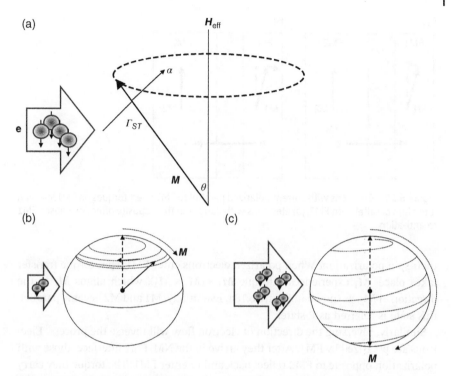

Figure 6.17 (a) Spin-polarized electrons exchange moments with the magnetization of opposite polarity, exerting a torque in a direction opposite to the damping. (b) When the current is below a threshold value, the precession angle θ first increases and then reaches a steady angle. (c) When the current is greater than a threshold value, the precession angle θ continues to grow beyond $\pi/2$ and magnetic reversal takes place.

$$J_{c0} = \frac{2e\alpha M_s t_F (H_k + H_D/2)}{\hbar P}. \tag{6.25}$$

6.4.3 Spin-Torque Transfer in a Spin Valve

Consider a spin valve structure (FM1/NM/FM2) as shown in Figure 6.18. Suppose that there is an angle θ between the magnetizations M_1 and M_2 of layers FM1 and FM2, respectively. Let us make FM2 much thicker, or pinned; thus, its magnetization M_2 is hard to switch. Free electrons in the NM electrode of FM2 (not shown in Figure 6.18) enter the film stack from the right side. When the electrons pass through the FM2 layer, FM2 acts as a spin filter. When leaving FM2, more free electrons are oriented to the direction of M2. Since the NM middle layer is very thin, thinner than the spin-relaxation length, the polarized electrons maintain

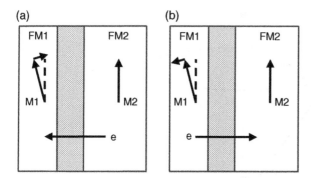

Figure 6.18 Electrons with current polarized parallel to FM2 exert torques on FM1, causing it to align parallel with FM2, (b) electrons with current in the opposite direction causes FM1 to anti-align.

their spin polarization. When polarized electrons enter FM1, spin-torque transfer takes place. M_1 experiences a torque $M_1 \times (M_1 \times M_2)$, which aligns M_1 to the direction of M_2, as shown in Figure 6.18a. Eventually, M1 and M2 become parallel; this state is referred as P-state.

Similarly, reversing the direction of electron flow will reverse the process. Electrons are polarized by FM1. After they arrive in the NM/FM2 interface, those with polarization opposite to FM2 reflect back and re-enter FM1. The torque they carry is $M_1 \times (M_1 \times (-M_2))$, which rotates M1 away from M2 and toward the opposite direction of M_2, as illustrated in Figure 6.18b. When M1 and M2 become anti-parallell, we refer this state as AP-state.

Thus, to switch the free layer of a spin valve from P-state to AP-state, the current is applied from the free layer into the pinned layer (electrons flow from the free layer to the pinned layer). To switch from AP- to P-state, current is applied from the free layer to the pinned layer (electrons flow from the pinned layer to the free layer).

A paper on the relation between spin-torque and spin-valve current can be found in [48].

6.4.3.1 Switching Threshold Current Density
Allowing for the partial polarization in the electron flux, the amplitude of the spin torque can be derived as follows [30]

$$|r| = \left(\frac{J_g(\theta)}{e}\right)\hbar, \tag{6.26}$$

where J is the current density, and e is electron charge. Note that J_g/e means the effective flux of the polarized electrons flowing from NM into FM1, and $\hbar$ is the (reduced) Plank constant. $g(\theta)$ is a function of the polarization factor P of the polarized electrons and the angle θ between polarized electron and M_1 of the FM1 layer, and it is equal to $[-4 + (1 + P)^3(3 + \cos\theta)/4P^{3/2}]^{-1}$. $g(\theta)$ vanishes at angle $\theta = 0$ and π, for which the component of transferred spin oriented orthogonal to the M_1 vanishes. Although the torque is theoretically zero when M_1 and M_2 are parallel or anti-parallel, in practice, the torque exists. The misalignment of the easy axis, as well as the thermal agitation at room temperature, causes the two M to deviate from perfect alignment. For small P, the transfer rate turns into ~$\sin\theta$.

The torque from the polarized electrons to M_1 changes its direction continually with time. At any instant in time, the torque pulls the magnetization away from the precession axis.

The torque is proportional to the flux of polarized free electrons entering FM1. As moment exchange takes place, the spin precession cone angle of FM1 increases until a new equilibrium cone angle is reached.

Consider a case with internal field $H = H_K$. The behavior of the precession cone angle is determined by the last two terms of Eq. (6.24) and can be derived as [49]

$$\frac{d\theta}{d(\gamma t)} = -\frac{1}{2}\alpha H_K \sin 2\theta - \left(\frac{\hbar J_g(\theta)}{eM_2 t_2}\right)\sin\theta. \tag{6.27}$$

6.4.3.2 Switching Time

The precession cone angle θ responds to the polarized current more quickly when the damping constant and magnetization are both small. Figure 6.19a shows the changes in FM1 precession cone angle as a function of normalized time for three current densities, $J3 < J2 < J1$, and the initial cone angle is 0.9π. H_K is in the direction of $\theta = \pi$. When the current density ($J3$) is small, the spin torque is insufficient to overcome the torque from damping, the cone angle gradually settles at $\theta = \pi$, and M_1 aligns to H_K. When the current density ($J2, J3$) is larger than a threshold, the cone angle shrinks (actually M_1 moves away from H_K), and M_1 eventually aligns to the direction of the spin moment of the incoming polarized electrons. A larger current accelerates the switching process. The switching pulse width requirement depends on the initial angle, which fluctuates under thermal agitation. A given write current pulse condition may or may not induce magnetization reversal. Thus, the switching is stochastic. It is an engineering task to reduce the write error rate, which is mainly accomplished by over-drive, i.e. to write with a current substantially larger than the threshold current.

Such transient behavior of the magnetization was verified by Koch on a Co/Cu/Co film stack [50]. The measurement result indicates that the time required to complete magnetic reversal is highly current dependent. There are two transition

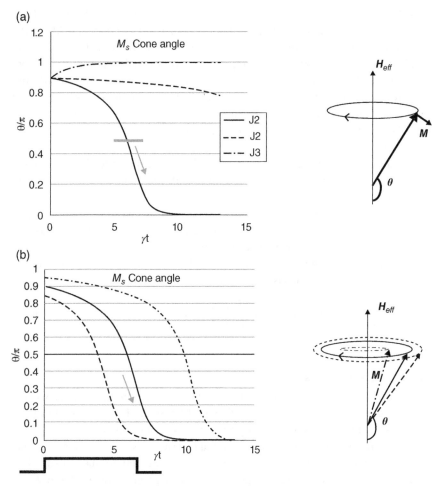

Figure 6.19 (a) Normalized precession cone angle (θ/π) between two FM1 and FM2 versus normalized transition time (γt). Solution of Eq. (6.27) with initial angle 0.9π. (b) Small variation in the initial angle changes the switching time. The switching is stochastic.

regimes. The first regime is the subnanosecond switching time τ at a current density over a critical value I_{c0}, in which

$$\tau^{-1} \approx \frac{P \cdot (\mu_B/e)}{m \cdot \ln(\pi/2\theta_0)}(I - I_{c0}), \quad \text{when} \quad I > I_{c0}. \tag{6.28}$$

where P is the polarization factor of the incoming electrons, θ_0 is the initial angle deviation from the easy axis $m = M_s V$, V is the volume of the free layer, and

$$I_{c0} = (\alpha/P)(2e/\hbar) \times m \times (H + H_K + 2\pi M_s) \tag{6.29}$$

is the zero-temperature threshold current for STT magnetic reversal. The second is a long switch time regime (>1nS) or "thermally activated" regime in which the switching current is below I_{c0},

$$\tau^{-1} \approx \tau_0^{-1} \exp\left[-\Delta_0(1 - I/I_{c0})\right], \quad \text{when} \quad I < I_{c0}, \tag{6.30}$$

or

$$I_C = I_{c0}\left[1 - \frac{1}{\Delta_0} \ln(\tau/\tau_0)\right], \tag{6.31}$$

where $\Delta_0 = E_b/k_B T$ is the normalized switching energy barrier between the AP- and P-state, and E_b is approximately equal to the anisotropy energy of the free layer $K_u V$ when all other forces are absent; τ is pulse width, and τ_0 is attempt time constant, which is typically 10^{-9} second. Note that $I_c = I_{c0}$ when $\tau = \tau_0$.

At room temperature, possessing thermal energy, the magnetization fluctuates around the easy axis. Each time, a write current pulse is applied to switch the magnetic moment, and the initial cone angle of the magnetization is slightly different. As a result, the switching time varies slightly. A given write pulse width may or may not be sufficiently long to switch the magnetization. Figure 6.19b illustrates this concept. This is one of the reasons that magnetization switching is stochastic.

The LLGS-based Macrospin model conveys a simplified picture of STT magnetization reversal under the injection of the spin current and/or spin-polarized current. In actuality, magnetization reversal may go through chaotic states before the reversal completes. Stohr's research team had recorded time-resolved images of the magnetization of a 150×100 nm^2 GMR structure [51]. They found that the magnetization reversal does not follow the Macrospin model; instead, the magnetization swirls and forms a vortex, and then the vortex drifts before reversal completes.

6.4.4 Spin-Torque Transfer Switching in Magnetic Tunnel Junction

In a spin valve, the voltage across the thin metal films of spin valve is negligibly small. As discussed in the previous section, its magnetization switching is analyzed based on spin valve current. The switching threshold currents are insensitive to switching direction, from AP-state to P-state or vice versa. The tunnel barrier of MTJ makes MTJ a high impedance device. Although its I–V relation is insensitive to the direction of current flow, its switching current threshold is not. The experimental MTJ switching data show that while the switching voltage is insensitive to the switching direction, the threshold switching currents are very different [52, 53]. For a typical MTJ with a TMR ratio of 100%, when the switching voltage of both polarities is the same, it means the switching current of two polarities differs by a factor of 2. Slonczewski and Sun had formulated a voltage-driven torque model for MTJ STT switching [54]. The model takes the spin polarization through the tunnel barrier and the voltage-dependent electron tunneling into consideration [55]. Deviating from previous models for spin valve [49, 56], the switching torque is now voltage-driven.

While the tunnel junction current is related to the voltage across the tunnel barrier through a *conductance dI/dV* relation, the torque is also related to the junction voltage as "*torkance*" $d\Gamma/dV$, where V is the voltage across the tunnel barrier. They formulate the current density across the MTJ as

$$J(V, \theta) = J_0(V)\left[1 + P_f P_r \cos(\theta)\right], \tag{6.32}$$

where θ is the angle between the magnetization M of the two ferromagnetic electrodes of the tunnel junction, and P_f and P_r are the polarization of free and reference layer, respectively, and are related to the tunneling magnetoresistance as $TMR = 2P_f P_r/(1-P_f P_r)$ and Julliere's polarization factor of the free and pinned layer of MTJ and

$$P_i = \frac{k_{i,+} - k_{i,-}}{k_{i,+} + k_{i,-}} \cdot \frac{k_0^2 - k_{i,+} k_{i,-}}{k_0^2 + k_{i,+} k_{i,-}}, \tag{6.33}$$

where $i = f, r, k_{i,+}$, and $k_{i,-}$ are the Fermi wave vector of the majority $(+)$ band and minority $(-)$ band, and k_0 is the imaginary wave vector in tunnel barrier. In MTJ, charge current and spin torque are connected through a factor $\hbar/2e$ (i.e. the ratio of spin current/charge current). The flow direction of the spin current is normal to the spin polarization direction. Thus, both the torque and current are bias voltage and angle θ dependent, in the form of $\sin\theta$.

Figure 6.20 shows an MTJ with pinned layer magnetization M_r and a free layer M_f. Electrons are injected from the pinned layer across the tunnel junction into the

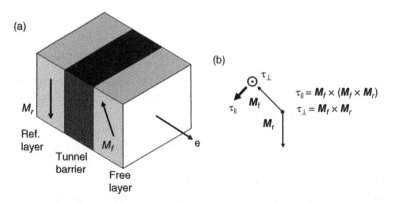

Figure 6.20 (a) Magnetic tunnel junction with electron enters the reference layer from the left and leaves from free layer on the right. The magnetization of the reference layer is not parallel to the free layer. (b) Two spin torque components acting on magnetization of the free layer M_f. One is $\tau_{\parallel} = M_f \times (M_f \times M_r)$ in the plane formed by two vectors M_f and the magnetization of reference layer M_r; the other is $\tau_{\perp} = M_f \times M_r$, which points to a direction perpendicular to the plane.

free layer. Each spin-polarized electron exerts two torques to the magnetization of the free layer: one is a torque $M_f \times (M_f \times M_r)$, which lies in the plane formed by $M_f \wedge M_r$, and is called in-plane torque ($\tau_{\parallel}$). The second torque is $M_f \times M_r$, which lies in a direction normal to the $M_f \wedge M_r$ plane, or out-of-plane, and is called the perpendicular or out-of-plane torque ($\tau_{\perp}$). It is also called the "field-like torque." Whereas $\tau_{\perp}$ in spin valve is negligible [50], $\tau_{\perp}$ in MgO-based MTJs can be a large fraction of $\tau_{\parallel}$ [53, 57, 58]. The Macrospin model of spin-torque transfer already includes $\tau_{\parallel}$ as the third term of Eq. (6.18). For a magnetic tunnel junction, it includes one more torque term, the out-of-plane torque term $\tau_{\perp}$.

6.4.5 Spin-Torque Ferromagnetic Resonance and Torkance

Within the Macrospin approximation, the dynamics of M of the precession layer of an in-plane film follows Landau-Lifshitz-Gilbert (LLG) equation with spin-torque terms,

$$\frac{dM}{dt} = -\gamma M \times H_{\text{eff}} + \alpha M \times \frac{dM}{dt} - \gamma \frac{\tau_{\parallel}(I, \theta)}{(M * Vol)} \hat{y} - \gamma \frac{\tau_{\perp}(I, \theta)}{(M * Vol)} \hat{x}, \quad (6.34)$$

where the third and fourth terms on the right side are in-plane and out-of-plane torque, respectively, both are functions of current and angle θ between M_f and M_r; the in-plane easy axis is in the z-direction, the in-plane hard axis is in the y-direction, and the x-axis points out of the film plane, as shown in Figure 6.20.

Spin-torque ferromagnetic resonance (ST-FMR) measurement [53, 57–63] was applied to study spin torque in MTJ. The difference between the FMR (Chapter 5) and the ST-FMR test is that the AC H-field used in the FMR test is replaced by a small AC current I_{AC}, across the two terminals of MTJ. When the AC frequency is near the resonance frequency of the magnetic layers, the magnetization in the free layer or both free and reference layers can be driven to precess, which modulates the angle θ between M_f and M_r. Since $R_{\text{MTJ}}(t)$ is a function of the instantaneous angle $\theta(t)$, the instantaneous terminal voltage $V(t) = (I_{\text{AC}}(t) \times R_{\text{MTJ}}(t))$ oscillates. From the magnitude and peak shape of the lowest-frequency normal mode, one can derive the in-plane and out-of-plane components of $d\tau/dV$, the *torkance* [64]. The in-plane torque is monotonic and changes sign as DC bias reverses. The out-of-plane spin torque is nearly quadratic. In an MTJ, the out-of-plane component is small but not negligible. The same conclusion was reached when the test was repeated using a network analyzer [65].

Figure 6.21 shows a measurement result. While $\tau_{\perp}$ is practically zero in spin-valves (FM/NM/FM), it can reach up to 30% of the in-plane torque $\tau_{\parallel}$ in magnetic tunnel junction in-plane [65]. As for the out-of-plane torque $\tau_{\perp}$, it can emulate the action of a field on the free layer magnetization M_f, which means that it can modify

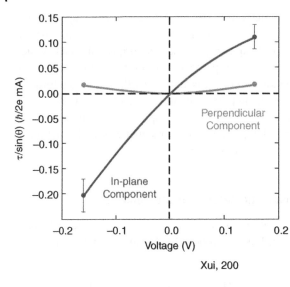

Figure 6.21 Measurement of bias *V* and angle dependence θ of spin torque in MTJ based on MTJ resistance at the resonance frequency. The extracted in-plane and out-of-plane torque is normalized to sin (θ) between the easy axis of free layer and the magnetization. The sample is a 90 nm diameter in-plane MTJ with a film stack [IrMn (6.1)/CoFe(1.8)/Ru/ CoFeB (2.0)]/ [MgOx]/ [CoFe(0.5)/ CoFeB(3.4)]/ [Ru(6.0)/Ta (3.0)/Ru(4.0)] [65].

the energy landscape seen by the magnetization. The current dependence of $\tau_\perp$ is generally more complex than $\tau_\parallel$.

6.5 Spin Current Interaction with Domain Wall

A magnetic domain wall in a ferromagnet is a transition region between domains of different magnetization orientation. Within the finite width of the domain wall, the orientation of the local magnetization rotates gradually from one edge to another edge of the transition region. Figure 6.22 illustrates a 180° domain wall,

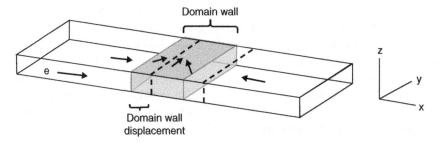

Figure 6.22 In-plane magnetized ferromagnetic film with Néel domain wall between two domains. The left domain is magnetized toward +x direction, the right one toward −x direction. Polarized electrons in the left domain flow toward the right domain. Once the polarized electrons flow into the domain wall, they exchange momentum with local magnetization.

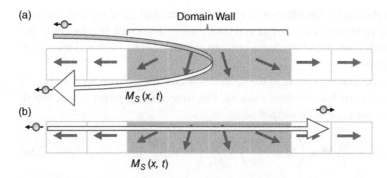

Figure 6.23 Two effects of an electron flow on a domain wall. (a) Electron reflection means the transfer of linear momentum to the domain wall. (b) "Adiabatic" transmission of electrons means the transfer of spin angular momentum [66].

two head-to-head domains separated by a Néel wall in an in-plane magnetized FM film.

Two ways the electron current may interact with the domain wall were proposed [66]. Figure 6.23 illustrates the concept. The first, called non-adiabatic transmission, is that conduction electrons act as a force on the position of wall "mass." The spatial gradient of the magnetization in the domain wall scatters the conduction electrons, and some are reflected [67, 68]. The reflected electrons transfer their *linear* momentum to the wall. This effect is proportional to the charge current and the wall "resistance" and, hence, is negligible except for very thin walls. The other is called "adiabatic" transmission. The traveling electron exchanges *angular momentum* (torque) with the local magnetization when they pass through the domain wall. The total spin angular momentum of the electrons and magnetization conserves. In other words, the magnetization in the wall absorbs the change of the electron spin, leading to a translational motion of the domain wall. This effect is the dominant one for thick walls.

6.5.1 Domain Wall Motion under Spin Current

Zhang and Li [69, 70] studied the response of the conduction electron spins in various spatial and time-varying magnetization $M(\mathbf{r},t)$ in the time-dependent semiclassical transport theory and formulated the dynamics of local magnetization with the LLG equation as

$$\frac{d\boldsymbol{M}}{dt} = -\gamma(\boldsymbol{H}_{\text{eff}} \times \boldsymbol{M}) + \frac{\alpha}{M_S}\left(\boldsymbol{M} \times \frac{d\boldsymbol{M}}{dt}\right) - \boldsymbol{\Gamma}_{\text{ST}}, \tag{6.35}$$

where M is the local magnetization (a function of time and location), and the third term on the right side is the torque carried by the spin current imposed onto the local magnetization, which includes both an in-plane component and an out-of-plane component.

For the case that current is in the x-direction and the domain wall motion is much slower than the electron velocity, the time-varying terms are dropped, and the spin torque can be simplified as

$$\Gamma_{ST} = -\frac{b_j}{M_S^2} M \times \left(M \times \frac{\partial M}{\partial x} \right) - \frac{c_j}{M_S} \left(M \times \frac{\partial M}{\partial x} \right) \tag{6.36}$$

In Equation (6.36), $b_j = JP\mu_B/(2eM_S)$, where J is the current density, μ_B is Bohr magneton, P its polarization rate, and $c_j = \xi b_j$, where ξ is the ratio of exchange relaxation time and electron spin-flip time constant, a dimensionless constant with value $\sim 10^{-2}$, which describes the degree of the nonadiabaticity between the spin of the non-equilibrium conduction electrons and local magnetization. The b_j and c_j terms are called "adiabatic" (or in-plane) and "non-adiabatic" (or out-of-plane) spin torque, respectively. While the adiabatic term is same as Slonczewski torque, the non-adiabatic term points in a direction orthogonal to the adiabatic one. For permalloy, the factor $\mu_B/(2eM_S)$ amounts to 7×10^{-11} m^3/C. The unit of b_j is m/S, which is velocity. Zhang and Li found that the initial speed of the domain wall is

$$v(0) = -\frac{1}{(1 + \alpha^2)} \left(\alpha \gamma H_{ext} W(0) + b_j + \alpha c_j \right), \tag{6.37}$$

while the terminal velocity of the domain wall is

$$v(\infty) = -\frac{1}{\alpha} \left(H_{ext} W(\infty) + c_j \right). \tag{6.38}$$

The terminal wall width $W(\infty)$ is slightly smaller than the initial wall width $W(0)$. While the adiabatic term (b_j) controls the initial velocity, the nonadiabatic term (c_j) controls the final velocity; the latter is inversely proportional to the damping constant α. In the absence of external field, the final velocity is comparable to the initial velocity, because α is much smaller than 1 since c_j/α is in same order as $b_j/(1 + \alpha^2)$. In a permalloy film (assuming $\alpha = 0.01 \sim 0.1$), this study predicted wall speed to be $6 \sim 60$ m/S, while Tatara estimated the domain wall speed can reach 250 m/S [67].

It is possible to drive the domain wall in thin film with perpendicular current injection in SV and in MTJ [71–75]. The polarized electrons from the reference layer exerts a torque to the domain wall on the free layer. The torque that drives the domain wall in the direction perpendicular to the electron motion is the out-of-plane torque. Indeed, the out-of-plane torque produces a magnetic field in the direction of the reference layer that has the proper symmetry to push the DW along

Figure 6.24 The average velocity of domain wall versus current density. In a FM film free of pinning site, the critical current density is proportional to the transversal anisotropy and wall width.

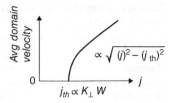

the plane of the free layer. This has been predicted in [75]. The in-plane torque can only slightly shift the domain wall by a few nanometers. While the out-of-plane torque amplitude is very small in metallic spin-valves, typically $c_J \ll 0.1\ b_J$, it has been shown experimentally that in magnetic tunnel junctions it can reach 30% of the in-plane torque.

6.5.2 Threshold Current Density

There is a threshold current density J_{th} below which the domain wall does not move (Figure 6.24) or the domain wall is "pinned"; J_{th} is usually also referred to as the "depinning" current density. The domain position is held by transversal anisotropy and by pinning sites [68, 76]. In an FM free of pinning sites, the critical current is proportional to the transversal anisotropy and the domain wall width. Once the current density exceeds the threshold, the average velocity takes the form of $\sqrt{(J)^2 - (J_{th})^2}$. The potential of a pinning site traps the domain wall, and it takes energy from the drive current to free the domain wall or to "de-pin" the domain wall [77]. The density of pinning sites affects both the threshold current density and the average domain-wall velocity. It is possible to speed up the average domain-wall velocity by the shape control that lowers the hard-axis coercivity, for example: the film thickness [76].

The domain wall velocity has been measured in wire-shaped devices with a simple permalloy (Py, $Ni_{81}Fe_{19}$) layer, spin valves, or MTJ film stack structures, in both current-in-plane and current-perpendicular-to-plane directions. All these experiments show current-induced domain wall motion (CIDM) action takes place. A wide range of the domain wall velocities were found, depending on the current flow directions. Table 6.1 shows a sample of experimental data. The de-pinning current density is shown from low-10^{10} to mid-10^{11} A/m^2.

More recently, current-induced domain wall motion is also found in asymmetric Pt/Co/Pt nanowire. The magnetization in Co is out-of-plane. An in-plane current, when above the de-pinning threshold, drives the domain wall. The polarized electrons not only result from the polarization of electron in the Co domain but also through the spin Hall effect (SHE) from the top and bottom Pt layers [82]. This structure provides additional engineering control of the properties of domain wall motion.

Table 6.1 Experimental data of current-induced domain wall motion. CIP: current-in-plane, CPP: current perpendicular to plane.

Sample	Current dir.	DW velocity (m/S)	J_{th} (10^{10} A/m^2)	Reference
Permalloy	CIP	3	70	[76]
SV CoO$_3$/Co7/Cu10/NiFe5/Au3	CIP	—	6.7	[78]
SV NiFe/Cu/Co	CIP	600	4	[79]
MgO - MTJ	CPP	500	6	[80]
MTJ [Co/Ni]$_n$/CoFeB	CIP	15–50	60–70	[81]

The fact that the velocity of the current-driven domain motion varies widely from the same to the opposite direction of the current flow; and the fact that the motion is heavily dependent on film stack material [82, 83], [84] suggests that in addition to the spin transfer torque of the conduction electrons, more mechanisms may be involved in the wall motion. More recent research shows that the SHE and Dzyaloshinskii-Moriya interaction (DMI) together play important roles in current-driven domain wall motion in various magnetic heterostructures [85], [86], [87], [88], [89]. This field remains active.

Homework

Q6.1 Due to spin Hall effect, when a current flows in a cylindrical conductor, the electron spins will accumulate at the lateral boundary. (a) Sketch the electron spin orientation on the surface. (b) Does the spin polarization reserve when the current direction is reserved?

A6.1 (a)

j

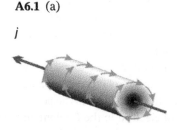

(b) From Eq. (6.7), yes, the spin current reserves if charge current is reversed.

Q6.2 What is the lateral distribution spin-up (spin-down) electrons in a rectangular conductor?

A6.2 For a stripe of with L, the spin-up electrons accumulate at y = 0, and spin down electrons accumulate at y = L. The lateral distribution of spin electron is the same as charge electron, a gradient starts from the edge.

$$n_\uparrow(y) = n_\uparrow(0) \exp\left(-\frac{y}{\lambda_s}\right); \; n_\downarrow(y) = n_\downarrow(L) \exp\left(-\frac{L-y}{\lambda_s}\right),$$ where λ_s is the spin flip length.

Q6.3 What is the spin current at a particular position in a conductor that all electrons are polarized in the z-direction and moving in the x-direction?

A6.3 (a) All carriers n are polarized in the z-direction, and $P = \dfrac{(n_\uparrow - n_\downarrow)}{(n_\uparrow + n_\downarrow)}$ $= 1 = P_z$. All carriers move with velocity v in the x-direction, and the spin current is the first term of Eq. (6.7) and $J_{i,j}^S = J_{x,z}^S = \hbar J_{x,z}^S = \hbar nvP = \hbar nv$.

Q6.4 Is there is a Hall voltage across a Hall bar when (a) an unpolarized current and (b) a polarized current flows in an NM stripe?

A6.4 (a) A typical Hall bar structure is shown in Figure Q6.4a. A unpolarized charge current flows in the x-direction, and it generates no Hall

Figure 6.Q4 (a) Electron in NM Hall bar, (b) FM inject electrons into NM Hall bar.

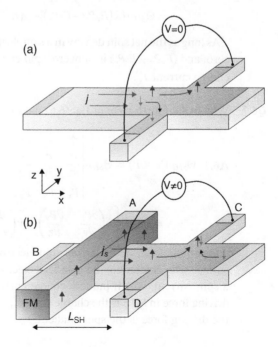

voltage, since electrons are made up of an equal number of spin-up electrons and spin-down electrons. (b) A spin-polarized current is generated with a structure, as shown in Figure Q6.4b. Applying a voltage across FM and NM, charge current injected from FM to NM is polarized. Since the number of spin-up electron is greater than those of spin-down, there is an Anomalous Hall voltage across the Hall bar.

Q6.5 What is the direction of spin current J^S for electrons with spin polarization in the −z direction. Refer to Figure Q6.4b.

A6.5 Since the FM injects electrons polarized in the −z direction and the elections flow in the x-direction, $J^S_{y,(-z)} = \left(\dfrac{\hbar}{2e}\right)\Theta_{SH}\left(-J^C_x\right)$. J^S flows in the −y direction. As more electrons accumulate in the −y side of the Hall bar, electrode D is more negative than C.

Q6.6 Why does the third term of Eq. (6.6) describe AHE?

A6.6 Let us expand the third term of Eq. (6.6) as

$$\Theta_{SH}\mu(\mathbf{E} \times P) = \Theta_{SH}\mu \begin{vmatrix} \hat{x} & \hat{y} & \hat{z} \\ E_x & E_y & E_z \\ P_x & P_y & P_z \end{vmatrix}$$

$$= \Theta_{SH}\mu\left\{\hat{x}\left(E_y P_z - E_z P_y\right) - \hat{y}\left(E_x P_x - E_z P_z\right) + \hat{z}\left(E_x P_y - E_y P_x\right)\right\}.$$

As long as the net spin density in a sample nonzero ($n_\uparrow - n_\downarrow \neq 0$), P is nonzero, and ($E_y P_z - E_z P_y$) is nonzero, spin current can induce an x-direction charge current J^C_x.

Q6.7 Why is the fourth term of Eq. (6.6) the inverse Spin Hall effect?

A6.7 $$\Theta_{SH}\mu(\nabla \times P) = \Theta_{SH}\mu \begin{vmatrix} \hat{x} & \hat{y} & \hat{z} \\ \dfrac{\partial}{\partial x} & \dfrac{\partial}{\partial y} & \dfrac{\partial}{\partial z} \\ P_x & P_y & P_z \end{vmatrix}$$

$$= \Theta_{SH}\mu\left\{\hat{x}\left(\dfrac{\partial P_z}{\partial y} - \dfrac{\partial P_y}{\partial z}\right) - \hat{y}\left(\dfrac{\partial P_z}{\partial x} - \dfrac{\partial P_x}{\partial z}\right) + \hat{z}\left(\dfrac{\partial P_y}{\partial x} - \dfrac{\partial P_x}{\partial y}\right)\right\}$$

Experimentally, net polarized spin electrons are excited by light on the surface. The gradient of the polarized electrons in y- and z-directions form a spin current, which induces charge current in the x-direction [9]. The driving force in SHE is the charge current, while in the inverse Hall effect, the driving force is the spin gradient.

Q6.8 Prove that while charge current changes sign in time reversal, spin current does not change sign.

A6.8 Both the charge current and the spin current change sign under space inversion (because spin is a pseudo-vector). In contrast, they behave differently with respect to time inversion: While the electric (charge) current changes sign,

$$J^C(-t) = en(V_{(-t)}) = -en(V_{(t)}) = -(env) = -J^C(t),$$

the spin current does not change sign, because spin, like velocity, changes sign under time inversion.

$$J^S_{x,z}(-t) = J^S_{-x,-z}(t)$$

Q6.9 Please sketch the spin potential and spin current as functions of x for the experimental setup of detecting ISHE.

A6.9

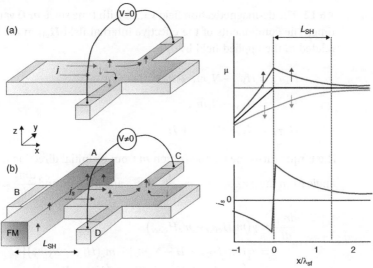

Q6.10 Assume a stream of free electron with current density of 1.6×10^6 A/cm², and each carries angular momentum $\frac{\hbar}{2}$. The electrons are all polarized such that $n = n_\uparrow$. What is the angular momentum of the ensemble?

A6.10 Let electron velocity be $0.8 \ 10^7$ cm/S. The spin angular momentum of the ensemble is $m = n_{\uparrow} \frac{\hbar}{2} = (J/ev)\frac{\hbar}{2} = 10^{18} \times 6.6 \times 10^{-34} / 6.28 \sim 10^{-10}$ (J S/m^3).

Q6.11 Rapid switching of magnetization of ferromagnetic thin film elements is important for spin electronics. Shall one choose a high damping or a low damping ferromagnet when switching the film with external field?

A6.11 A precessing magnetization around z-axis can be expressed as

$$M(t) = M_z \hat{z} + M_r(t)\hat{r}, \qquad (Q6.2.1)$$

where $\hat{r}$ is the radial direction normal to the $\hat{z}$ axis. $M_z \cong$ constant and M_r circles in a plane normal to the z-axis. The first term of Eq. (1a) gives the period of each precessing cycle of $t \sim \dfrac{2\pi}{\gamma H}$. The send term gives the decay of M_r due to damping, and $M_r \propto e^{-\lambda_g t}$, and decays to $0.1 \times$ by $\lambda_g t = 3$, or $t = 3/\lambda_g$.

A large damping reduces the number of precession cycle.

Q6.12 Verify the solution of Eq. (6.4) is Eq. (6.6).

A6.12 The de-magnetization field varies with time since $m(t)$ varies with time. The components of the effective internal field H_{eff} in the film are related to the applied field by

$$H_{\text{eff},x} = H_{k,x} - N_x m_x + h_0 \, e^{j\omega t}, \qquad (Q3.1a)$$

$$H_{\text{eff},y} = H_{k,y} - N_y m_y, \qquad (Q3.1b)$$

$$H_{\text{eff},z} = H_{k,z} - N_z m_z + H. \qquad (Q3.1c)$$

The torque turns the magnetization m from its initial direction at a small angle from the z-axis; thus, $m_x, m_y \ll m_z$, and $m_z \approx m$, and $\dfrac{dm_z}{dt} \approx 0$. Then,

$$\frac{dm_x}{dt} = \gamma\left(m_y H_{\text{eff},z} - m_z H_{\text{eff},y}\right)$$
$$= \gamma\left[\left(m_y(H_{k,z} + H - N_z m_z) - m_z(H_{k,y} - N_y m_y)\right)\right]$$
$$\approx \gamma\left[\left(m_y(H_{k,z} + H - N_z m) - m(H_{k,y} - N_y m_y)\right)\right]$$

thus,

$$\frac{dm_x}{dt} = \gamma\left[(m_y H_{k,z} - m H_{k,y})\right] + \gamma\left[(H - (N_z - N_y)m)m_y\right] \approx \gamma A m_y, \qquad (Q3.2a)$$

here, $A = H - H_{k,y} - (N_z - N_y)m$

$$\frac{dm_y}{dt} = \gamma\left[(m_x H_{k,z}) - m(H_{k,x} + h_0 e^{j\omega t})\right]$$

$$+ \gamma[(H - (N_z - N_x)m)m_x] \approx \gamma B m_x + C,$$

$\qquad\qquad\qquad\qquad\qquad\qquad\qquad\qquad$ (Q3.2b)

here, $B = H - H_{k,x} - (N_z - N_x)m,$ $\qquad\qquad\qquad$ (Q3.3a)

and $C = m(H_{k,x} + h_0 e^{j\omega t}).$ $\qquad\qquad\qquad\qquad$ (Q3.3b)

Since the excitation is $h_0 e^{j\omega t}$, the solution of m_x and m_y take the time dependence of $e^{j\omega t}$, and Eq. (Q3.2a) and Eq. (Q3.2b) can be written as

$$j\omega m_x = \gamma A m_y,$$ $\qquad\qquad\qquad\qquad\qquad\qquad$ (Q3.4a)

$$j\omega m_y = \gamma B m_x + C.$$ $\qquad\qquad\qquad\qquad\qquad$ (Q3.4b)

Solve Eq. (Q4a) and Eq. (Q4b), and one gets

$$m_x = \frac{\gamma B h_0}{1 - \dfrac{\omega^2}{\gamma^2 AB}},$$ $\qquad\qquad\qquad\qquad\qquad$ (Q3.5a)

and $m_y = \dfrac{-j\omega}{\gamma A} m_x.$ $\qquad\qquad\qquad\qquad\qquad\qquad$ (Q3.5b)

From Eq.(Q3.5a), the resonant frequency is at $\omega = \omega_0 = \gamma\sqrt{AB}.$ (Q3.6a)

Q6.13 Prove that a static magnetic field cannot exchange energy with magnetic moment. What about a time-varying magnetic field?

A6.13 A static field can excite magnetic moment and cause it to precess. The energy level of $\boldsymbol{m}$ is $\boldsymbol{m} \cdot \boldsymbol{H} = mH\cos\theta$. As long as angle θ does not change, the energy of $\boldsymbol{m}$ does not change. One easy way to view the physics is that when $\boldsymbol{m}$ precesses over one cycle and back to its initial position, there is no energy exchange. The time-varying magnetic field can induce energy exchange, either from the time varying magnetic field (or torque) to the magnetic moment or from the transfer of magnetic moment to lattice as damping.

Q6.14 Calculate the Oersted field in a circular shape MTJ as a function of distance from the center. The diameter of the MTJ is 60 nm, with a uniformly distributed current of 300 μA. What is the highest field in the MTJ, and where is it located?

A6.14 The Oersted field in a circular MTJ is $H = jr/2$, where J is current density and r is the distance from center. At the center, $H = 0$. The maximum field at the edge of MTJ is 478 A/m, or ~ 6 Oe. It is too small to switch the magnetization. Since 300 µA is the typical current that switches a STT-mode MTJ, spin torque is more efficient than field.

Q6.15 The Internal field in a Co is 10^4 Tesla. When an electron enters Co film with spin polarization at an angle θ from the magnetization of Co film, at a ballistic speed of 10^6 m/S, how long a distance does it travel when its moment completes 1 cycle of precession?

A6.15 The Larmor precession frequency $\omega = \gamma H_{eff}$, and the precession time of 1 cycle $= 2\pi/(\gamma H_{eff})$. For $\gamma = 1.76 \times 10^{11}$ rad/ST, it takes 3.57×10^{-9} sec to precess 1 cycle, during which the electron travels ballistically over a distance of 3.57 nm. Within this distance, the electron spin interacts with the moment of the Co film. Its transversal component of angular momentum is absorbed by the precession mechanism.

Q6.16 The spin-torque s acting on a magnetization of M is $M \times (M \times s)$. When M and s are not collinear, but having an angle of θ and $\theta \neq 0$ or π. How does the anti-damping term in Eq. (6.4) depend on θ?

A6.16 The angle dependence of anti-damping term $M \times (M \times s)$ can be rewritten as $sM^2 \sin\theta\ \mathbf{u}$, where $\mathbf{u}$ is a unit vector in the direction normal to M and s.

Q6.17 Eq. (6.8) shows that spin-torque transfer efficient is independent of the area of the device, but the thickness of the ferromagnet. What is the implication of scaling?

A6.17 The critical switching current scales with MTJ area. For the first order, this is correct. As discussed in Chapter 7, the smaller the MTJ, the more it behaves like the Macrospin model.

References

1 Hall, E.H. (1879). On a new action of the magnet on electric currents. *J. Math.* 12 (3): 287.

2 Hall, E.H. (1887). On the rotational coefficient in nickel and cobalt. *Philo. Mag.* 12 (74): 157.

3 Nagaosa, N., Sinova, J., Onoda, S. et al. (2010). Anomalous Hall effect. *Rev. Mod. Phys.* 82: 1539.

4 Kato, Y.K., Myers, R.C., Gossard, A.C., and Awschalom, D.D. (2004). Observation of the spin Hall effect in semiconductors. *Sci.Exp. Science* 306 (5703): 1910–1913. https://doi.org/10.1126/science.1105514.

5 Hirsch, J.E. (1999). *Phys. Rev. Lett.* 82: 1834.

6 Smit, J. (1958). The spontaneous Hall effect in ferromagnet. *Physica* 24 (1–5): 39.

7 Berger, L. (1970). Side-jump mechanism for the Hall effect in ferromagnetics. *Phys. Rev. B* 2 (11): 4559.

8 Guo, G.Y., Murakami, S., Chen, T.-W., and Nagaosa, N. (2008). Intrinsic Spin Hall Effect in Platinum: First-Principles Calculations. *Phys. Rev. Lett.* 100: 096401 – Published 3 March; also see [Mizukami, 2001] S. Mizukami, Y, Ando, T. Miyazaki, *J. Mag. Mater.* 226, 1640 (2001).

9 Bakun, A.A., Zakarcheya, B.P., Rogarchev, A.A. et al. (1984). Observation of surface photocurrent caused by optical orientation of electron in a semiconductor, *Pos'ma Ah, Eksp, Teor. Fiz.* (*1984*) [*Sov. Phys*]. *JETP Lett.* 40: 1293.

10 Xiao, D., Chang, M.-C., and Niu, Q. (2010). Berry phase effects on electronic properties. *Rev. Mod. Phys.* 82: 1959.

11 Hoffmann, A. (Oct 2013). Spin Hall effects on metals. *IEEE Trans. Magn.* 49 (10): 5172.

12 Dyakonov, M.I. (2012). Spin Hall Effects, arXiv:1210.3200

13 Dyakonov, M.I. and Khaetskii, A. (2008). Spin Hall effects, *Spin Physics in Semiconductor,* ser. In: *Solid State Sciences,* vol. 157 (ed. M.I. Dyakonov), 211–243. New York: Springer.

14 Tedrow, P.M. and Meservey, R. (1973). Spin polarization of electrons tunneling from films of Fe, Co, Ni, and Gd. *Phys. Rev. B* 7 (1) https://doi.org/10.1103/PhysRevB.7.318.

15 Lassailly, Y., Drouhin, H.-J., van der Sluijs, A.J. et al. (1994). *Phys. Rev. B* 50: 13054.

16 Strijkers, G.J. (2001). *Phys. Rev. B* 63: 104510.

17 Huang, S.X., Chen, Y.T., and Chien, C.L. (2008). *Appl. Phys. Lett.* 92: 242509.

18 Feng, Y.P. et al. (2017). Prospects of spintronics based on 2D materials. *WIREs Comput. Mol. Sci.* 7: e1313. https://doi.org/10.1002/wcms.1313.

19 Baoli Liu, Junren Shi Wenxin Wang, Hongming Zhao, Dafang Li, Shoucheng Zhang, Qikun Xue, Dongmin Chen (2006). Experimental Observation of the Inverse Spin Hall Effect at Room Temperature, Cornell University, arXiv:cond-mat/0610150

20 Omori, Y., Auvray, F., Wakamura, T. et al. (2014). Inverse spin Hall effect in a closed loop circuit. *Appl. Phys. Lett.* 104: 242415.

21 Kimura, T., Otani, Y., Sato, T. et al. (2007). Room Temperature Reversible Spin Hall Effect. *Phys. Rev. Lett.* 98: 156601.

22 Johnson, M. and Silsbee, R.H. (1985). Interfacial charge-spin coupling: injection and detection of spin magnetization in metals. *Phys. Rev. Lett.* 55 (17): 1790.

23 Valenzuela, S.O. and Tinkham, M. (2004). Spin-polarized tunneling in room-temperature mesoscopic spin valves. *Appl. Phys. Lett.* 85 (24): 5914. https://doi.org/10.1063/1.1830685.

24 Valenzuela, S.O. and Tinkham, M. (2006). Direct electronic measurement of the spin Hall effect. *Nature* 442 (7099): 176.

25 Isasa, M., Villamor, E., Hueso, L.E. et al. (2015). Temperature dependence of spin diffusion length and spin Hall angle in Au and Pt. *Phys. Rev. B* 91: 024402.

26 Batley, J.T., Linfield, E.H. et al. (April 2015). Spin relaxation through Kondo scattering in Cu/Py lateral spin valves. *Phys. Rev. B* https://doi.org/10.1103/PhysRevB.92.220420 arXiv.org > cond-mat > arXiv:1504.07515.

27 Dyankonov, M.I. and Perel, V.I. (1971). Current-induced spin orientation of electrons in semiconductors. *Phys. Lett.* 35A (6).

28 Dyakonov, M.I. (2007). Magnetoresistance due to edge spin accumulation. *Phys. Rev. Lett.* 99: 126601.

29 Berger, L. (1996). Emission of spin waves by a magnetic multilayer traversed by a current. *Phys. Rev. B* 54: 9353.

30 Slonczewski, J.C. (1996). Current-driven excitation of magnetic multilayers. *J. Magn. Magn. Mater.* 159: L1.

31 Kittel, C. (2004). *Introduction to Solid State Physics*, 3e. Ch. 16. Wiley.

32 Tserkovnyak, Y., Brataas, A., and Bauer, G.E.W. (2002). Enhanced Gilbert Damping in Thin Ferromagnetic Films. *Phys. Rev. Lett.* 88: 117601.

33 Tserkovnyak, Y., Brataas, A., Buer, G.E.W., and Halperin, B.I. (2005). Non-local magnetization dynamics in ferromagnetic heterostructures. *Rev. Mod. Phys.* 77 (4): 1375.

34 Mosendz, O., Pearson, J.E., Fradin, F.Y. et al. (2010). Quantifying spin Hall angles from spin pumping: Experiments and Theory. *Phys. Rev. Lett.* 104: 046601.

35 Watanabe, S., Hirobe, D., Shiomi, Y. et al. Generation of megahertz-band spin currents using nonlinear spin pumping. *Sci. Rep.* 7: 4576. https://doi.org/10.1038/s41598-017-04901-4.

36 Mizukami, S., Ando, Y., and Miyazaki, T. The study on ferromagnetic resonance linewidth for NM/80NiFe/NM (NM=Cu, Ta, Pd and Pt) films. *Jpn. J. Appl. Phys.* 40: 580. http://iopscience.iop.org/article/10.1143/JJAP.40.580.

37 Lee, H., Wen, L., Pathak, M. et al. (2008). Spin pumping in Co56Fe24B20 multilayer systems. *J. Phys. D. Appl. Phys.* 41 (215001).

38 Boone, C.T., Nembach, H.T., Shaw, J.M., and Silva, T.J. (2013). Spin transport parameters in metallic multilayers determined by ferromagnetic resonance measurements of spin-pumping. *Journal of Applied Physics* 113: 153906.

39 Otani, Y. and Kimura, T. (2011). Manipulation of spin currents in metallic systems. *Phil. Trans. R. Soc. A* 369: 3136–3149. https://doi.org/10.1098/rata2011.0010.

40 Heinrich, B., Tserkovnyak, Y., Woltersdorf, G. et al. (2003). Dynamic exchange coupling in magnetic bilayers. *Phys. Rev. Lett.* 90 (18): 187601.

41 Berger, L. (1984). Exchange interaction between ferromagnetic domain wall and electric current in very thin metallic films. *J. Appl. Phys.* 55: 1954.

42 Tsoi, M., Jansen, A.G.M., Bass, J. et al. (1998). Excitation of a magnetic multilayer by an electric current. *Phys. Rev. Lett.* 80: 4281.

43 Meyers, E.B., Ralph, D.C., Katine, J.A. et al. (1999). Current-induced switching of domains in magnetic multilayer devices. *Science* 285 (6): 867. www.sciencemag.org.

44 Grollier, J., Cros, V., Hamzic, A. et al. (2001). *Appl. Phys. Lett.* 78 (23): 509.

45 Ralph, D.C. and Stiles, M.D. (2008). Spin Transfer Torques. *J. Magn. Magn. Mater.* 320: 1190–1216.

46 Fert, A., Cros, V., George, J.-M. et al. (2004). Magnetization reversal by injection and transfer of spin: experiments and theory. *J. Magn. Magn. Mater.* 272–276 (3): 1706.

47 Sun, J.Z. (2000). Spin-current interaction with a monodomain magnetic body: A model study. *Phys. Rev. B* 62: 570.

48 Slonczewski, J.C. (2002). Currents and torques in metallic magnetic multilayers. *J. Magn. Magn. Mater.* 247: 324–338.

49 Slonczewski, J.C. (1989). Conductance and exchange coupling of two ferromagnets separated by a tunneling barrier. *Phys. Rev. B* 39: 6995.

50 Koch, R.H., Katine, J.A., and Sun, J.Z. (2004). Time-Resolved Reversal of Spin-Transfer Switching in a Nanomagnet. *Phys. Rev. Lett.* 92 (8): 088302.

51 Acremann, Y., Strachan, J.P., Chembrolu, V. et al. (2006). Time-Resolved Imaging of Spin Transfer Switching: Beyond the Macrospin Concept. *Phys. Rev. Lett.* 96: 217202.

52 Jan, G., Wang, Y.-J., Moriyama, T. et al. (2012). High spin torque efficiency of magnetic tunnel junctions with MgO/CoFeB/MgO free layer. *Appl. Phys. Exp.* 5: 093008.

53 Luc Thomas, Guenole Jan, Son Le, Santiago Serrano-Guisan, Yuan-Jen Lee, Huanlong Liu, Jian Zhu, Jodi Iwata-Harms, Ru-Ying Tong, Sahil Patel, Vignesh Sundar, Dongna Shen, Yi Yang, Renren He, Jesmin Haq, Zhongjian Teng, Vinh Lam, Paul Liu, Yu-Jen Wang, Tom Zhong, and Po-Kang Wang (2014). Probing magnetic properties of STT-MRAM devices down to sub-20 nm using Spin-torque FMR. *IEEE Digest of IEDM*. IEEE. pp. 17–849.

54 Slonczewski, J.C. and Sun, J.Z. (2007). Theory of voltage-driven current and torque in magnetic tunnel junctions. *J. Magn. Magn. Mater.* 310: 169–175. publisher: Elsevier (Holland).

55 Meservey, R. and Tedrow, P.M. (1994). Spin-polarized electron tunneling. *Phys. Rep.* 238 (4): 173–243.

56 Slonczewski, J. (2005). *Currents, torques, and polarization factors in magnetic tunnel junctions. Phys. Rev. B* 71: 024411.

57 Sankey, J.C. et al. (2008). Measurements of the spin-transfer-torque vector in magnetic tunnel junctions. *Nat. Phys.* 4: 67–71.

58 Kubota, H. et al. (2008). Quantitative measurement of voltage dependence of spin-transfer torque in MgO-based magnetic tunnel junctions. *Nat. Phys.* 4: 37–41.

59 Fuchs, G.D., Sankey, J.C., Pribiag, V.S. et al. (2007). Spin-torque ferromagnetic resonance measurements of damping in nanomagnets. *Appl. Phys. Lett.* 91: 062507.

60 Deac, A.M. et al. (2008). Bias-driven high-power microwave emission from MgO-based tunnel magnetoresistance devices. *Nat. Phys.* 4: 803–809.

61 Wang, C., Cui, Y.-T., Sun, J.Z. et al. (2009). Bias and angular dependence of spin-transfer torque in magnetic tunnel junctions. *Phys. Rev. B* 79: 224416.

62 Skowronski, W., Czapkiewicz, M., Frankowski, M. et al. (2013). Influence of MgO tunnel barrier thickness on spin-transfer ferromagnetic resonance and torque in magnetic tunnel junctions. *Phys. Rev. B* 87: 094419.

63 Matsumoto, R., Chanthbouala, A., Grollier, J. et al. (2011). Spin-torque diode measurements of MgO-based magnetic tunnel junctions with asymmetric electrodes. *Appl. Phys. Exp.* 4: 063001.

64 Wang, C., Cui, Y.-T., Katine, J.A. et al. (2011). Time-resolved measurement of spin-transfer-driven ferromagnetic resonance and spin torque in magnetic tunnel junctions. *Nat. Phys.* 7: 496. http://www.nature.com/naturephysics.

65 Xue, L., Wang, C., Cui, Y.-T. et al. (2012). Network analyzer measurements of spin transfer torques in magnetic tunnel junctions. *Appl. Phys. Lett.* 101: 022417.

66 Berger, L. (1984). Exchange interaction between ferromagnetic domain wall and electric current in very thin metallic films. *J. Appl. Phys.* 55: 1954.

67 Tatara, G. and Kohno, H. (2004). Theory of current-driven domain motion: spin transfer versus momentum transfer. *Phys. Rev. Lett.* 92: 086601.

68 Tatara, G., Takayama, T., Kohno, H. et al. (2006). Threshold Current of Domain Wall Motion under Extrinsic Pinning, β-Term and Non-Adiabaticity. *J. Phys. Soc. Jpn.* 75: 064708.

69 Zhang, S. and Li, Z. (2004). Domain-wall dynamics driven by adiabatic spin-transfer torques. *Phys. Rev. B* 70: 024417.

70 Zhang, S. et al. (2004). Roles of non-equilibrium conduction electrons on the magnetization dynamics of ferromagnets. *Phys. Rev. Letts* 93 (12).

71 Ravelosona, D. et al. (2006). Domain Wall creation in nanostructures driven by a spin-polarized current. *Phys. Rev. Lett.* 96: 186604.

72 Rebei, A. and Mryasov, O. (2006). Dynamics of a trapped domain wall in a spin-valve nanostructure with current perpendicular to the plane. *Phys. Rev. B* 74: 014412.

73 Lou, X., Gao, Z., Dimitrov, D.V., and Tang, M.X. (2008). Demonstration of multilevel cell spin transfer switching in MgO magnetic tunnel junctions. *Appl. Phys. Lett.* 93: 242502.

74 Boone, C.T. et al. (2010). Rapid Domain Wall motion in permalloy nanowires excited by a spin-polarized current applied perpendicular to the nanowire. *Phys. Rev. Lett.* 104: 097203.

75 Khvalkovskiy, K.V. et al. (2009). High domain wall velocities due to spin currents perpendicular to the plane. *Phys. Rev. Lett.* 102: 067206.

76 Yamanouchi, A., Chiba, M., Matsukura, D. et al. (2006). Velocity of domain-wall motion induced by electrical current in the ferromagnetic semiconductor (Ga, Mn) As. *Phys. Rev. Lett.* 96: 096601; A. Yamaguchi, T. Ono, S. Nasu, K. Miyake, K. Mibu; T. Shinjo, Phys. Rev. Lett. 92, 077205 (2004); A. Yamaguchi, S. Nasu, H. Tanigawa, T. Ono, K. Miyake, K. Mibu, and T. Shinjo, *Appl. Phys. Lett.* 86, 012511 (2005).

77 Z. Li, J. He and S. Zhang (2005). Effects of spin current on ferromagnets, arXiv: Cond.-mat/0508735v1

78 Lim, C.K., Devolder, T., Chappert, C. et al. (2004). Domain wall displacement induced by subnanosecond pulsed current. *Appl. Phys. Lett.* 84: 20. spin valve.

79 Uhlíř, V., Pizzini, S., Rougemaille, N. et al. (2010). Current-induced motion and pinning of domain walls in spin-valve nanowires studied by XMCD-PEEM. *Phys. Rev. B* 81: 224418.

80 Metaxas, P.J., Sampaio, J., Chanthbouala, A. et al. (2013). High domain wall velocities via spin transfer torque using vertical current injection. *Sci. Rep.* 3: 1829. https://doi.org/10.1038/srep01829.

81 Sampaio, J., Lequeux, S., Metaxas, P.J. et al. (2013). Time-resolved observation of fast domain-walls driven by vertical spin currents in short tracks. *Appl. Phys. Lett.* 103: 242415.

82 Haazen, P.P.J., Murè, E., Franken, J.H. et al. (2013). Domain wall depinning governed by the spin Hall effect. *Nat. Mater. Lett.* https://doi.org/10.1038/ NMAT3553.

83 Grollier, J., Chanthbouala, A., Matsumoto, R. et al. (2011). Magnetic domain wall motion by spin transfer. *C. R. Phys.* 12: 309.

84 Yang, S.-h., Ryu, K.-S., and Parkin, S. (2015). Domain-wall velocities of up to 750 m s^{-1} driven by exchange-coupling torque in synthetic antiferromagnets. *Nat. Nanotechnol.* 10: 221–226.

85 Ajejas, F., Křižáková, V., de Souza Chaves, D. et al. (2017). Tuning domain wall velocity with Dzyaloshinskii-Moriya interaction. *Appl. Phys. Lett.* 111: 202402.

86 Herrera Diez, L., Voto, M., Casiraghi, A. et al. (2019). Enhancement of the Dzyaloshinskii-Moriya interaction and domain wall velocity through interface intermixing in Ta/CoFeB/MgO. *Phys. Rev. B* 99: 054431.

87 Yoshimura, Y., Kim, K.-J., Taniguchi, T. et al. (2016). Soliton-like magnetic domain wall motion induced by the interfacial Dzyaloshinskii–Moriya interaction. *Nat. Phys.* 12: 157–161.

88 Ryu, K.-S., Thomas, L., Yang, S.-H., and Parkin, S. (2013). Chiral spin torque at magnetic domain walls. *Nat. Nanotechnol.* 8: 527–533.

89 Emori, S., Bauer, U., Ahn, S.-M. et al. (2013). Current-driven dynamics of chiral ferromagnetic domain walls. *Nat. Mater.* 12: 611–616.

7

Spin-Torque-Transfer (STT) MRAM Engineering

7.1 Introduction

Spin-torque transfer magnetoresistive random-access memory (STT-MRAM) is far more energy efficient than field MRAM, which was briefly described in Chapter 5 and extensively in [1]. Their characteristics are very different. This chapter describes the memory operation and the performance of STT-MRAM. As a nonvolatile memory, the key performance metrics are memory access performance, namely, READ and WRITE; data retention performance; and device reliability performance, especially the limit of write endurance cycle.

In Section 7.2, we discuss energy barriers as a function of a magnetic tunnel junction (MTJ) film stack and device structure. The thermal energy barrier dictates the data retention performance, while the switching energy barrier dictates the switching current threshold. In Section 7.3, we focus on the switching properties. The discussion starts from a simple uniform magnetization reversal model, called the Macrospin model, in which the magnetization of the entire MTJ free layer magnetization is assumed to precess in unison under spin current. Thus, the free layer can be represented by a single magnetization vector. It precesses under the injection of anti-damping spin current until reversal takes place. From this model, the stochastic switching properties can easily be understood. The stochastic switching properties are summarized as a performance metric, the write error rate (WER) as a function switching voltage (or current). From there, we further discuss certain frequently observed switching abnormality. This section closes with a more complex magnetization reversal process. It is called the domain mediated magnetization reversal process.

Then, we discuss two STT-MRAM device reliability issues: tunnel barrier degradation and data retention. Section 7.4 covers a MgO tunnel barrier degradation model. Section 7.5 covers the relation between the thermal energy barrier and the data retention time performance at the chip level. We introduce the bit-level and

Magnetic Memory Technology: Spin-Transfer-Torque MRAM and Beyond,
First Edition. Denny D. Tang and Chi-Feng Pai.
© 2021 The Institute of Electrical and Electronics Engineers, Inc.
Published 2021 by John Wiley & Sons, Inc.

the chip-level characterization methods. The methods allow us to extract the thermal stability factor of MTJ and to project the data retention time of an STT-MRAM array.

In Section 7.6, we discuss the 1 MTJ-1 transistor (1M-1T) MRAM cell design and scaling. Both the STT switching efficiency and the properties of CMOS transistor must be considered in the design, since the cell properties are partly determined by the transistor. For example, the cell size and the operating voltage and current, etc., are determined by the CMOS technology platform node. Scaling MTJ requires lower cell currents. The film magnetic properties need to be adjusted accordingly. In addition to the adjustment of the free layer, the pinning strength of the pinned-layer is also an important subject.

Section 7.7 covers the SPICE model for memory chip-level circuit simulation. A comprehensive MTJ model includes an LLG equation solver, so that it reflects accurate transient I–V characteristics and predicts accurate WER. Section 7.8 discusses test chips and testing methodologies for weeding out weak bits in a memory chip. Test chip is indispensable during technology development. Weak bits include bits that switch unreliably or abnormally and have low read and write margin, short data retention time, and short life expectancy.

7.2 Thermal Stability Energy and Switching Energy

MTJ is a binary resistor, having two resistance states. The *thermal stability energy barrier* $E_{b,therm}$ of an MTJ is the energy required to switch the MTJ state when MTJ is in idling. It determines the data retention of MRAM cells. The *STT switching energy barrier* $E_{b,STT}$ is the energy required to STT-write MRAM cells. It determines the MRAM write critical current. They do not have to be the same, depending on the direction of the easy axis of the free layer.

The magnetic moment of the free layer is $m = M_s V$, where V is the volume of the free layer. The thermal stability energy barrier is the energy required to rotate the moment 90°. Thus,

$$E_{b,therm} = \frac{1}{2} \int_0^{\pi/2} (M_s V) H_{\text{eff}} \cdot (\cos \theta) d\theta = \frac{1}{2} M_s V H_{\text{eff}}, \tag{7.1}$$

where θ is the angle between moment and easy axis, and $H_{\text{eff}} = H_K + H_D$ and H_D is the demag field.

For in-plane MTJs, the shape anisotropy H_D of the free layer dominates H_{eff}; thus,

$$E_{b,therm} \cong \frac{1}{2} M_s V H_D = \frac{1}{2} M_S^2 V (N_y - N_x), \tag{7.2}$$

where N_x and N_y are in-plane demagnetizing factors. When the MTJ dimension (I, w) is much larger than the free layer thickness t, the out-of-plane demagnetizing factors, N_z, are much larger than N_x and N_y. And $(N_y - N_x) \ll N_z$. The film shape keeps the easy axis (H_{eff}) in plane.

During STT switching, the anti-damping torque of the injected spin current raises the precession cone angle. Its effort is hindered by the strong out-of-plane demag field. Thus, the switching energy barrier for STT switching of the in-plane MTJ is

$$E_{b,STT} \approx \frac{1}{2} N_z M_S^2 V \tag{7.3}$$

and $N_z \gg N_x, N_y$, thus, $E_{b,therm.} < E_{b,STT}$ for in-plane STT MTJs.

For out-of-plane MTJs, the out-of-plane interfacial anisotropy is stronger than the crystalline anisotropy and in-plane shape anisotropy ($4\pi M_S$) combined. The net anisotropy is out of plane. At room temperature, magnetic moment precesses around the out-of-plane H_{eff}. During STT switching, the anti-damping torque of the spin current is counteracted by a weaker in-plane demag field, which can be ignored. The energy barrier for switching is approximately the same as that of thermal energy barrier and is

$$E_{b,STT} \approx E_{b,therm} = \frac{1}{2} A \left(K_i - 4\pi t M_S^2 \right), \tag{7.4}$$

where A is the area and t is the thickness of the free layer.

Furthermore, it is convenient to express as a stability energy as thermal stability factor Δ, where $\Delta = E_{b,therm}/k_B T$, the ratio of MTJ thermal energy barrier and thermal energy.

Since the switching current threshold is proportional to the switching energy barrier, the switching efficiency (defined as Δ/I_{c0}, thermal stability factor/critical switching current) of the out-of-plane MTJ is higher. Figure 7.1 illustrates this difference.

The previous analysis is based on the Macrospin model, in which the magnetization of the entire free layer is assumed to precess in unison. Exceptions may happen when (i) the free layer area is large, and (ii) the free layer is very thin and the film exchange constant is weak, such that domain wall is small and narrow, and domain nucleates before switching completes. This will be discussed in Section 7.3.6.3.

Before we leave this section, we should discuss how the switching energy barrier of MTJ compares to those of volatile memories, such as SRAM and DRAM. After all, MRAM will be competing against volatile memory.

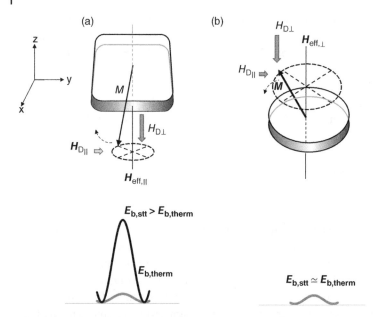

Figure 7.1 Switching energy barrier and the thermal stability energy of (a) in-plane and (b) out-of-plane MTJ. The demagnetization field in the out-of-plane free layer is much smaller since the MTJ diameter is much larger than the MTJ thickness. The demag field hinders the effort of anti-damping torque τ_{stt} of spin current, more so in in-plane MTJ than in perpendicular MTJ. The in-plane switching energy barrier is larger than the thermal stability energy barrier.

7.3 STT Switching Properties

Two important switching parameters are the switching energy barrier and the switching threshold current density. The switching behavior is highly dependent on the pulse width of the switching current. When the write pulse is many orders longer than the precession period, typically less than 1 ns, the switching can be successfully accomplished partially with the help of magnetization thermal fluctuations. The long write pulse regime is called thermal regime. In this regime, the longer write pulse is, the higher probability a precessing magnetization may flip. The switching current can be below the critical current. On the other hand, when the write pulse is in the same order of magnitude as the magnetization precession period, the switching relies on the anti-damping action of spin current. It is called precession regime. To achieve faster switching, more spin current is needed to grow the precession cone in a shorter time. Figure 7.2 shows the dependence of the write current as a function of the write pulse width. The border between regimes is around a few nanoseconds.

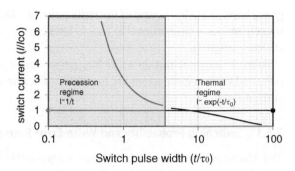

Figure 7.2 MTJ switch-current as a function of the current pulse width. The horizontal line is the critical current I_{c0}. τ_0 is the characteristic time related to the precession cycle period.

The critical current density that induces magnetization reversal has been derived,

$$J_{c0} = \frac{2e\alpha M_s t_F (H_K + H_D/2)}{\hbar P}, \qquad (7.5)$$

based on a single-domain model, which is called the Macrospin model [2]. For in-plane MTJ, H_K is bulk anisotropy, and H_D is shape anisotropy, which dominates the total anisotropy. The switching threshold current is

$$I_{c0} = \frac{\alpha}{P}\frac{2e}{\hbar}\mu_0 H_D M_s V/2, \quad \text{where V is the volume of the free layer.} \quad (7.6a)$$

For perpendicular MTJ, the perpendicular interfacial anisotropy H_K dominates. Thus, the switching threshold current I_0 (also called switching critical current) is

$$I_{c0} = \frac{\alpha}{P}\frac{2e}{\hbar}\mu_0 H_K M_s V. \qquad (7.6b)$$

Both equations are in SI unit. At room temperature, the critical current can be expressed in terms of thermal stability factor as

$$I_{c0} = \frac{\alpha}{P}\frac{2e}{\hbar}2\Delta k_B T. \qquad (7.7)$$

For $\Delta = 60$, $I_{c0} \approx \frac{\alpha}{P} 1.5\,\text{mA}$.

Switching threshold drops as temperature increases, because both H_K and M_S drop with temperature. It is expected that the critical current drops as the ambient temperature rise and also if the resistance-area (RA) product is higher due to self-heating. When a write current is applied, the MTJ tunnel barrier heats up. Larger-diameter MTJ heats more than a smaller one. Thus, one expects that the switching current density is higher when switching a smaller-diameter MTJ.

To switch magnetization with a short current pulse, higher current is required to speed up the growth rate of the precession cone angle such that the cone angle grows beyond $\pi/2$ by the end of the write pulse [3]. The switching time is approximately inversely proportional to $(I_w - I_{c0})$, where the write current I_w exceeds the critical current I_{c0}.

To switch a magnetization with a long current pulse, thermal fluctuation and the polarized current both act on the magnetization, and switching can take place at current $I < I_{c0}$.

7.3.1 Switching Probability and Write Error Rate (WER)

The Macrospin model, which excludes magnetization thermal fluctuation, predicts the deterministic switching property. When thermal noise is included in consideration, the switching property is no longer deterministic, but stochastic, since the initial magnetization jitters, and the cone angle is nonzero and is random [4–7]. Other mechanisms may also contribute to the stochastic switching property. The following analysis considers only the thermal noise.

Thermal energy causes the magnetization to jitter around the easy axis (H_{eff} direction). The cone angle θ between the magnetization and the easy axis distributes over a small but finite range. The range is narrower when temperature is low (less thermal energy) and when the anisotropy is strong (the thermal stability factor is large); in other words, the magnetization jitters closer to the easy axis. Nonetheless, each write event starts at a random angle.

Figure 7.3 shows the statistical distribution of the magnetization angle as a function of normalized time τ for a normalized current ι, where τ is write pulse width

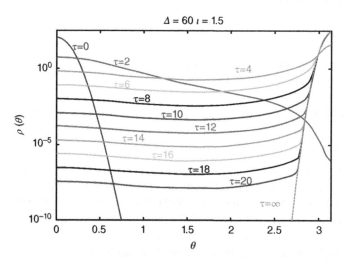

Figure 7.3 The distribution of magnetization precession angle θ of an MTJ with energy stability factor. The normalized write current is $\iota = 1.5$. $\Delta = 60$ [7].

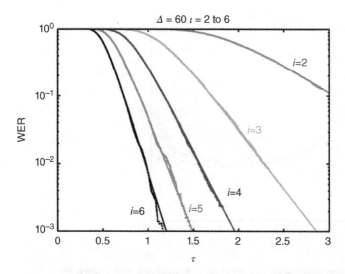

Figure 7.4 Write error rate (WER) versus write pulse width of an MTJ with thermal stability factor $\Delta = 60$, $\iota = I/I_{c0}$, and for effective field $\mu_0 H_K = 1$ T, the corresponding precession period $\tau \sim 10^{-10}$ sec. Solid lines are approximated solution. Points are Fokker-Plank solution [7].

t_{pw} normalized with precession cycle period $\left(\tau = \dfrac{\alpha\gamma\mu_0 H_K}{1 + \alpha^2} t_{pw}\right)$ and normalized current $\iota = \dfrac{I}{I_{c0}}$. At time zero ($\tau = 0$), no current is applied, the magnetization jitters around $\theta = 0$, with a 3-sigma spread of cone angle ~0.5 rad. This sigma corresponds to the MTJ thermal stability factor $\Delta = 60$ at room temperature. Statistically speaking, once a spin current of magnitude 1.5x of the switching threshold ($\iota = 1.5$) is injected, the 100% of the cone angle reaches π in four precession cycles ($\tau = 4$). The reversal completes.

As a consequence of stochastic switching behavior, there is a finite probability that a given write pulse, defined as pulse width and current, will fail to switch. The failure rate is called WER. Butler further calculated the WER versus τ at various normalized current ι. Figure 7.4 shows the calculated numerically WER [7]. The thermal stability factor is $\Delta = 60$.

For write current below the critical write current, the spin current cannot grow the cone angle beyond $\pi/2$ by itself. But, thermal fluctuation can assist during precession. Eventually, the MTJ free layer flips. Figure 7.5 illustrates this point. Initially ($\tau = 0$), the magnetization fluctuates around 0, the easy axis; there is a small probability that the precession cone angle θ is greater than 0. After write current is turned on, the STT torque pulls θ away from easy axis. Even when the write current is below critical current ($\iota = 0.5$), the cone angle continues to spread.

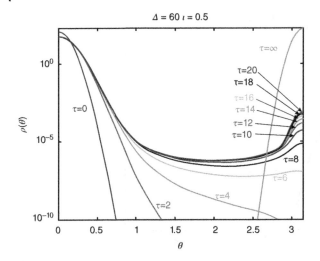

Figure 7.5 Cone angle distribution as a function of normalized time τ when the normalized write current $\iota = 0.5$ is applied.

Nonetheless, as long as the write current stays on, eventually majority θ reaches $\pi/2$, and the magnetization switches.

In the thermal regime (long switching pulse width), $J < J_{i0}$. The switching probability can be approximated as [3, 6]

$$P_{sw} = 1 - \exp\left\{-\frac{t_{PW}}{\tau_0}\exp\left[-\Delta\left(1 - \frac{I}{I_{c0}}\right)\right]\right\}, \tag{7.8a}$$

where t_{PW} is the pulse width and I_{c0} is the critical write current.

In the short pulse regime (<10 ns), the reversal process cannot be considered thermal activation over an energy barrier. In this regime, the required current density must be larger than the critical current density. The switching probability is expressed as follows [2]

$$P_{sw} = \exp\left\{-4\Delta\exp\left[-\frac{2t_{PW}\left(\frac{I}{I_{c0}} - 1\right)}{\tau_D}\right]\right\}, \tag{7.8b}$$

where

$$\tau_D = \frac{1}{\alpha\gamma\mu_0 H_K} \tag{7.9}$$

is the characteristic time associated with the magnetic moment precession cycle time (for example, for $\alpha = 0.01$ and $\mu_0 H_K = 1$T, then $\tau_D = 3.5$ ns).

These two equations provide an approximated description of switching probability. WER measurement data from hardware deviates substantially from Eq. (7.8). One possibility is that the MTJ temperature varies with applied voltage (and current).

Another is that MTJ does not switch ideally as described in the Macrospin model. In this respect, WER characterization becomes routine in MRAM development.

Although the switching critical current is well-defined, there has not been a unified definition of the switching threshold among experimental publications. One finds from the literature that it is becoming a common practice to call the write current at the 50% write success rate as the threshold current. That is not good enough for product development. No one can use a product that fails 50% of the time. In industry, "write current," rather "critical current," is more commonly accepted. Typically, WER should be at least than 10^{-9}. Thus, the memory cell write current I_w must be much larger than I_{c0}. I_w is a current that the gating transistor of an MRAM cell must supply to the MTJ such that it guarantees successful write.

Although previous switching threshold analyses are based on current density, in practice, it is more convenient to measure switching threshold voltage, like the threshold voltage of a transistor. From here on, this book will quote threshold in either current or voltage.

The WER can be experimentally measured by repeatedly switching the MTJ between the AP- and P-state with a particular pulse width. The write voltage is proportional to the product of MTJ resistance and area, RA, since

$$V_W = I_W \cdot R = (J_W \cdot A) \cdot R = J_W \cdot (RA),$$

where V_W and I_W are write voltage and write current, respectively. Lowering the RA of MTJ will lower the write voltage while keeping the write current the same. The write bit error rate (WER) is

$$\text{WER} = 1 - P_{SW} \tag{7.10}$$

Figure 7.6 shows a measured WER as a function of write current for different write pulse width. The data shows a gradual increase of write current as pulse

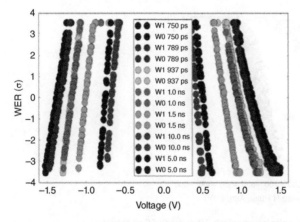

Figure 7.6 Actual WER measurement of an out-of-plane MTJ down to sub-ns write pulse width. The transition from long to short write pulse is more gradual [8].

width is shortened, which is the most interesting to the MRAM industry. Empirically, the WER of long write pulse in the range of 50–100 ns can be reasonably fitted with a *complement error function*, WER $= 1-P_{SW} = erfc(V_W)$, where V_W is write voltage, error function is the integration of a normal distribution function from the center to the edge, $erfc(z) = \int_z^\infty \exp{(-x^2)}dx$. In this case,

$$z = \frac{V_W - V_{W50}}{\sqrt{2}\sigma_{temporal}},$$

where V_{W50} is the switching threshold voltage (WER $= 0.5$) and $\sigma_{temporal}$ is the temporal standard deviation of the WER.

A wider write pulse gives rise to a steeper decay in the WER, or the $\sigma_{temporal}$ is smaller. On the other hand, for shorter pulses, the decay is slower than a normal function and the WER drops slower than those of long pulse. Error rate measurements are required to make projections of the switching window size for chip-level design. It is possible that more transition mechanisms are involved.

Nonetheless, for practical device applications, the WER should reach below 10^{-9}. When an array of MTJs is considered, the distribution of WER from MTJ to MTJ should also be considered. Figure 7.7 shows the variation of V_W of 64 MTJs of an array. The spatial sigma at a given WER value should be included in the consideration of MRAM design. The spread is highly correlated to the MTJ integration process.

7.3.2 Switching Current in Precessional Regime

Experimental studies (e.g. Figure 7.6) show that the WER of short pulse (precessional regime) does not follow the prediction of the Macrospin model [9]. The

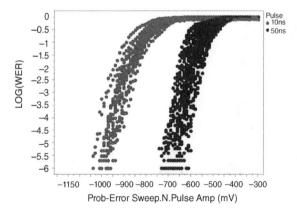

Figure 7.7 WER of 64 MTJs write pulse width 10 ns (red), 50 ns (blue).

reason is that the energy barrier is a fixed value of a function of free-layer structure and material parameter. Its value depends on whether domain is nucleated during magnetic reversal. When magnetic reversal involves domain nucleation, the switching energy barrier is effectively smaller than that when the domain is not nucleated (i.e. the entire free layer magnetization reverses in one stroke). Since the energy barrier involving with domain nucleation is smaller, it is more likely that most of the switching does take place with domain nucleation. The switching is mediated by domain wall propagation and DW energy. The switching current at a given WER can be expressed as a function of precession period and write pulse width [10, 11] involving an effective energy barrier as

$$I_W(P_{WER}) = \frac{M_s V e}{P g \mu_B} \left[\frac{2}{\tau_D} + \frac{1}{t_{PW}} \ln \left(\frac{\pi^2 \Delta}{4 P_{WER}} \right) \right], \tag{7.11}$$

where $\Delta = H_K M_S V/2$ (the uniform switching energy barrier), V is the free layer (FL) volume, P is polarization, μ_B is the Bohr magneton, and g is the Landé factor. The characteristic timescale τ_D is given by $\tau_D = 1/(\gamma \alpha H_K)$, with γ the gyromagnetic ratio and H_K the anisotropy field. And t_{PW} is the write pulse width. Eq. (7.11) has important consequences for the design of STT-MRAM devices. First, the write current is proportional to the free layer moment, $M_S V$, irrespective of pulse length or WER. Second, the write current "floor" in the long pulse limit is proportional to $1/\tau_D$, i.e. to α and H_K. Third, the current increase in $1/t_{PW}$ at short pulse lengths depends primarily on the free layer magnetic moment, rather than anisotropy or damping. In this precessional switching regime (short pulse), both low magnetic moment and low damping are necessary for low switching current at deep error rate regime using nanosecond long write pulses. It is illustrated in Figure 7.8 in the range of 1–10 ns.

7.3.3 Switching Delay of an STT-MRAM Cell

In a 1M-1T MRAM cell, the cell current shows a step function when MTJ switches. At a given write current, the switching delay is measured from the beginning of the write pulse to the time the MTJ switches state. Figure 7.9 illustrates the switching delay ($t_2 \rightarrow t_3$, $t_6 \rightarrow t_7$). The delay varies over a range (solid line and dashed line) for a single cell from event to event. As long as the write pulse is long enough to cover the distribution of the delay, WER will be low to nil.

7.3.4 Read Disturb Rate

Since a current is applied to an MTJ to READ the cell data, there is a finite chance the cell is written. Thus, if the unwanted switching occurs during read, it is called a read disturb. The read disturb rate (RDR) = 1−WER.

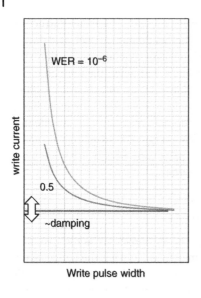

Figure 7.8 Write current versus write pulse width in the precessional switching regime for WER = 0.5 and 10^{-6}.

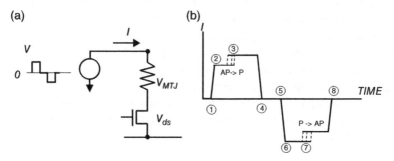

Figure 7.9 (a) A memory cell with an MTJ in series with a transistor. The transistor is on, and the voltage source (V) cycles between positive and negative. (b) The current through the MTJ changes as the MTJ switches state, from AP to P at t_3, and the current reversesas MTJ switches from P to AP at time t_6. The delay time varies from event to event.

Butler had calculated the RDR as a function of read pulse width and MTJ thermal stability factor. Figure 7.10 shows the calculated result of $\Delta = 60$ [7].

7.3.5 Switching Under a Magnetic Field – Phase Diagram

In the thermal switching regime, the external magnetic field in the direction of the easy axis assists (or retards) the STT switching threshold [12–23]. The field alters the switching energy barrier in the same manner as in the field MRAM in the

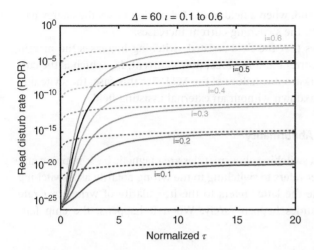

Figure 7.10 Shows a way to project the RDR from measured WER data [7].

single-domain analysis. Combining the field term and the current term together, one obtains the switching energy barrier as

$$\Delta = \Delta_0 \left(1 - H/H_{c0}\right)^2 (1 - I/I_{c0}), \tag{7.12}$$

where $\Delta_0 = E_{b,STT}/k_B T$. Figure 7.11 shows the measured switching current in the presence of the easy-axis magnetic field. When the MTJ is injected with a write current to switch it from P to AP state, an external field in the AP direction lowers the energy barrier $\Delta = \Delta_0 \left(1 - |H|/H_{c0}\right)^2$. As a result, the switching current is

Figure 7.11 Phase diagram of switching field and write current. H_{ext} is the external field, H_{offset} is the offset of the R-H loop, H_{c0} is the critical (switching) field without current injection, I_{c0} is the critical STT switching current, I_{bd} is the breakdown current, and the switching energy barrier under zero field and current is 56 $k_B T$. Reproduced from [12], with the permission of AIP Publishing.

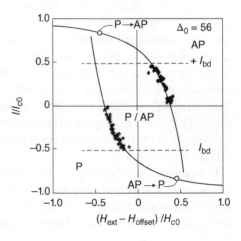

reduced. On the other hand, when a field in the P direction raises the energy barrier $\Delta = \Delta_0 (1-|H|/H_{c0})^2$, the switching current increases.

Figure 7.11 also shows that the field-assisted switching increases the margin between the switching current and the breakdown current, I_{bd}, which is the current that will cause a catastrophic failure of the MTJ. For this case, the switching current exceeds the breakdown without an external assisting field.

7.3.6 MTJ Switching Abnormality

Two frequently observed switching abnormalities are magnetic back-hopping and "ballooning." The former refers to switching to the wrong polarity as a switching current is applied, while the latter refers to the irregularity of write-error rate behavior. The abnormalities happen rarely. When it happens, the chip fails at WRITE.

7.3.6.1 Magnetic Back-Hopping

When an MTJ receives a polarized current in the direction that attempts to switch the free layer to the same state (e.g. from AP- to AP-state), the MTJ should not switch. However, it has been observed experimentally that the MTJ may switch to the opposite state (in this case, P-state) and then switch back (to AP state) repeatedly. Such a phenomenon is called *back-hopping*. It happens frequently at currents much greater than the write critical current, with a finite probability, mostly in the samples with low anisotropy [14, 15]. Figure 7.12 shows back-hopping events. A voltage pulse is applied across the MTJ. The transient waveform is recorded. Figure 7.12a shows three traces of the transient waveform: the applied voltage pulse such that $V_{MTJ} = 0.8$ V starting at $t = 0$ for all three events. Take the dark trace as an example; one observes the MTJ switches resistance state around $t = 15$ ns. The medium dark trace behaves similarly; the MTJ changes resistance at $t = 120$ ns. In both events, the MTJ does not change state after the initial switch. The light trace shows the initial switching at $t = 210$ ns and follows with a telegraph-noise-like waveform across the MTJ at $t = 605$ ns, indicating that the MTJ hops back and forth between states. That is back-hopping. One back-hopping event occurs in three traces at $V_{MTJ} = 0.8$ V over 700 ns. At $V_{MTJ} = 0.9$ V, back-hopping events take place more often, six times. The back-hopping events happen randomly in time [16]. Figure 7.12a also tells us that MTJ switching is stochastic: The latency of the first switching event is random, spreading from 15 to 210 ns in this particular sample.

The cause of back-hopping has been elusive for quite some time. It has been attributed to the out-of-plain torque $M \times S$ carried by the injecting electrons. Unlike the in-plane torque [17, 18], the out-of-plane torque is proportional to the square of the bias voltage. Further experimental studies indicated that back-

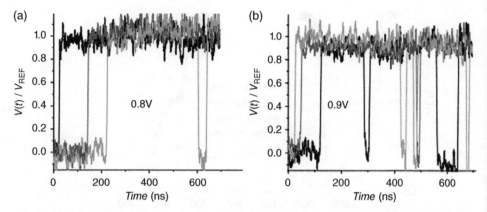

Figure 7.12 Transient switching signals of a step long write pulse to MTJ for write voltage = 0.8 (a) and 0.9 V (b), with three traces in a different color for each write voltage. At 0.8 V write pulse, the initial switching takes place at a different time (from 15 ns to 210 ns). One of the three traces (in the lightest color) shows state reversal (back-hopping) around 605 ns. At 0.9 V write pulse voltage, the initial switching time is shortened to within ~105 ns. Back-hopping becomes more frequent and can be found in all three traces. The back-hopping event is becoming more frequent at higher bias. *Source:* Reproduced from [16], with the permission of AIP Publishing. It has been pointed out that the MTJ bits with a high tendency of back-hopping can be detected with a phase diagram [16]. MTJ bits with negative slope in the phase diagram, i.e. a retarding field that can reduce the MTJ switching current, tend to exhibit back-hopping at high bias voltage (or current).

hopping is correlated to the instability in the synthetic pinned layer. After the free layer is switched, a high writing current can even cause the reference layer to switch in the case of a synthetic antiferromagnetic reference layer, and part of the reference layer can even undergo complex spin flip dynamics [19]. Back-hopping occurs in both in-plane and out-of-plain MTJ. For the out-of-plane MTJ, please refer to Section 7.6.3.

7.3.6.2 Bifurcation Switching (Ballooning in WER)

As one increases the switching voltage (or current), one expects the WER to decrease. However, one may find that the WER of some bit cells may decrease at a slower rate or even stop decreasing for a range of switching voltages and then continue to decrease normally after the switching voltage exceeds certain value. This change in WER is referred to as ballooning. Figure 7.13 illustrates such abnormal switching phenomenon. Ballooning starts at about 400 mV in the figure and ends at about 500 mV. It happens in a small percentage of samples. It appears in both switching directions. An external field can shift the trigger point and magnitude; thus, ballooning is magnetic in nature [14].

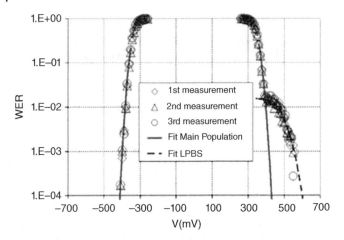

Figure 7.13 Abnormal bifurcation switching at low probability [15].

7.3.6.3 Domain Mediated Magnetization Reversal

Recall that the domain wall width is $DW_{\text{width}} \approx \pi \sqrt{\dfrac{A_{\text{ex}}}{K_{\text{eff}}}}$. In the high H_K perpendicular MTJ, the domain wall (DW) width can be narrower than the MTJ diameter, especially in the very thin free layer in which the wall energy is small. DW nucleation during STT switching becomes possible. In such situation, the magnetic reversal process composes two distinct phases: First the nucleation of domain wall occurs; then a propagation phase happens, in which the domain wall sweeps through the MTJ. In the region that the polarization of injected electrons is co-linear with the magnetization, the spin torque is nil. Once the domain nucleates, the spin torque of the polarized electrons acts only on a fraction of the MTJ, not the entire MTJ. Thus, the switching efficiency drops. Domain propagation results in a much slower switching speed for complete reversal. The domain wall propagation mechanism also makes the switching more sensitive to the pinning sites that originates from random defects of the magnetic and electrical properties of the devices. Long-life domain wall (>100 uS) in the free layer under spin current can place an MTJ in intermediate resistance states [8]. The intermediate resistance states under STT current is observed when an MTJ is at low bias, but not observed at high bias. (Figure 7.14).

7.4 The Integrity of MTJ Tunnel Barrier

It has been proven that the MTJ for field MRAM is reliable [20]. The situation for STT-MRAM is different. When STT-MRAM is in write operation, an electric field of the order of 0.5V/nm is applied across the tunnel barrier, and a current density

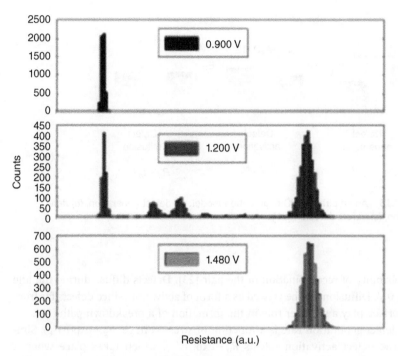

Figure 7.14 STT switching from P- to AP-state. No switching at 0.9 V (top), intermediate states are observed in STT switching at 1.2 V (middle), but not at 1.48 V (bottom) [8].

in the high-10^6 A/cm^2 range flows through the MTJ. Although the junction does not break down immediately under such stress condition, the junction integrity degrades and eventually breaks down after prolong write stress cycles [16]. The breakdown of MgO limits the number of write cycle of MRAM. Thus, it is necessary to understand the time-dependent dielectric breakdown (TDDB) property of the tunnel junction and to project the MTJ life expectancy.

7.4.1 MgO Degradation Model

The MgO degradation model is depicted in Figure 7.15. When electrons inject from the bottom electrode (BE) interface and tunnel through MgO into top electrode (TE), its energy is released at the interface of top electrode and generates defects. Defects can be attributed to Frenkel pairs of oxygen interstitial (O^{2-}) and vacancies VO^{2+} [21]. In the case of bipolar voltage (current) stress, the reverse stress voltage activates the defects, displacing the interstitial O^{2-} and vacancy VO^{2+}, reducing

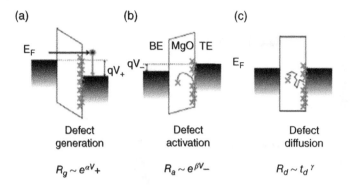

Figure 7.15 An empirical MgO degradation model: (a) defect generation, (b) defect activation, (c) defect diffusion [23].

the probability of recombination of the pair [23]. Defects diffuse during the idle time at 0 V. Diffusion can be viewed as a form of activation, since defects far from the interface play a stronger role in the formation of a breakdown path [22].

The defect generation rate is a function of stress voltage, $R_g \sim \exp(\alpha V_+)$. Similarly, the defect activation rate is $R_a \sim \exp(\beta V_-)$, which takes place when a reverse voltage is applied to the MTJ. The defect diffusion rate is $R_d \sim t_d^{\gamma}$, where α, β, and γ are constants; and V_+, V_-, and t_d are positive stress voltage, negative stress voltage, and diffusion time, respectively. Since each rate is different, the buildup of the defect depends on the stress waveform and the idle time (no stress voltage). It also depends on whether the stress voltage is the unipolar- or bipolar-voltage. Bipolar stress (write "1," "0," "1," "0," ....), which happens to be the actual use condition [24, 25], degrades the MgO junction faster than the unipolar voltage stress. For a bottom-pin MTJ, stress electrons entering MgO dielectric from the top MgO-CoFeB interface (WRITE "1," AP-state) degrades MgO slower than those from the bottom interface (WRITE "0," P-state). The observation suggests that the quality of the top MgO-CoFeB interface may be different from that of the bottom interface. If indeed the defects are generated at the electron exiting interface, as illustrated in Figure 7.15, then the bottom MgO-CoFeB interface is more resistant to stress. Qualitatively, the Carboni stress model [23] can explain the differences.

The rate increases at higher temperatures, and the MgO fail rate increases at higher ambient temperatures [22]. Experimental study [26] shows that lower-*RA* MTJ tends to outlast the higher-*RA* MTJ at the same switching write-error rate. It is due to self-heating. The trend is consistent with the temperature dependency of the MTJ degradation model.

7.5 Data Retention

Data retention performance is one metric of nonvolatile memory. As mentioned at the beginning of this chapter, the probability of a bit switching, or flipping state, when the bit is in idling, is determined by a thermal energy barrier $E_{b,therm}$. Once the thermal energy barrier is determined, one can predict the data retention. We will start the discussion from bit-level determination method to the chip-level determination method.

For an out-of-plane MTJ, $E_{b,therm} = E_{b,STT}$. Thus, one can determine $E_{b,therm}$ based on the STT switching properties. For in-plane MTJ, $E_{b,therm} < E_{b,STT}$. We cannot use this method. We will show in Section 7.5.3 that one can determine the thermal energy barrier with the baking method.

For many nonvolatile memory applications, the required data retention time is in the order of years. Accelerated testing is commonly practiced to predict the data retention time. Two commonly used acceleration methods are as follows: (1) adding an aiding field, and (2) raising the ambient temperature by baking. No current is applied to the MTJ junction in the latter test; thus, one obtains $E_{b,therm}$.

7.5.1 Retention Determination Based on Bit Switching Probability

The most popular method for determining $E_{b,therm}$ is based on Eq. (7.8a, b). The energy barrier is determined from the slope of the switching voltage (current) versus the write pulse width on a semi-log plot. Figure 7.16 shows the write voltage versus write pulse width for a group of MTJ bit cells. The absolute value of the switching voltage is different from bit to bit; however, the slope is essentially the same. From the slope, one obtains $E_{b,STT}$. Notice that the slope is not a constant; it varies with pulse width and is smaller at longer pulses. Besides, since the switching current heats up the MTJ, the MTJ is at different temperature when

Figure 7.16 Voltage at 50% switching probability versus switching pulse width for a group of MTJ of ~60 nm diameter [27] (Copyright (2012) The Japan Society of Applied Physics).

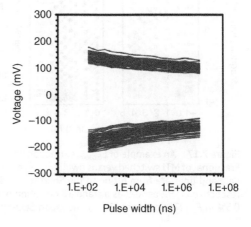

measured at different write currents. That adds uncertainty in the determination of $E_{b,STT}$, based on Eq. (7.8a 2b). This method is frequently used when the number of samples is limited.

For a chip of many bits, there is a distribution of values for the thermal energy barrier. The method in the next sections is required to characterize the data retention time of the chip. The bits at the lower end of the energy barrier distribution fail first, while those at the higher end fail later.

Due to that fact that switching probability is a double exponential function of the energy barrier, the sigma of switching energy barrier $\sigma(E_b)$ of an MTJ array cannot be ignored.

7.5.2 Energy Barrier Determination Based on Aiding Field

This test is based on a *magnetization decay model* [28]. It examines the time-to-magnetic reversal of a large number of MTJs under a magnetic field below the switching coercivity, and no current is applied to the MTJ. Thus, the MTJ is at constant temperature throughout the testing. The result is more reliable than the one in Section 7.5.1. The switching energy barrier reduction under an external aiding field is

$$\Delta = \Delta_0 \left(1 - H/H_{c0}\right)^2, \text{where } \Delta_0 = E_b/k_B T.$$

An aiding field enhances the switching rate, thus reducing the time required to switch, or bit flip. The $E_{b,therm}$ determined from this method corresponds to the data retention time when the MTJ is idling (under a field). Fig. 7.17 illustrates the energy barrier extraction.

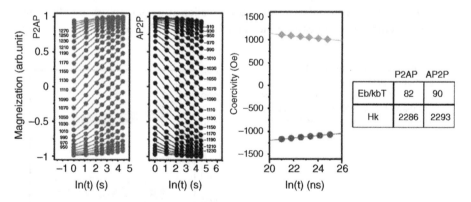

Figure 7.17 An example of $E_{b,therm}$ extraction based on the magnetization Decay (Sharrock) model. Fractional of MTJ switched versus dwell time in an applied field (from 840 Oe to 1400 Oe). Points are data and lines are from fitting to the Sharrock model. Each branch of switching is fitted to its respective E_b, H_K, as well as a standard deviation of approximately 10% in the anisotropy field and 0.5% in E_b [27] Copyright (2012) The Japan Society of Applied Physics.

7.5.3 Energy Barrier Extraction with Retention Bake at Chip Level

Due to that fact that the switching probability is a double exponential function of the energy barrier, the sigma of the switching energy barrier $\sigma(E_b)$ of an MTJ array cannot be ignored. It is expected that there is a distribution of $E_{b,therm}$ among bits in the memory array. The retention bake method described here allows the determination of both the mean and sigma of thermal energy barrier $E_{b,therm}$.

This method is simply to accelerate the magnetic reversal or retention failure by raising the sample temperature. It is also called the retention bake method, which gives $E_{b,therm}$. For an individual bit, the switching probability at no write current is

$$P_{sw}(t_{bake}) = 1 - \exp\left\{ - \frac{t_{bake}}{\tau_0} \exp\left[- \frac{E_{b,therm}}{k_B T} \right] \right\}.$$

If thermal barrier of every bit is the same, then, the array FailBit count is

$$\text{FailBit}(t_{bake}) = N P_{SW}(t_{bake}), \tag{7.13}$$

where t_{bake} is bake time, and N is the total number of test bits. By counting the FailBit at several bake temperatures and bake times, one can extract the effective energy barrier and calculate the retention time at the operation temperature.

However, in a memory array, the energy barrier $E_{b,therm}$ is distributive, and Eq. (7.13) can be obtained by integrating the FailBit over the distribution of thermal energy barrier, as

$$\text{FailBit}(t_{bake}) = N \int_0^{t_{bake}} \left[\int P_{SW}(E_{b,therm}) dE_{b,therm} \right] dt. \tag{7.14}$$

The data retention time of bits at the low end of the $E_{b,therm}$ distribution is shorter; they flip first. As shown in Figure 7.18a, assuming that the distribution is Gaussian, the FailBit cluster from the low end of the $E_{b,therm}$ distribution at the early stage of the baking then spreads toward the center of the distribution as baking progresses.

First integrate over the energy barrier distribution; then integrate over bake time. The FailBit/N versus bake time is illustrated in Figure 7.18b. The slope of the FailBit count versus bake time (t_{bake}) is very sensitive to the sigma of $E_{b,therm}$. Two MTJ arrays with the same mean $E_{b,therm}$ but different sigma behave very differently.

By taking data from multiple retention bakes, the $E_{b,therm}$ mean and sigma can be extracted. Figure 7.19 shows the data and extraction. From this set of retention bake data, one can extrapolate the data retention FailBit count of a chip.

The effective thermal energy barrier of MTJ bit arrays can be expressed as $E_{b,therm}^{eff} = E_{b,therm}^{mean} - \sigma(E_{b,therm})^2/2$ [29]. The second term is the sigma of thermal energy barrier of bits in the array.

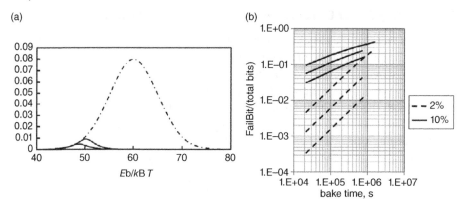

Figure 7.18 (a) The distribution of FailBits for an array of bits having thermal stability distribution in the form of a Gaussian (dot-dash). Once baked, bits with a lower thermal barrier flip first. As bake time increases, the FailBit population spreads from the lower edge toward the center of the distribution. Two different retention bake times (two solid lines) shows how the FailBits spread. (b) The retention bake FailBit versus bake time at three bake temperatures for two arrays of the same $E_{b,therm}$ mean but different sigma (2% and 10%). Notice the slope changes.

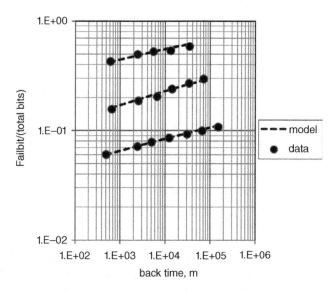

Figure 7.19 Retention bake data (solid diamonds) and fitting (dashed lines) at three different temperatures (85C, 115C, 150C). Extracted stability factors are 66, 53, and 39 for 85, 115, and 150C. The sigma of thermal stability Δ of the array is 25% [31].

This method works well when the total number of FailBit is small, less than 10%. The thermal barrier of P-to-AP and AP-to-P can be determined.

In case of long-time (or higher temperature) retention bake, some flipped bits (FailBits) may flip back to their initial state. The apparent FailBit rate slows down. The analysis should include bits that flip back. Appendix A covers the flip back analysis [30].

7.5.4 Data Retention Fail at the Chip Level

An STT-MRAM chip can lose data in many ways. While the chip is idling, the data integrity is dictated by data retention. When the chip is in operation, data integrity is dictated by WER and read disturb rate. Since each cell is individually written by the cell current, the STT-MRAM cell array does not suffer from the half-select disturb problem as in the field MRAM cells (Chapter 5).

Consider the switching energy barrier $E_{b,STT}$ between the two states of the MTJ. During the WRITE operation, the $E_{b,STT}$ is reduced to ~0 when written by injecting a large write current I_W. During the read operation, the $E_{b,STT}$ is reduced to $E'_{b,STT} = (1 - I_R/I_{c0}) E_{b,STT}$, where I_R is the read current (Figure 6.28). In the case of out-of-plane MTJ, $E_{b,STT} = E_{b,therm}$.

The probability of one data error occurring in a STT MRAM chip over a period of time is defined as follows:

$$\sum_{i=1}^{bits} \sum_{j=1}^{3} P_{ij}\tau_{ij} = 1, \tag{7.15}$$

where i and j are number of bits and state of the bit cell ($j = 1$: standby, 2: READ, 3: WRITE), respectively; P_{ij} is the probability of magnetization reversal of cell i and in state j, and similarly, τ_{ij} is the time duration of cell i and in state j. Thus,

$$\sum_{i=1}^{bits} (P_{is}\tau_{is} + P_{ir}\tau_{ir} + P_{iw}\tau_{iw}) = 1 \tag{7.16}$$

where subscript s is for standby, r is for READ, and w is for WRITE.

7.6 The Cell Design Considerations and Scaling

We discuss the engineering of one-MTJ-one-transistor (1M-1T) STT MRAM cell in this section. The key parameters of a memory are cell size, access latency, power dissipation, data retention time, and endurance cycle. The property matching between the MTJ and the transistor is important in determining the memory performance. For example, the cell size is not determined by the MTJ, but also the transistor and the metal pitch of the CMOS. Today, in order to supply enough current to switch the MTJ, the size of a planar CMOS transistor is much larger than the MTJ. But, if a planar FET is replaced by a FinFET, or a Gate-all-around (GAA)-FET, the situation can be very different. Cell scaling properties determines the longevity of the memory technology. Only those that can continue to scale in density will last in this competitive memory market.

7.6.1 STT-MRAM Bit Cell and Array

Figure 7.20 shows the 1M-1T STT-MRAM cell (a) and a 2×2 array (b). To READ an MTJ resistance state of a bit cell, the word line (WL) of the selected row is activated, and the FET conducts. Current in the selected bit line (BL) flows through the MTJ and transistor to the SL, which is grounded. The BL voltage is compared to a reference. To WRITE a bit cell, activate the WL, raise the selected BL voltage, and ground the SL; this state allows the write current to flow from Bl to SL to WRITE "0" (P-state) or vice versa to WRITE "1" (AP-state). Figure 7.20c shows a generic layout of a 2x2 MRAM cell array. MTJ is a small fraction of a MRAM cell.

We will discuss the cell design consideration in the next few sections.

7.6.2 CMOS Options

The CMOS process evolved from simple transistors into logic CMOS and memory (DRAM) CMOS. The logic CMOS in general emphasizes speed performance, and the transistor drain-to-source current at saturation ($I_{ds, sat}$) is larger and thus can drive more current per unit gate width. The logic transistor is usually leaky when OFF. The memory CMOS is denser in layout, but the current driving capability is lower. The DRAM transistor leakage is exceptionally low.

For MRAM, the desirable CMOS process should offer transistors with a large current driving capability and a dense layout. The leakage is less important;

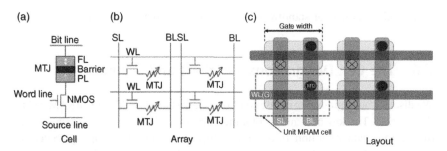

Figure 7.20 (a) 1M-1T MRAM cell, (b) a 2×2 cell array. (c) 2×2 cell array layout.

Figure 7.21 Switching current of MTJ versus MTJ diameter (assume = CMOS node) for two MTJ switching current densities (1 and 4 MA/cm^2). The shaped area is the current provided by minimal size FET current versus CMOS nodes. The FET current is assumed to be ~0.5 $I_{ds,\ sat}$.

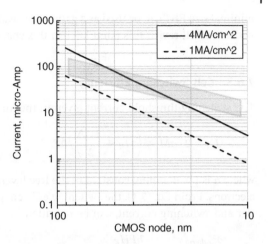

however, it should be better than logic transistor. Unfortunately, such a dense-layout and high current-drive transistor is not readily available.

As transistor is scaled down, another design consideration is the voltage compatibility issue. The voltage supply of the scaled transistor at sub-30 nm nodes is V_{dd} ~ 1.2–0.8 V. The 1M-1T STT cell is required to adjust MTJ RA (resistance area product) to satisfy $V_{MTJ} = RA \times J_w$ ~ ½ V_{dd}.

Figure 7.21 shows although the switching current is in the μA range, the write current density in MTJ is 4MA/cm^2 (solid line) and 1MA/cm^2 (dotted line). And the switching current is therefore proportional to the square of MTJ diameter, while the current driving capability of the cell transistor is proportional to the gate width. Foundry minimum size transistors at various nodes are in the shaded band. (Here the foundry minimal size transistor current is taken as half of the FET $I_{ds,\ sat}$; since the gating FET is in series with an MTJ, its source-drain voltage will not reach the saturation.)

At larger dimension nodes, the MTJ area is large, and its switching current is larger than the current drive of the minimal size transistor. When scaling down, the MTJ area shrinks faster than the minimal gate width of the FET, and the minimal size FET can provide enough switching current.

A typical logic foundry provides minimal transistor size is about 30-50 F^2 at that node. The size of the MTJ is far smaller than the gating FET in the 1M-1T cell. Unless the foundry tailors the transistor process for better memory cell density, the MRAM built-in logic CMOS process is hard to reach the density of DRAM and NAND.

Clearly, reducing the write current while keeping other performance parameters unchanged is the most important task in the development of STT-MRAM technology for better cell density. When the write current density is 4 MA/cm^2, the minimal size FET can be used as a gating transistor only when @ 20 nm or smaller node. We will revisit this point later in this chapter.

7.6.3 Cell Switching Efficiency

The switching threshold current given in the Macrospin model is

$$J_{c0} = 2 \, e\alpha M_S t_F \left(H_K + \frac{H_D}{2} \right) / \hbar P, \tag{7.17}$$

where α is the damping constant of the free layer, and P is the polarization of the electrons. From Eq. (7.1), the switching efficiency, defined as thermal energy barrier and switching current, can be rewritten as

$$\frac{E_{b,therm}}{I_{c0}} = \frac{\hbar P H_K}{4 e \alpha (H_K + 2\pi M_s)}. \tag{7.18}$$

Eq. (7.18) clearly shows the reduction of damping constant α [32, 33], and the improvement of the polarization coefficient P is important. The out-of-plane MTJs are preferred for their better $E_{b,therm}/I_{c0}$, since its switching energy barrier is the same as the thermal stability energy barrier.

α and P are related to the choice of the film material and the capping material to the free layer of the MTJ stack. MgO capping provides the lowest effective damping, $\alpha = 0.004$ [34]. Materials with strong perpendicular anisotropy such as multilayer $[Co/Pt]_n$, etc., are not a good material for free layer due to the high damping. Damping is also dependent on the size of the free layer, multi-eigenmode excitation happens in larger size (~100 nm) free layers where the damping at the center is different from the edge of the free layer [35, 36].

The high polarization of a Heusler alloy material [36, r2] is ideal for free layer material; unfortunately, most Heusler alloys require high temperature annealing to form the desired crystal order. The integration into the MTJ is nontrivial. So far, little work has been done with Heusler alloys [37]. CoFeB of various compositions remains the workhorse material in industry until a better material is found. For more details on these materials, see [38].

We should also keep in mind that data retention time is also important when we address the switching current reduction. Data retention time relates to the switching efficiency in that both rely on Eq. (7.18). New film material, film stack configuration, and MTJ structures are explored, such as thermal-assist STT-MRAM [39], nano-current-channel structure [40], and double-spin-filter structure [41].

Macrospin model does not predict the dependence of switching efficiency. The switching efficiency on the MTJ area. Nonetheless, test results [27, 42, 43] show that as MTJ of the same film stack is reduced in size, the efficiency rises. For the out-of-plane MTJ with a diameter of 50 nm, Δ / I_{c0} can be as high as ~2–3.

From an energy consumption point of view, the switching efficiency is

$$\frac{E_{b,therm}}{E_{SW}} = \frac{E_{b,therm}}{I_{SW}} \frac{1}{[V_{SW} \times t_{PW}]},$$

where subscript SW stands for switching.

Reducing the damping constant of the free layer is the most direct way to lower the write current without affecting the data retention property. As shown in Section 6.7, the effective damping constant of the MTJ free layer is affected by a normal metal capping layer to the free layer. Normal metal with short spin diffusion length increases the effective damping of the free layer. The effective damping constant is reduced to 0.004 [11, 34], as shown in Figure 7.22a, and that further improves the switching efficiency E_b/I_{c0}.

Improving switching efficiency indirectly improves the MTJ endurance. The smaller write current stresses the tunnel barrier to a lesser degree, and thus the lifetime of the tunnel barrier is longer.

7.6.4 Cell Design Considerations

As a memory, the most important performance metric is memory cell density. The next important metrics are access performance, such as cell access power and access time, and then device endurance. Since MRAM is nonvolatile, data retention is also a metric. The specifications of data retention time and device endurance are product application dependent. For example, for system on a chip (SoC) embedded cache applications, the energy barrier design target is seconds

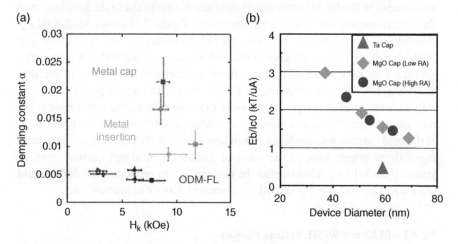

Figure 7.22 (a) Effective damping of 30 nm MTJ with three kinds free layer (Ta contact, metal insertion, optimized damping/moment [ODM]) structures, (b) switching efficiency. One important measure of merit of STT-MRAM technology is E_b/I_{c0} in units of $k_BT/\mu A$ [11].

[43], while for cold storage applications, the target is typically 3–10 years. Similarly, the MTJ endurance should be higher in cache, lower in cold storage.

Design is a multiparameter optimization and trade-off process. Here we suggest a design optimization flow to deal with the complex trade-offs of the MTJ film/structure parameters for each generation of CMOS technology. The parameter optimization process can be simplified by starting from the retention requirement. Once MTJ switching efficiency Δ/I_w is known, memory cell size and memory access performance will automatically fall into place.

7.6.4.1 WRITE Current and Cell Size

From Δ and Δ/I_W, cell current is determined; thus, the gate width of gating transistor dimension is determined. For MTJ designed for shorter data retention time, its thermal stability factor Δ is smaller; thus, the WRITE current is smaller, which requires a smaller size gating transistor. Therefore, the cell is smaller. The net result is that the chip size is smaller. We will discuss the array density later in Chapter 9.

7.6.4.2 READ Access Performance and *RA* Product of MTJ

In the READ access, the on-chip sense amplifier senses the MTJ bit resistance by reading the bit line voltage. A larger TMR ratio will accelerate the bit line voltage development, thus reducing the sense amp signal development time. However, the sense amp contributes only a portion of the total READ access delay, and the TMR contribution to the access delay is not significant; rather, the contribution of resistance spread to the READ error rate is significant. Due to the finite distribution of the MTJ resistance, the bits on a bit line spread. Figure 7.23 shows the R_p and R_{ap} distribution. A measure of the goodness of a technology is $DR = (R_{p0} - R_{ap0})/[\sigma(R_p) + \sigma(R_{ap})]$, where subscript 0 stands for average value and σ is sigma. The spread is a measure of the MTJ stack patterning technology. The spread of the transistor resistance further degrades the READ margin and access performance.

In addition to read margin, one should examine read access performance: the read disturb rate (RDR) and sensing time delay. The former sets the upper limit of the read current, and the latter sets in lower limit of the read current. For sense signal development time of the order of 5 ns, the typical cell current must be greater than 5–10 µA. That implies the resistance-area product of the MTJ should scale down as well. This point will be discussed further in Section 7.6.5.

7.6.4.3 READ and WRITE Voltage Margins

Read and write margins can be derived from the WER and RDR, as shown in Figure 7.24. A well-designed MTJ provides a well-separated voltage margin that ensures 10^{-9} error rate.

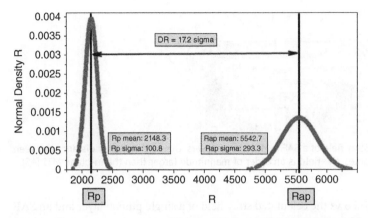

Figure 7.23 An illustration of R_p and R_{ap} distribution on a chip. The ratio of $(R_{ap0}-R_{p0})/[\sigma(R_{ap}) + \sigma(R_p)]$ = 17.2, where subscript 0 means average value and σ is sigma [44].

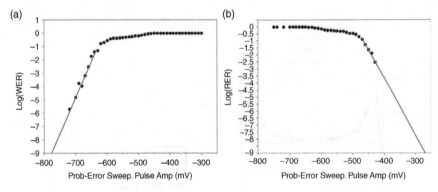

Figure 7.24 An illustration of extracting the operating WRITE voltage and READ voltage. (a) From write-error-rate (WER), one may extract the WRITE voltage and V_W ~ -800 mV for WER <10^{-9}. (b) From RDR = 1- WER, one extracts the READ voltage, which is V_R ~ -250 mV at RDR <10^{-9}. The margin is ~550 mV.

7.6.4.4 Stray Field Control for Perpendicular MTJ

The edge pole of the SAF pinned layer of an in-plane MTJ forms a good flux closure at the film edges. The stray field is confined locally at the edge of the pinned layer, and a well-balanced SAF spreads no net stray field to the free layer. In an out-of-plane FM film, the poles are on the surface of the pinned layer, rather than at the side edge. Figure 7.25 depicts the magnetic flux from the SAF pinned layer. The stray field spreads into the free layer and affects the R-H loop and the stability of the MTJ.

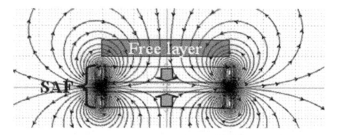

Figure 7.25 Stray field of a SAF p-MTJ pinned layers. Due to the shorter distance between the magnetic poles, the field is an order of magnitude larger than the in-plane MTJ [45].

Figure 7.26 shows the calculated stray field of a single pinned layer and an SAF pinned layer. The stray field of a single pinned layer is twice of an SAF pinned layer. The magnitude of the stray field is larger at the edge of the MTJ and smaller at the center, in both the vertical and lateral directions. As the MTJ diameter reduces, the edge field encroachment toward the center of MTJ is more severe.

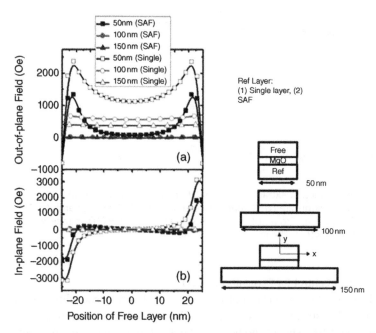

Figure 7.26 Calculated (a) out-of-plane and (b) in-plane stray field as a function of free layer position. The diameter of the free layer is fixed at 50 nm, and the pinned layer is with the size of 50 nm, 100 nm, and 150 nm. The open symbols and the close symbols are the results of p-MTJ with a single and SAF pinned layer, respectively [45].

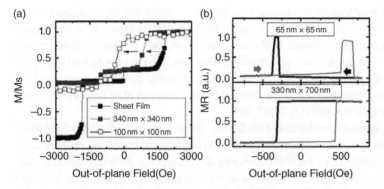

Figure 7.27 (a) Magnetization reversal from +H to −H of p-MTJ sheet film (black dot) and pattern array (blue dot and green dot). The anti-parallel coupling field in SAF pinned layer decreases with a reducing MTJ size. (b) R-H loops of p-MTJ with different sizes. Although the large MTJ shows a good R-H loop, the small MTJ fails due to the failure of RKKY coupling of the SAF pinned layers [45].

For an MTJ of 50 nm diameter having a $[Co/Pt]_n$ multilayer as the SAF pinned layer, the stray field at the edge of the pinned layer is >1 K Oe. It not only offsets the R-H loop of the MTJ; it also compromises the RKKY coupling of the SAF pinned layer and thus the stability of the pinned layer. The control of the stray field becomes not only a design issue, but also a scaling issue [45].

Figure 7.27a compares the major loop M-H scan of the film and two different sized MTJs made of the same MTJ film stack. The degradation to the MTJ coercivity becomes more severe as one scales down the MTJ size. In the small diameter MTJ, the stray field encroaches to the center of the MTJ, and the pinning of the SAF pair fails, while in the large MTJ, the SAF pair holds up well (Figure 7.27b).

One engineering solution to the stray field is to form MTJs with a step-etch structure. By extending the pinned layer edge outward, moving the pinned layer edge away from the free layer edge, the stray field diminishes, and a well-balanced MTJ R-H loop can be obtained [45]. Another way to overcome the pinning failure is to raise the RKKY coupling and to raise the coercivity of the pinned layer [46].

7.6.4.5 Suppress Stochastic Switching Time Variation Ideas

One may refer to Eq. (6.3); the Slonczewski spin torque is $M_F \times (M_F \times M_P)$, where the subscript F and P refer to the free layer and pinned layer, respectively. When magnetization in the free layer and the reference layer is collinear, the torque is zero. One way to start the STT precession is with thermal agitation of the free layer so that the free layer is no longer collinear to the reference layer. However, this random thermal agitation leads to the uncertainty in the switching time and switching current.

Another way to start precession is to lift the in-plane free layer magnetization slightly out-of-plane with a perpendicular (orthogonal) polarizer. Now its magnetization is no longer collinear to the reference layer, even with no applied current (and vice versa, for an out-of-plane MTJ). Thus, once the spin polarized current is injected, STT action starts from nearly the same initial condition, instead of the random condition dictated by the thermal noise. This class of STT-MRAM is called *orthogonal spin transfer* (OST) [47, 48].

The experimental evidence of benefit from an MTJ with an orthogonal polarizer appears insignificant. Subsequent studies [7] confirm experimental results (insignificant benefit), and a careful balance of the orthogonal polarizer and the magnetic shape anisotropy is critical to achieve a good WER [49], [50]. Another similar idea is to shorten the switching with a *precessing* orthogonal polarizer [51].

7.6.5 The Scaling of MTJ for Memory

The cost of memory technology continues to fall. CMOS technology continues to scale. The MTJ magnetic properties of the 1M-1T MRAM cell should be adjusted to achieve cell size reduction while maintaining the performance. The objective of MRAM scaling is to find a set of simple rules to raise the cell density without major redesign while maintaining its performance. Here, the order of scaling consideration is (i) density and (ii) performance.

From a density consideration, MRAM scaling follows CMOS technology scaling. Since the MTJ is integrated into a CMOS circuit to form the basic storage bit element, the 1M-1T cell, timing-wise, MTJ scaling synchronizes with CMOS scaling. The density of the 1M-1T MRAM cell is dictated by the gating transistor size, which in turn is a function of MTJ switching current. That is why reducing switching current has been the focus of the MTJ design.

CMOS scaling is characterized by the minimal lithographic feature size F. The minimal size MTJ is F^2, and the minimal switching current is $I_{MTJ} = J_W F^2$, where J_W is the MTJ write current density. In the course of CMOS scaling, the drain current density provided by each generation of FET has been maintained reasonably close to $J_{ds,sat} = 0.5$ mA/μm (for low-speed transistor) -1 mA/μm (for high-speed transistor), where $J_{ds,sat}$ is the saturate drain-source current density in units of current per gate width. Thus, scaling down the size of the MTJ cuts down the write current. That, in turn, shrinks the transistor gate width and therefore the cell size. Nonetheless, the cell size is still limited by the gate width of the transistor defined by the write current, as shown in Figure 7.21.

Today, Si foundries migrate their technology from the planar FET platform to the 3D finFET platform starting in the 20-nm node. The finFET gate width is no longer along the in-plane direction, but the fin height, along the out-of-plane direction. The current of finFET is proportional to the fin height. A minimal-layout transistor can carrier unlimited current, in principle. When finFET is deployed for MRAM applications, the switching current is no longer a cell size scaling issue. It is only an MTJ reliability issue.

7.6.5.1 In-Plane MTJ

The data retention energy barrier of in-plane MTJ is dominated by the shape anisotropy. The typical MTJ shape aspect ratio is in the range of $l/w = 2$–3, while the narrow size dimension w is limited by the technology feature size F, and l is the length. And $(l, w) \gg t$, the thickness of the free layer. Since the shape anisotropy for data retention is $(N_y - N_x)lwt M_S^2 \geq 60k_BT$, the long side is in the y-direction. To maintain constant $E_{b,therm}$, the free layer thickness t should be raised as t $(AR = 3) = 23.41w^{-0.728}$ and $t(AR = 2) = 29.76w^{-0.744}$. Figure 7.28 shows the relation between t and w that satisfies $E_{b,therm} = 60k_BT$. When MTJ $w < 10$ nm, the film thickness t is comparable to w. For a CoFeB free layer on an MgO tunnel barrier, the interface contributes out-of-plane anisotropy K_i; thus, the in-plane shape anisotropy competes against the out-of-plane anisotropy in such small dimension MTJs. To keep the easy axis in-plane, the condition $(N_z - N_y)M_S^2 \geq K_i/t$ should be satisfied. At $w = 10$ nm or below, the free layer thickness is approaching the MTJ width, the easy axis begins to tilt out-of-plane with the significant help from the shape anisotropy. This makes it harder to stabilize the magnetization in the in-plane position for w below 10 nm [52].

The experimental study of in-plane MTJ scaling has shown that the thermal stability of in-plane MTJ is marginal at the 28 nm node, due to the difficulty of

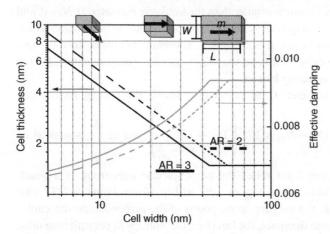

Figure 7.28 Constant $E_{b,therm}$ scaling of in-plane MTJ. Free layer thickness increases as MTJ size shrinks [52] Copyright (2011) The Japan Society of Applied Physics.

lowering the switch current I_{SW} and maintaining the shape anisotropy in small-dimension MTJs in current workhorse material CoFeB [53].

As shown in Figure 7.28, the width/length aspect ratio, n, of the in-plane MTJ has been kept between 2.5 ~ 3.5 to maximize the shape anisotropy, while minimizing the MTJ area and, thus, the write current. So, the minimal MTJ area is ~0.78×n F^2, where F is the minimal lithographic feature size of the manufacturing process. On the other hand, the minimal out-of-plane MTJ is a circle of diameter, F, and its area is 0.78 F^2. In addition, at the same retention performance level, the switching energy barrier $E_{b,STT}$, not the thermal stability barrier $E_{b,therm}$, of the in-plane MTJs is larger than that of the out-of-plane MTJs (Section 7.2). A larger transistor is needed to supply switch current to in-plane MTJ cells. Due to these two reasons, the minimal memory cell area of the in-plane MTJ is always larger than that of an out-of-plane cell.

7.6.5.2 Out-of-Plane (Perpendicular) MTJ

To maintain constant data retention time as we scale the MTJ diameter, $E_{b,therm}$ should be kept constant. Its thermal stability energy is a function of interfacial, shape, and bulk (crystalline) anisotropy,

$$E_{b,therm} = AK_{eff} = A\big(K_i + 0.5(N_x - N_z)M_s^2 t + K_b t\big), \qquad (7.19)$$

where the first term is the interfacial anisotropy, the second term is the volume shape anisotropy, and the third term is bulk anisotropy. A and t are the area of the MTJ and the effective thickness of the free layer, respectively; N_x and N_z are demag factor of the free layer in the in-plane and out-of-plane directions, respectively; t is free layer thickness, and M_s is the free layer magnetization. For the MTJ of free layer thickness, t is much smaller than the free layer diameter, D, $N_x \sim 0$, and $N_z \sim 1$; thus, $(N_x - N_z)$ is negative. A negative shape anisotropy is in-plane, which reduces the net perpendicular anisotropy K_{eff}. Therefore, one must keep the free layer t very thin, typically 1.3 nm, so that K_{eff} is positive (perpendicular) for the MTJ with the MgO tunneling barrier.

When the MTJ diameter D is sufficiently small, the demag factors can be approximated as $(N_x - N_z) \sim 1/2 \left(1 - \dfrac{3}{1 + \dfrac{4t}{\sqrt{\pi D}}} \right)$ and changes sign when

$t/D > 0.89$. A thick free layer raises the perpendicular anisotropy of a small-diameter MTJ. As illustrated in Figure 7.28, with a 10 nm diameter MTJ with 10 nm thick free layer, the magnetization points in the perpendicular direction.

Between 10 and 40 nm diameter, the interfacial anisotropy K_i per unit area must be raised as the MTJ diameter is reduced so that AK_i is roughly constant. One way to raise the unit area is the use of an MgO cap to the free layer. The MgO/CoFeB

(a)   (b)

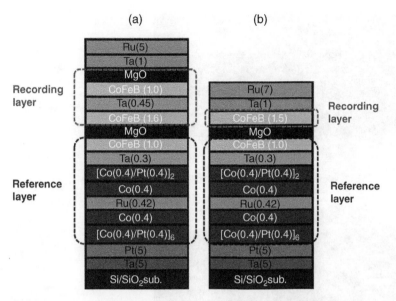

Figure 7.29   (a) Scaled MTJ with dual-MgO interface free layer MTJ, (b) single-MgO MTJ [60].

interface provides stronger out-of-plane anisotropy than the CoFeB/Ta interface. A dual-MgO MTJ film stack is shown in Figure 7.29, and a TEM cross-section is shown in Figure 7.30. The composite CoFeB/Ta/CoFeB free layer can absorb B into Ta during post-deposition anneal and effectively reduces the M_s of the free layer [71]. Having two MgO/CoFeB interfaces reduces the damping coefficient [27] and raises the effective K_{eff}. Table 7.1 compares the MTJ performance of MTJ wth MgO and Ta as the free layer cap. The RA of MTJ with MgO cap is higher, and the switching current density is lower. Active research is ongoing in search for higher interfacial anisotropy K_i. First-principle calculations [54–57] and experimental attempts have been made [58, 59].

It should be noted that although the sigma of thermal energy barrier $\sigma(E_{b,therm})$ is not a scaling parameter, its importance cannot be ignored. The data retention performance of a chip of many bits exponentially depends on that. It was pointed out that the $\sigma(E_{b,therm})$ of the in-plane MTJ has been found experimentally consistently larger than that of the out-of-plane MTJ in various laboratories [30].

Experimental studies of MTJ scaling in the teens of nanometer CMOS nodes [42, 46, 53, 62], discovered many new engineering issues related to MTJ design. Figure 7.31 shows the experimental data of the $E_{b,therm}$ from two development teams [42, 27]. For the MTJ with a diameter smaller than 40 nm, the thermal stability factor is ~F^2 as indicated by the line. F is feature size and is the MTJ diameter

Figure 7.30 A dual-MgO MTJ. *Source:* Reproduced from [61], with the permission of AIP Publishing.

Table 7.1 Compares the MTJ performance characteristics of two free layer film stacks: CoFeB/Ta(cap) and CoFeB/MgO (cap) [27].

	Ta cap		MgO cap (Hi-*RA*)	
RA ($\Omega\,\mu m^2$)	7.0		12.1	
Resistance (kΩ)	2.4	3.7	5.0	7.3
Diameter^a (nm)	59	63	54	45
I_{c0} (μA)	104	62	52	38
V_{c0} (mV)	295	298	310	326
J_{c0} (MA/cm^2)	3.8	2.0	2.3	2.4
E_b ($k_B\,T$)	45	91	90	89
Efficiency ($k_B\,T/\mu$A)	0.43	1.5	1.7	2.3

^{a} Estimated.

Figure 7.31 Energy barrier of dual-MgO out-of-plane MTJ. Line $E_b \sim AK_i$. MTJ data: square [27], diamond [42]. The line is E_b. $\propto$ diameter^2.

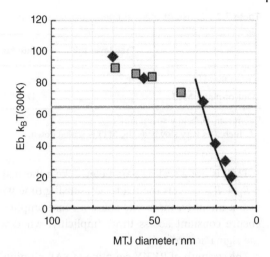

in this case. For those with a diameter greater than 40 nm, the thermal stability does not follow the F^2 relationship. Rather, it increases slowly.

The thermal energy barrier scaling F^{-2} rule applies only to MTJs with diameters smaller than the ~40 nm. It was originally attributed to "subvolume" excitation [63] in in-plane MTJs. Such deviation is also found in perpendicular MTJs. Domain formation limits the thermal stability. The dependence of $E_{b,therm}$ of domain-wall mediated switching is $F^{0.67}$ [33].

There is a limit of interfacial anisotropy, 1.46 mJ/m^2, one can obtained from the Fe—O (O of MgO) interface. To scale down to a 10-nm diameter, a different approach has been proposed. With reference to Figure 7.28, when the free layer thickness and diameter is in the same order, shape anisotropy can offer perpendicular anisotropy, as shown in Eq. (7.19). This approach is called *perpendicular shape anisotropy*, and MTJ based on this anisotropy is called *PSA*-MTJ. An 8-nm diameter PSA-MTJ was built having a stack syAF/MgO/CoFeB(2 nm)/W (0.2 nm)/NiFe(30 nm)/Ta/Ru. The 30-nm thick NiFe is a low damping, low sensitivity to film stress material to thicken the free layer, and the W dusting decouples the fcc NiFe from to 001 CoFeB. The thermal stability factor is ~80, and the switching current is 5 µA ($J = 1\ 10^7$ A/cm^2) [70] for STT write current. Since MgO junction degradation at this current may be too high and the MgO endurance is an issue, one may use the SOT write scheme.

The thermal energy barrier is proportional to the anisotropy of the free layer of the out-of-plane MTJ. Table 7.2 column 2 (Interfacial K_i) shows that one needs to raise the free layer interfacial anisotropy K_i to keep the total free layer anisotropy ~$K_i\,D^2$ constant (thus the data retention time) as the diameter of the MTJ scales down. The interfacial anisotropy of 19.7 k Oe has been obtained [11]. With such surface anisotropy, scaling to a 30-nm diameter MTJ is possible.

Table 7.2 Constant thermal energy barrier scaling.

	Free layer thk	Interfacial K_i	V_w	$I_w{}^{(b)}$
In-plane MTJ	$w^{-0.75}$	–	const	$J_w\, wl$
Out-of-plane MTJ	–	$D^{-2\,(a)}$	const	$J_w\, D^2/4$

^{a} diameter, D, below 40 nm and
$^b J_w$ increases very slowly as the MTJ is scaled down.

TMR is not a scaling parameter. Like a thermal energy barrier, the ratio TMR/$\sigma(R_p + R_{ap})$ affects the read-back sensing time window. As more bits are packed into a memory, more bits share a sense amplifier. This ratio must be raised to ensure constant access time. Implicitly, when we scale parameters, we adjust the sigma accordingly.

The strength of RKKY coupling of SAF pinning layer of the perpendicular MTJ should be scaled up for a smaller-diameter MTJ. As pointed out in Figure 7.27, the spread of the edge stray field from the edge to the middle of the MTJ weakens the stability of overall pinned layer pinning strength more severely in smaller-diameter MTJs than in larger MTJs. Under the STT WRITE operation, a highly polarized current not only switches the free layer, but also disturbs the weakened pinned layer.

For more MTJ scaling from device point of view, see [29, 64]. On the other hand, from a memory circuit READ and WRITE functionality point of view, additional constraints should be considered. A scaled memory must maintain a bit-error-rate performance, both write-error rate (WER) and read-disturb rate (RDR), below chip error specification. Readers are referred to the scaling analysis given in Appendix B of this book (courtesy of T. Sunaga).

7.7 MTJ SPICE Models

MTJ is a magnetic device. An equivalent electrical circuit model is required for memory designers to design an MRAM chip. The more comprehensive the model, the more the circuit designer can optimize the MRAM performance.

7.7.1 Basic MTJ Equivalent Circuit Model for Circuit Design Simulation

The design of MRAM memory chips requires a circuit model that describes the electrical behavior, not the magnetic behavior, of the MTJ so that current design tools are able to simulate the magnetic memory circuitry.

The generic MTJ circuit model consists of two resistors, a switch and a capacitor (Figure 7.32a). R_p is the parallel-state resistor, and $dR + R_p = R_{ap}$. Both are voltage

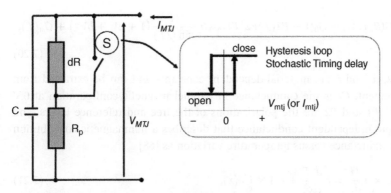

Figure 7.32 A generic equivalent circuit of an MTJ circuit model. A switching threshold and static hysteresis are built into the switch macro.

V_{MTJ} dependent. The capacitor is the effective junction capacitor across the tunnel barrier. The switch mimics the switching properties of the MTJ, such as the hysteresis loop, switching threshold, and switching latency. When the switch S is open, the MTJ is in AP-state; when it is closed, it is in P-state. All parameters can be extracted from MTJ testing data at the device level or the film level (capacitor) [65, 66]. This compact SPICE model is a static model and does not capture the stochastic magnetization dynamic behavior of the MTJ. The compact model is simple and is sufficient for DC design point assessment.

7.7.2 MTJ SPICE Circuit Model with Embedded Macrospin Calculator

To accurately simulate the transient behavior of an MRAM cell in a circuit, the MTJ SPICE model should describe dependency of the stochastic switching error rate and latency on the write pulse width and current. The MTJ SPICE model embedded with a Macrospin calculator in either Verilog-A or C language overcomes the deficiencies of compact SPICE models [67]. It offers write current dependence stochastic switching delay and, thus, WER and read disturb rate, and their temperature dependence.

Internal nodes are added to the compact MTJ SPICE model to simulate the magnetization dynamics, while the external nodes simulate the MTJ electrical behavior. An MTJ is modelled by nine nodes, consisting of four external nodes and five internal nodes. The internal nodes describe the three components of the magnetization (m_x, m_y, m_z) and the MTJ temperature T. The temperature is calculated from V^2/R of the MTJ. The MTJ conductance depends on the angle θ between the magnetization of the two FM layers, the bias voltage (V) applied across the MTJ, and the temperature. The resulting total tunneling conductance can then be written as

$$G(\theta, T, V) = G_0[1 + P1(T)P2(T)\cos\theta]\frac{\lambda T}{\sin(\lambda T)}\left(1 + aV + bV^2\right) + G_{si}(T),$$

$$(7.20)$$

where λ, a, and β are material-dependent constants and can be extracted from measurement. G_0 is the conductance in parallel magnetic configuration at 0 V and 0 K, $P1$ and $P2$ are the polarizations of the free and reference layers, and G_{si} is spin-independent conductance that describes a nonmagnetic contribution to the conductance versus temperature variation as [68]

$$C \times \rho \frac{dT}{d\tau} = k\frac{d^2T}{dx^2} + RA \times J^2\delta(x),$$

$$(7.21)$$

where C is specific heat, ρ is volume density of material, k is the thermal conductivity, RA is resistance-area product, J is current density, and δ is Dirac distribution; the barrier is at $x = 0$; and time τ and space coordinate x are independent variable.

Adding the Macrospin calculator and random initial magnetization cone angle θ, the model can provide a reasonable description of the STT-switching, such as the stochastic switching behavior of the MTJ, and allow a circuit designer to predict the WER. Figure 7.33b shows the switching latency, which is a function of write

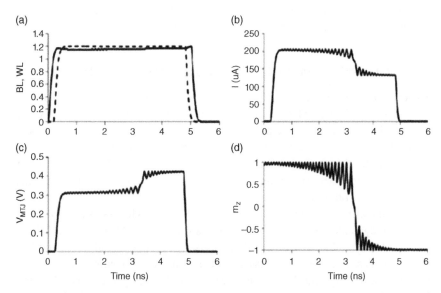

Figure 7.33 Verilog-A simulation of the switching transient of a 1M-1T cell. Magnetization switching dynamics is included. The waveforms of (a) WL and BL, (b) I_{MTJ}, (c) V_{MTJ}, and (d) z-component of M. The step in current and voltage waveform across MTJ reflects the switching delay [69].

pulse. By overlapping the current responses of the voltage pulse train, the model shows the variations of switching latency. In this example, if the write pulse is shorter than 3 ns, some attempts to write will fail.

The inputs of this model are external magnetic field and voltage across the MTJ. Thus, the model is applicable to field-mode switching or STT-mode switching.

7.8 Test Chip, Test, and Chip-Level Weak Bit Screening

In various stages of the MRAM technology development, test chips are designed to extract device-level parametric and chip-level dynamic performance parameters, such as READ and WRITE cycle time. For Gb-level product development, a tens Mb-level test chip is a reasonably good choice. It provides a good sampling of the problems that may occur in the product. In general, two kinds of bit failure must be dealt with. They are the hard failure bits and soft failure bits, FailBit. Hard FailBits fail permanently. Soft FailBits fail intermittently. Soft FailBits are marginal bits and sometimes lumped into the group of weak bits. Weak bits can be bits of small sense margin when READ or with marginal or abnormal WRITE behaviors, bits of short data retention, bits of lower WRITE endurance, etc. Each kind of FailBit must be identified and dealt with in a different way.

A test chip can be used to map the hard FailBits, to create statistics of soft FailBit distribution, and to classify bit failure modes. Such information from the test chip serves as

- An early feedback to the development effort so that the cell margin is improved.
- An input for determining the strength of error-correction code for handling the soft errors, and the least number of redundant rows and columns for replacing hard FailBits.
- An input for calibrating MTJ SPICE model for product chip design.

In early stages of technology development, test chips are built to provide early information. The information is channeled into the product design data base. Gradually, product and test chips are placed on same mask and built together. By the time the product is mature, the test chip is removed from the mask.

One challenging testing task is to construct a test flow to identify soft FailBits. For a typical product, the failure rate of the MRAM bit must be 10^{-6} (1 part per million, or ppm) or below; it seems very low, but for a Gb-level product, error will happen too often to bear. At this level, FailBits are hard to locate, and the failure event occurs so rarely and takes a lot of test time to capture. A good test chip should be designed to offer FailBit information in a short test time. Memory testing is a costly step in production, and engineers must manage the test time and screen out the marginal bits on a chip. Imagine that each bit requires 1 μS to test. Testing a 1Gb would cost 1000 seconds (~1/3 of an hour), a prohibitively long time.

Notice that the digital chip–level testing methodology, such as a pass-or-fail test, is much faster than analog chip testing methodology. The analog test methodology relies on the A-D converter in the memory tester, which is very slow. For example, to measure the value of the bit current of each bit (an analog value), one may not need to bring the bit current out of the chip into a tester and to measure the current with the A-D converter of a tester. One may also do the same current measurement with a comparator circuit on the test chip. An on-chip sense amplifier with a tunable reference will serve this purpose well. By adjusting the reference level, one can find out the range of reference current in which the bit fails to be read correctly. From the reference, the value of bit current can be extracted. A tunable sense amp can be used to map out the sense margin of each bit in an array.

Bit failure could be caused by circuit problems, rather than magnetic problems. This section will cover only a subset of the chip testing that addresses the magnetic problems. Circuit-induced issues will not be addressed (we assume that the sense amp is perfect, e.g. no offset).

Before we start, we introduce a few test notations. Each describes a test procedure. {W1}↑: writes the entire bit array "1," and the arrow ↑indicates writing from the lowest bit address to the highest address; similarly, ↓ is from the highest bit address to the lowest address. The same style applies to {W0}, {R1}, and {R0}. {W1}^{n} stands for repeating the write "1" step n times.

When a bit is expected to store "1" and the readout is not "1" but "0," the bit is a failed bit, a FailBit. Although when a test identifies a FailBit, it does not tell us why: whether the bit is not previously written successfully so the bit stores incorrect value or read out incorrectly, or data was written successfully but later corrupted due to other reasons before read, or due to read-disturb. When a FailBit is found, the address of the FailBit is recorded for further analysis.

Let us examine the bit screening with the purpose of looking for (i) read/write marginal bits, (ii) short data retention bits, and (iii) low endurance bits in the array. Item (i) can be detected at the wafer level, and (ii), (iii) usually cannot. Endurance in general is tested offline to predict the failure rate for a given product life.

These four bit-screening tests are discussed next.

7.8.1 Read Marginal Bits

In general, most screening methods are based on worsening the read condition of the bit cell, such as (a) shifting the test chip sense amp reference cell resistance to reduce sense margin, (b) shortening the sensing time, or (c) collapsing the MTJ junction voltage by adjusting the BL or WL voltages. That implies the WL voltage, the BL voltage, and the reference or sense amp of test chip are tunable during testing. Raising the test temperature or adjusting bit line voltage supply will also reduce the TMR and, thus, the sense margin.

A read margin test procedure involves a basic test sequence of {W1}, {R1}, {W0}, and R{0} and its variations. The test chip is set to test at various I_w, I_R, and timing of the read condition. To construct the RDR sequence of {W1*}, $[\{R1\}]^n$ is performed, where {R1} is performed with read current in the opposite direction of {W1*}. Here, "*" of the first test step means to write with a large write current to make sure that every bit in the array is successfully written to state "1". It can be skipped if {R1} does not disturb the bit content. {R1} is performed at the read current of interests. The sequence is repeated n times, and the end result is the RDR. To reach RDR = 10^{-6}, n = 10^6 at each read current. Similarly, one repeats the procedure on the "0" state. To save test time, the test is usually terminated early, and the rate is extrapolated to the desired error level.

Sometimes, a test chip is designed to equip with built-in-self-test (BIST) logic circuitry, also called and "engine," to execute the test once the chip receives a command from tester. That reduces the signal traffic between the test chip and tester and, thus, shortens the test time.

7.8.2 Write Marginal Bits

In general, a WRITE fail is caused by insufficient write current and/or write pulse width. It would be straightforward to screen out the write margin bits. There are many ways to test; one test sequence is $[\{W1^*\}. \{R1\}, \{W0\}, R\{0\}]^n$. {W0} is executed at the write current of interests. By extracting the statistics of a successful WRITE of a given write condition, one can construct the write-error-rate (WER). Some bits suffer from "ballooning" fail (Section 7.3.4.2); these bits require extensive testing to identify. Since it is highly related to the wafer processing, finding out the existence of such bits is important.

7.8.3 Short Retention Bits

As described in Section 7.5, data retention time is a function of thermal energy barrier, or more fundamentally, the net bit anisotropy of an MTJ, be it in-plane or out-of-plane. Bit anisotropy is affected by many factors, such as film uniformity, cell area variation, etch residue, etc. We have discussed in this chapter that the retention performance of a chip is a function of both the mean and sigma of thermal stability factor. These two parameters can be extracted from the aiding-field procedure (Section 7.5.2) and the thermal bake procedure (Section 7.5.3). These accelerated tests provide thermal stability data in a short test time. The tests are conducted off-line in packaged chips.

Since anisotropy of each bit is not readily measurable at the bit level, one may correlate the retention time with coercivity of the MTJ bits. Anisotropy and coercivity are correlated. And bit coercivity is readily measurable. The conventional R-H loop test methodology provides bit coercivity information. But this test is slow, due to the fact that the rise time and fall time of an electromagnet on a tester is slow and due to the fact that the different magnitude of field is applied to each bit many times to form the bit R-H loop. From the loop, the coercivity is extracted.

A fast chip level coercivity test procedure is shown in Figure 7.34a. The test procedure starts out with a reset field H* applied to the entire array $\{W0\}_H$. The magnet pole is larger than the test chip. So, its field is sufficient to WRITE the entire test chip in one shot. Then apply a field of magnitude H1 to do $\{W1\}_H$, and it is followed by $\{R1\}$. [$\{W0^*\}_H$, $\{W1\}_{H1}$, $\{R1\}$] completes the first cycle of the test procedure. Initially, H1 is small, say, well below H_c, no bit in the array flips, and every bit remains in "0" state. Or #$\{R1\}$ bit = 0. As one increments n, one raises the field. When Hn field is close to H_c, some weak bits start to flip. The number of "1" bits increases. (See Figure 7.34b). This procedure is repeated with an increment in the field of $\{W1\}_{Hn}$ until every bit in the array flips to "1." By taking the digital differentiation of Figure 7.34b, one gets Figure 7.34c, the H_c distribution of the array bits, which may be correlated to the thermal stability factor. It also identifies the location of short-retention bits in the array for bit screening purposes.

7.8.4 Low Endurance Bits

In publications, there have been plenty of bit endurance studies based on accelerated life expectancy tests (Section 7.4). The time-dependent-dielectric-breakdown (TDDB) test methodology for dielectric is well established. Consistent conclusions

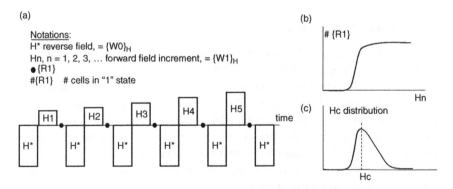

Figure 7.34 (a) Schematics of test flow of chip-level bit coercivity test, (b) accumulated flip bit count, (c) Hc distribution.

are: (1) MgO tunnel barrier degrades under bipolar stress faster than unipolar stress, and (2) MgO degrades faster in higher temperature. An MTJ film stack design guideline for low breakdown fail [26] has been given. All these are offline evaluations. All such tests require long test time.

The subject of identifying the low endurance bits in a memory array is rarely discussed in publications. It is desirable to screen off the weak bits that pass at the product testing stage and fail earlier than majority bits in product life. The methodology either is not being practiced or is believed to be highly sensitive trade secret or very sensitive.

Nevertheless, establishing a bit endurance methodology is essential. A test chip should be designed to provide an efficient block stress function to accelerate this time-consuming test for identifying the endurance failure mode and for learning the bit failure syndromes prior to the catastrophic failure. We believe that given enough effort, an endurance screening methodology will be developed.

It is inevitable that every product chip will have unscreened weak bits that pass the manufacturing test. One can minimize but cannot totally eliminate the weak bits. These weak bits cause soft fail in the product life. The remedy is to add error-correction code (ECC) bits into the array. The penalty of having ECC bits in the array is the reduction of effective bit density and the increase in data access latency. An endurance test chip function should give the fail bit distribution and the ECC strategy.

Homework

Q7.1 Compare two 16 Gb MRAM chips, one with thermal stability factor $\Delta = 70$, $\sigma = 10\%$, and the other $\Delta = 60$, $\sigma = 8\%$. The chips are idle at room temperature. What is the probability of bit retention fail of these two chips?

    A7.1 The variation of thermal stability factor of the MTJ array comes from variation of the MTJ bit area and etching-induced damages and re-deposition. In case 1, the 10% sigma of energy barrier, statistically, ~2 bits will suffer retention fail in one second, 1900 bits fail in one hour, and so on. In case 2, the sigma is 8%, and the retention fail is in same order and slightly less severe (see Figure Q7.1).

    Here, ignoring the etching-induced energy barrier variation, we estimate sigma from the variation of the MTJ bit size. The thermal energy barrier is proportional to the MTJ diameter. For a 70-nm diameter MTJ bit, if the sigma of the bit diameter is 4 nm, the area sigma is 11%. For a 40-nm MTJ bit with sigma of 2-nm, the area sigma is 10%. The demand

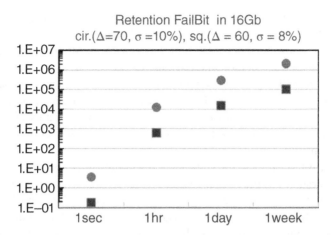

Figure 7.Q1 Two 16 Gb STT-MRAM chips. Their thermal energy (mean, sigma) barriers are different. One is (mean, sigma) = $70 k_B T$ and 10% sigma. And the other $60\ k_B T$ and 8% sigma. The tail bits of the lower end of the energy distribution fail first. For example, 3 bits fail in 1 second in the first chip, while less than 1 bit fails in the second chip. With tighter distribution in energy barrier, although the mean is lower, fewer retention fail bits are projected.

to the photolithography and etching tool and processing uniformity is very strict.

For this reason, it is believed that a large capacity MRAM would need to be scrubbed (refreshed) periodically, like a DRAM. The difference is that MRAM is scrubbed less often. Thus, the energy barrier of a large capacity MRAM should be further raised for cold store application.

Q7.2 Take the best switching efficiency data point in Figure 7.22b $\left(\dfrac{E_b}{I_{co}} = 3\ (k_B T/ \mu A) \right)$. Estimate the ratio of $(E_b\ /\ E_{SW})$ for an MTJ with $E_b = 60\ k_B T$, or 2.4e-19 J. From a typically operating point, $I_{SW} \sim 2.5 \times I_{c0}$, and $V_{SWJ} \sim 0.5$ V, pulse width of 20 ns,

A7.2 $E_b = 60\ k_B T \rightarrow I_{c0} = 20\ \mu A$. We come up $E_{b,therm}/E_{SW} \sim 4.8 \times 10^{-7}$. It takes $\sim 10^7 \times$ thermal barrier energy to switch a bit with STT mechanism.

Q7.3 In a DRAM cell, binary data is stored as electron charge in a capacitor. For a typical design, 30fC electron charge. When the charge is absent in the bit cell capacitor, data = "0," and when 30fC is stored, data is "1." In an MRAM, the direction of magnetic moment in a ferromagnet is the data.

When the moment of the free layer points in the same direction as the pinned layer in an MTJ, the MTJ resistance is in low state, the cell stores "0," and vice versa. What is the equivalent electron charge stored in a volume of ferromagnet Fe? The unit lattice of Fe crystal is body-centered cubic (*bcc*). The lattice constant is 2.87A, and the magnetic moment of the unit is 2.2 Bohr Magneton (μ_B).

A7.3 When the moment in a Fe lattice reverses, the net moment change is $2 \times 2.2\,\mu_B$. One can view that as 2×2.22 electrons swapping between majority band and minority band in each Fe unit lattice. The equivalent electron charge is 2×2.2 e, where e is the electron charge. Let us consider a Fe of volume V $= 10\,\text{nm} \times 10\,\text{nm} \times 3\,\text{nm}$ as an example; its volume is large enough to have sufficient thermal stability as the storage element of a memory bit cell when interfacial K_i is 2 erg/cm^2. It contains V/(unit lattice cell) $=$ 12,700 unit cells. Thus, the equivalent charge is $12700 \times (2 \times 2.2\text{e}) = 9\text{fC}$. If the charge takes 20 ns to supply, the equivalent current is $i = \frac{Q}{t} = 0.5\ \mu A$. The equivalent energy barrier is ½ Km $= 80$ $k_B T$. This homework exercise illustrates that the energy of capacitive data storage in the volatile DRAM is in same order as nonvolatile magnetic data storage. Nonvolatility and low active power can co-exist. Although fundamentally the two are comparable, in practice DRAM requires much less energy than MRAM to switch. There is a lot of room to improve MRAM (courtesy of H. Yoda, private communication).

References

1 Tang, D.D. and Lee, Y.J. (2010) *Magnetic Memory*, Chapter 5. Cambridge Press.
2 Sun, J.Z. (2000). Spin-current interaction with a mono domain magnetic body: a model study. *Phys. Rev. B* 62 (1): 570–578; also see J. Z. Sun, Spin angular momentum transfer in current-perpendicular nanomagnetic junctions, *IBM J. Res. Develop.*, 50, no. 1, pp. 81–100, Jan. 2006.
3 Koch, R.H., Katine, J.A., and Sun, J.Z. (2004). Time-resolved reversal of spin-transfer switching in a nanomagnet. *Phys. Rev. Lett.* 92 (8): 088302.
4 Tomasello, R., Puliafito, V., Azzerboni, B., and Finocchio, G. (2014). Switching Properties in Magnetic Tunnel Junctions with Interfacial Perpendicular Anisotropy: Micromagnetic Study. *IEEE trans on Magnetics* 50 (7) https://doi.org/10.1109/TMAG.2014.2307280.
5 C. K. A. Mewes and T. Mewes(2014). Matlab Based Micromagnetics Code M^3. http://bama.ua.edu/~tmewes/Mcube/Mcube.shtml (accessed July 11, 2020).

6 Yagami, K., Tulapurkar, A.A., Fukushima, A., and Suzuki, Y. (2005). Estimation of thermal durability and intrinsic critical currents of magnetization switching for spin-transfer based magnetic random access memory. *J. Appl. Phys.* 97: 10C707.

7 W. H. Butler, Tim Mewes, Claudia K. A. Mewes, P. B. Visscher, William H. Rippard, Stephen E. Russek, and Ranko Heindl, Switching distributions for perpendicular spin-torque devices within the macrospin approximation, *IEEE Trans. On Magentics, VOL. 48, NO. 12,* p.4684, DECEMBER 2012

8 Guenole Jan, et al (2016). Achieving Sub-ns switching of STT-MRAM for future embedded LLC applications through improvement of nucleation and propagation switching mechanisms. *IEEE Symposium on VLSI Technology* Digest of Technical Papers, IEEE, DOI:10.1109/VLSIT.2016.7573362.

9 Thomas, L., Jan, G., Zhu, J. et al. (2014). Perpendicular spin transfer torque magnetic random access memories with high spin torque efficiency and thermal stability for embedded applications (invited). *J. Appl. Phys.* 115: 172615.

10 He, J., Sun, J.Z., and Zhang, S. (2007). Switching speed distribution of spin-torque-induced magnetic reversal. *AIP, J. of APPL Phys.* 101: 09A501.

11 Luc Thomas, Guenole Jan, Santiago Serrano-Guisan, Huanlong Liu, Jian Zhu, Yuan-Jen Lee, Son Le, Jodi Iwata-Harms, Ru-Ying Tong, Sahil Patel, Vignesh Sundar, Dongna Shen, Yi Yang, Renren He, Jesmin Haq, Zhongjian Teng, Vinh Lam, Paul Liu, Yu-Jen Wang, Tom Zhong, Hideaki Fukuzawa, and PoKang Wang, STT-MRAM devices with low damping and moment optimized for LLC applications at 0x nodes, *IEEE Dig. Of IEDM 2018*, paper 27.03

12 Higo, Y., Yamane, K., Ohba, K. et al. (2005). Thermal activation effect on spin transfer switching in magnetic tunnel junctions. *Appl. Phys. Lett.* 87: 082502.

13 Inokuchi, T. et al. (2006). Current-induced magnetization switching under magnetic field applied along the hard axis in MgO-based magnetic tunnel junctions. *Appl. Phys. Lett.* 89: 102502.

14 T. Min, et al., Back-hopping after spin torque transfer induced magnetization switching in magnetic tunneling junction cells, *Dig. Of MMM 2008*, DB-02 Oct 2008.

15 T. Min, Q. Chen, R. Beach, G. Jan, C. Horng, W. Kula, T. Torng, R. Tong, T. Zhong, D. Tang, P. Wang, M. Chen, J.Z. Sun, J. K. Debrosse, D. C. Worledge, T. M. Maffitt, W. J. Gallagher, A Study of Write Margin of Spin Torque Transfer Magnetic Random Access Memory Integrated with CMOS Technology, Joint MMM-Intermag Conference paper, AA-05, (2009)

16 Sun, J.Z., Gaidis, M.C., Hu, G. et al. (2009). High-bias backhopping in nanosecond time-domain spin-torque switches of MgO-based magnetic tunnel junctions. *J. Appl. Phys.* 105: 07D109.

17 S-C Oh2009 Se-Chung Oh, Seung-Young Park, Aurélien Manchon, Mairbek Chshiev, Jae-Ho Han, Hyun-Woo Lee, Jang-Eun Lee, Kyung-Tae Nam, Younghun Jo, Yo-Chan Kong, Bernard Dieny, and Kyung-Jin Lee, Bias-voltage dependence of perpendicular spin-transfer torque in asymmetric MgO-based magnetic tunnel junctions, *Nat. Phys.*, 2009 DOI: https://doi.org/10.1038/NPHYS1427

18 Devolder, T., Kim, J.-V., Garcia-Sanchez, F. et al. (2016). Time-resolved spin-torque switching in MgO-based perpendicularly magnetized tunnel junctions. *Phys. Rev. B* 93: 024420.

19 P. Wang, G. Jan, Thomas, A. Wang, T. Zhong, T. Torng, Y. Lee, H. Liu, J. Zhu, S. Le, S. Serrano-Guisan, R. Tong, J. Haq, J. Teng, D. Shen, R. He and V. Lam (2017). Development of STT MRAM for embedded memory applications. INSPEC Accession Number: 17100831, DOI: 10.1109/INTMAG.2017.8007930, IEEE International Magnetics Conference (INTERMAG). INSPEC Accession Number: 17100831.

20 Akerman, J., Brown, P., Shu, M.J. et al. (2004). *IEEE Trans. Device Mat. Reliab.* 4 (3): 428.

21 Uberuaga, B.P. et al. (2004). *Phys. Rev. Lett.* 92: 115505.

22 Stathis, J.H. (1999). *J. Appl. Phys.* 86: 5757.

23 Carboni, R., Ambrogio, S., Chen, W. et al. (2016). Understanding cycling endurance in perpendicular spin-transfer torque (p-STT) magnetic memory. *Dig. of IEDM*: 516.

24 Kan, J.J., Park, C., Ching, C. et al. (2017). A study on practically unlimited endurance of STT-MRAM. *IEEE Trans. Electron. Devices* 64 (9): 3639–3646.

25 [T. Sunaga, private communication]

26 Jian Zhu, Yuan-Jen Lee, Huanlong Liu, Son Le, Jodi Iwata-Harms, Sahil Patel, Ru-Ying Tong, Vignesh Sundar, Santiago Serrano-Guisan, Dongna Shen, Renren He, Jesmin Haq, Zhongjian Jeffrey Teng, Vinh Lam, Yi Yang, Yu-Jen Wang, Tom Zhong, Luc Thomas, Hideaki Fukuzawa, Guenole Jan and Po-Kang Wang, (2018). Comprehensive reliability study of STT-MRAM devices and chips for Last Level Cache applications at 0x nodes," *IEDM 2018, MRAM Poster Session*, IEDM.

27 Jan, G., Wang, Y.-J., Moriyama, T. et al. (2012). High spin torque efficiency of magnetic tunnel junctions with MgO/CoFeB/MgO free layer. *Appl. Phys. Exp.* 5: 093008.

28 Sharrock, M.P. (1994). Time dependence of switching fields in magnetic recording media (invited). *J. Appl. Phys.* 76 (10).

29 Luc Thomas, Guenole Jan, Son Le, Yuan-Jen Lee, Huanlong Liu, Jian Zhu, Santiago Serrano-Guisan, Ru-Ying Tong, Keyu Pi, Dongna Shen, Renren He, Jesmin Haq, Zhongjian Teng, Rao Annapragada, Vinh Lam, Yu-Jen Wang, Tom Zhong, Terry Torng and Po-Kang Wang (2015). Solving the Paradox of the Inconsistent Size Dependence of Thermal Stability at Device and Chip-level in Perpendicular STT-MRAM, *IEEE Dig. Of IEDM 2015*, IEEE, paper 26.4.

30 K. Tsunoda, M. Aoki, H. Noshiro, Y. Iba, S. Fukuda, C. Yoshida, Y. Yamazaki, A. Takahashi, A. Hatada, M. Nakabayashi, Y. Tsuzaki, and T. Sugii, Area Dependence of Thermal Stability Factor in Perpendicular STT-MRAM Analyzed by Bi-directional Data Flipping Model, *IEEE Dig. Of IEDM 2014.* P.486

31 [Private communication, Courtesy to Industrial Technology Research Institute]

32 Shaw, J.M., Nembach, H.T., and Silva, T.J. (2011). Damping phenomena in Co90Fe10/Ni multilayers and alloys. *Appl. Phys. Lett.* 99: 012503. https://doi.org/ 10.1063/1.3607278.

33 Thomas, L., Jan, G., Le, S., and Wang, P.-K. (2015). Quantifying data retention of perpendicular spin-transfer-torque magnetic random access memory chips using an effective thermal stability factor method. *Appl. Phys. Lett.* 106: 162402. https:// doi.org/10.1063/1.4918682.

34 E. Kitagawa, S. Fujita, K. Nomura, H. Noguchi, K. Abe, K. Ikegami, T. Daibou, Y. Kato, C. Kamata, S. Kashiwada, N. Shimomura, J. Ito, and H. Yoda, Impact of ultra low power and fast write operation of advanced perpendicular MTJ on power reduction for high-performance mobile CPU, *IEEE Dig. Of IEDM,* p.677, (2012]

35 Tom Silva, Hans Nembach, Justin Shaw, Brian Doyle, Kaan Oguz, Kevin O'brien, and Mark Doczy, Characterization of Magnetic Nanostructures for STT-RAM Applications by use of Macro-and Micro-scale Ferromagnetic Resonance, http:// www.nist.gov/document-1229 (accessed July 11, 2020).

36 Nembach, H.T., Shaw, J.M., Boone, C.T., and Silva, T.J. (2013). Mode- and size-dependent landau-lifshitz damping in magnetic nanostructures: evidence for nonlocal damping. *Phys. Rev. Lett.* 110: 117201.

37 Kubota, T., Tsunegi, S., Oogane, M. et al. (2009). Half-metallicity and Gilbert damping constant in Co2FexMn1 − xSi Heusler alloys depending on the film composition. *Appl. Phys. Lett.* 94: 122504.

38 Dieny, B. (2016). *Introduction to Magnetic Random-Access Memory.* Wiley.

39 Prejbeanu, I.L., Bandiera, S., Alvarez-Herault, J. et al. (2013). Thermally assisted MRAMs: ultimate scalability and logic functionalities. *J. Phys. D: Appl. Phys.* 46: 074002.

40 Meng, H. and Wang, J.P. (2006). Composite free layer for high density magnetic random access memory with lower spin transfer current. *Appl. Phys. Lett.* 89: 152509.

41 Huai, Y., Paul P, Nguyen (2004). Magnetic element utilizing spin transfer and MRAM devices using the magnetic element. US Patent 6,714,444.

42 S. Ikeda, H. Sato, H. Honjo, E. C. I. Enobio, S. Ishikawa3, M. Yamanouchi, S. Fukami, S. Kanai3, F. Matsukura, T. Endoh and H. Ohno (2014). Perpendicular-anisotropy CoFeB-MgO based magnetic tunnel junctions scaling down to 11 nm. *IEEE Dig. Of IEDM.* IEDM. p. 796.

43 Jog Adwait Jog, Asit K. Mishra Cong Xu, et al. (2011). Cache Revive: Architecting Volatile STT-RAM Caches for Enhanced Performance in CMPs. *Intel Technical Report CSE-11-010.*

44 [Private communication, courtesy of Applied Material]

45 Sheng-Huang Huang, Ding-Yeong Wang, Kuei-Hung Shen, Cheng-Wei Chien, Keng-Ming Kuo, Shan-Yi Yang, Yung-Hung Wang (2012). Impact of stray field on the switching properties of perpendicular MTJ for Scaled MRAM, *IEEE Dig of IEDM*.IEDM. paper 29.2

46 Woo Chang Lim, Y. J. Lee, J. M. Lee, W. K. Kim, J. H. Kim, K. W. Kim, K. S. Kim, Y. S. Park, H. J. Shin, S. H. Park, J. H. Kim, J. H. Jeong, M. A. Kang, Y. H. Kim, W. J. Kim, S. Y. Kim, Y. C. Cho, H. L. Park, H. S. Ahn, J. H. Park, S. C. Oh, S. O. Park, S. Jeong, S. W. Nam, H. K. Kang, E. S. Jung (2013). Enhancement of switching margin by utilizing superior pinned layer stability for sub-20 nm perpendicular STT-MRAM" *Symposium on VLSI Technology Digest of Technical Papers*, VLSI. Paper 6.3

47 D. Bedau, D. Backes, H. Liu, J. Langer, P. Manandhar, A. D. Kent (2011). Orthogonal Spin Transfer MRAM. *IEEE Dig. Of 56th Annual Magnetism & Magnetic Materials Conference (MMM).* IEEE. p. 165.

48 G. E. Rowlands, T. Rahman, J. A. Katine, J. Langer, A. Lyle, H. Zhao, J. G. Alzate, A. A. Kovalev, Y. Tserkovnyak, Z. M. Zeng,6 H. W. Jiang,6 K. Galatsis,5 Y. M. Huai,7P. Khalili Amiri, K. L. Wang, I. N. Krivorotov, and J.-P. Wang, Deep subnanosecond spin torque switching in magnetic tunnel junctions with combined in-plane and perpendicular polarizers, *Appl. Phys. Lett.* 98, 102509, (2011)

49 Se-Chung, O., Park, S.-Y., Manchon, A. et al. (2009 https://doi.org/10.1038/nphys1427). Bias-voltage dependence of perpendicular spin-transfer torque in asymmetric MgO-based magnetic tunnel junctions. *Nature Physics* 5 (12): 898–902.

50 Mejdoubi, A., Lacoste, B., Prenat, G., and Dieny, B. (2013). Macrospin model of precessional spin-transfer-torque switching in planar magnetic tunnel junctions with perpendicular polarizer. *Appl. Phys. Lett.* 102: 152413.

51 US Patent 10553787, (2020), Mustafa Michael Pinarbasi, Michail Tzoufras, Bartlomiej Adam Kardasz.

52 Devolder, T. (2011). Scalability of magnetic random access memory based on an in-plane magnetized free layer. *Appl. Phys. Express* 4: 093001.

53 Y. Kim, S. C. Oh, W. C. Lim, J. H. Kim, W. J. Kim, J. H. Jeong, H. J. Shin, K. W. Kim, K. S. Kim, J. H. Park, S. H. Park, H. Kwon, K.H. Ah, J. E. Lee, S. O. Park, S. Choi, H. K. Kang, C. Chung (2011). Integration of 28 nm MJT for 8 ~ 16Gb level MRAM with full investigation of thermal stability. *Dig. Of VLSI Tech. Symp.* VLSI. p. 210.

54 Yang, H.X., Chshiev, M., Dieny, B. et al. (2011). First-principles investigation of the very large perpendicular magnetic anisotropy at Fe|MgO and Co|MgO interfaces. *Phys. Rev. B* 84: 054401.

55 Shouzhong Peng, Mengxing Wang, Hongxin Yang, Lang Zeng, Jiang Nan, Jiaqi Zhou, Youguang Zhang, Ali Hallal, Mairbek Chshiev, Kang L. Wang, Qianfan Zhang, Weisheng Hao (2015). Origin of interfacial perpendicular magnetic anisotropy in MgO/CoFe/metallic capping layer structures. *Scientific reports*, 5, 18173.

56 Manchon, A. and Zhang, S. (2008). Theory of nonequilibrium intrinsic spin torque in a single nanomagnet. *Phys. Rev. B* 78: 212405.

57 Koon, N.C., Williams, C.M., and Das, B.N. (1991). Giant magnetostriction materials. *J. Magn. Magn. Mater.* 100: 173–185.

58 Tao, B.S., Li, D.L., Yuan, Z.H. et al. (2014). Perpendicular magnetic anisotropy in $Ta/Co_{40}Fe_{40}B_{20}/MgAl_2O_4$ structures and perpendicular $CoFeB/MgAl_2O_4/CoFeB$ magnetic tunnel junction. *Appl. Phys. Lett.* 105: 102407.

59 Luc Thomas, Guenole Jan, Santiago Serrano-Guisan, Son Le, Jodi Iwata-Harms, Jian Zhu, Yuan-jen Lee, Huanlong Liu, Ru-ying Tong, Sahil Paten, Vignesh Sundar, Dongna Shen, Jesmin Haq, Yi Yang, Jeffrey Teng, Renren He, Vinh Lam, Paul Liu, Tom Zhong, Allen Wang, Terry Thorng, Po-kang Wang, High perpendicular anisotropy in sub-30 nm MRAM devices measured by spin-torque ferromagnetic resonance (2017). *IEEE Digest of International Magnetics Conference (INTERMAG), 2017*, Dublin. DOI: https://doi.org/10.1109/INTMAG.2017.8007703.

60 S. Ikeda, H. Sato, H. Honjo, E. C. I. Enobio, S. Ishikawa, M. Yamanouchi, S. Fukami, S. Kanai, F. Matsukura, T. Endoh and H. Ohno, Perpendicular-anisotropy CoFeB-MgO based magnetic tunnel junctions scaling down to 1X nm., *IEEE Digest of IEDM paper 32.2*, p.796, (2014)

61 Thomas, L., Jan, G., Zhu, J. et al. (2014). Perpendicular spin transfer torque magnetic random access memories with high spin torque efficiency and thermal stability for embedded applications (invited). *J. Appl. Phys.* 115: 172615.

62 H.K. Yoda, S. Fujita, N. Shimomura, et.al., Progress of STT-MRAM Technology and the Effect on Normally-off Computing Systems, *Digest of IEEE IEDM*, p.239, (2012)

63 Sun, J.Z., Robertazzi, R.P., Nowak, J. et al. (2011). Effect of subvolume excitation and spin-torque efficiency on magnetic switching. *Phys. Rev. B* 84: 064413.

64 Amiri, P.K., Alzate, J.G., Cai, X.Q. et al. (2015). Electric-field-controlled magnetoelectric RAM progress, challenges, and scaling. *IEEE Trans. Magn.* 51 (11): 3401507.

65 Mukherjee, S.S. and Kurinec, S.K. (Sep. 2009). A stable SPICE macro-model for magnetic tunnel junctions for applications in memory and logic circuits. *Trans. Magn.* 45 (9): 3260–3268.

66 Harms, J.D., Ebrahimi, F., Yao, X., and Wang, J.-P. (Jun. 2010). SPICE macromodel of spin-torque-transfer-operated magnetic tunnel junctions. *IEEE Trans. Electron Devices* 57 (6): 1425–1430.

67 Guo, W., Prenat, G., Javerliac, V. et al. (2010). SPICE modeling of magnetic tunnel junction written by spin-transfer torque. *Journal of Physics D: Applied Physics*, IOP Publishing 43 (21): 215001. http://hal.archives-ouvertes.fr/hal-00569612.

68 Shang Chang He, N., Janusz, J., Ronnie, M., and Jagadeesh, S. (1998). *Physics Rev. B.* 58: R2917.

69 Lim, H., Lee, S., and Shin, H. (2014). Unified Analytical Model for Switching Behavior of Magnetic Tunnel Junction. *IEEE ELECTRON DEVICE LETTERS* 35 (2): 193, IEEE.

70 Perrissim, N., Lequeux, S., Strelkov, N. et al. (2018). A highly thermally stable sub-20 nm magnetic random-access memory based on perpendicular shape anisotropy. *Nanoscale* (25) https://doi.org/10.1039/CBNR06365A.

71 Honjo, H., Ikeda, S., Sato, H. et al. (2016). Improvement of thermal tolerance of CoFeB–MgO perpendicular-anisotropy magnetic tunnel junctions by controlling boron composition. *IEEE Trans. Magn.* 42, 3401104: 7.

8

Advanced Switching MRAM Modes

8.1 Introduction

This chapter describes three promising new modes of magnetoresistive random-access memory (MRAM) cells that are currently under research and development at the time of this manuscript preparation. All are based on magnetic tunnel junction (MTJ) technology, and all promise very fast access time. The MTJ is switched in a way similar but different from the pure spin-transfer mechanism. The first kind is called current-induced domain-wall motion mode (CIDM) MRAM, and the MTJ free layer is switched by moving the domain wall position in free layer with current. The second kind is called spin orbit torque (SOT) mode MRAM, sometimes also called spin Hall effect (SHE) mode MRAM. Its MTJ free layer is switched by a spin current from a strong spin-orbit coupling material, such as Pt, W, and Ta. The third kind is called precession-toggle mode MRAM.

Both CIDM memory cell and SOT memory cell are configured in 1M-2T (1 MTJ, 2 transistors) cell topology. The read terminal and write terminal are separated. MTJ serves as read sensor; during memory read access, only a small read current flows through the tunnel barrier. The large write current does not. Subnanosecond write pulse can be applied to shorten the write access time and not affect the cell MTJ endurance performance. And the MTJ resistance-area product (RA) is no longer restricted by the power supply voltage of the scaled complementary metal-oxide-semiconductor (CMOS) circuits. A larger RA means thicker MgO tunnel barrier, which eases the tunnel barrier manufacturing. In addition, higher-sense (read) voltage can be obtained without suffering from READ disturb issue of spin-transfer torque magnetic random-access memory (STT-MRAM); thus, the read access performance is also improved. They are better than STT-MRAM in these respects.

Magnetic Memory Technology: Spin-Transfer-Torque MRAM and Beyond,
First Edition. Denny D. Tang and Chi-Feng Pai.
© 2021 The Institute of Electrical and Electronics Engineers, Inc.
Published 2021 by John Wiley & Sons, Inc.

Their drawbacks are: (i) the 1M-2T cell size is larger, and (ii) the write current density, and thus the total current consumption, is larger than STT-MRAM. Since larger cell transistors are required to supply the large write current, the power consumption of the cell is larger than the STT-MRAM cell. Active academic research into better SOT material is ongoing to overcome the drawbacks.

The precession-toggle mode MRAM is based on the voltage-control magnetic anisotropy (VCMA) effect. It is also called magneto-electric RAM (MeRAM) [37]. It is a 1M-1T cell and is written with unipolar write pulse. The development of precession-toggle cell is in its infant stage. It requires precision control of the write pulse width, accurate down to a fraction of precession cycle, or subnanosecond. Nonetheless, the concept is appealing, because the switching time is half a magnetization precession cycle, the fastest among all MRAMs.

The VCMA effect can be deployed in other modes of MRAM to improve access performance. For example, it is also used to select bits in a multibit-word SOT MRAM cell during WRITE. A voltage is applied to the MTJ to selected bits of a word and temporarily lowers the anisotropy of selected bits in the word during WRITE so that only the selected bits are written while the rest bits in the word are not.

8.2 Current-Induced-Domain-Wall Motion (CIDM) Memory

It has been well known that an external H-field induces movement of the domain wall in ferromagnet (FM). Here, the idea is to induce domain wall movement with a current in FM, without external H-field. The advantage of current-induced domain-wall (DW) motion is that the action is localized within the current path, only the selected cell. Thus, the notorious half-select cell disturbance issue of the field MRAM cell array (Chapter 5) does not happen in the CIDM MRAM cell array. Besides, the energy efficiency of the CIDM MRAM cell is much higher than the field MRAM. For a 5 nm thick 100 nm wide FM film, the threshold current of the CIDM MRAM cell can be in the order of 100−200 μA, about two orders lower than the field-MRAM switching current.

Two cell configurations will be discussed: the single domain-wall cell and the multiple domain-wall cell, the Racetrack. The single-bit cell can be accessed as random-access memory, while the Racetrack is a sequential-access memory with extremely high bit density.

8.2.1 Single-Bit Cell

Figure 8.1 shows a schematic of a single-bit domain wall cell. Circuit-wise, it is a 1M-2T, four-terminal memory cell. The MTJ is in top-pin configuration. The two ends of the free layer are extended and magnetically pinned by two hard magnets. The two hard magnets are of opposite polarity; thus, there is a domain wall in the free layer. The two ends of the free layer are electrically connected to the complementary data lines DL and DL/ through the hard magnets and transistors. The domain wall in the free layer is driven by current. The cell state is determined by the final position of the DW, which in turn is determined by the direction of the write current in the free layer. During the WRITE operation, WL is raised, transistors are turned on, a write current flows through the free layer, and its direction is set by the voltages of DL and DL/. During the READ operation, all WL and BL voltage are raised, two transistors are "on," and the read current flows from bit line (BL) through MTJ to DL and DL/, both grounded.

Micromagnetic simulations reveal that the CIDM mechanism of an in-plane FM is less efficient than an out-of-plane one. Figure 8.2 compares the threshold write current density of in-plane and perpendicular magnetized film. The lowest CIDM write current density is found in thin PMA film and is between 10^7 and 10^8 A/cm^2. In a CIDM cell with (Co/Ni) laminated film as the free layer and (Co/Pt) lamination as the pinned and reference layer, the DW movement speed is expected to be ~50 m/S, indicating nanosecond-range switching time for a 50-nm wide MTJ is

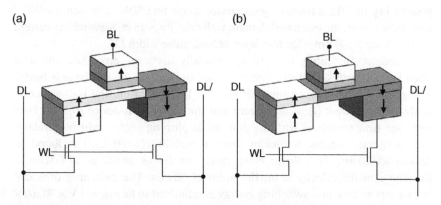

Figure 8.1 A single-bit domain wall cell. (a) When the DW is driven to the right-side, MTJ is in P-state. (b) When DW is driven to the left side, the MTJ is in AP state. MTJ acts as a read-out device. The write current flows through the FM stripe, and does not flow through MTJ tunnel barrier.

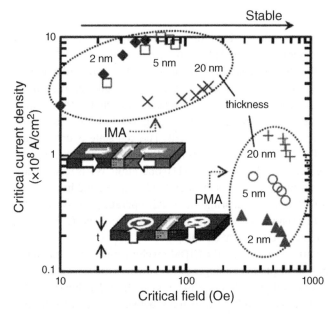

Figure 8.2 Micromagnetic simulation of CIDM threshold in-plane write current density Jc in thick in-plane (IMA) and perpendicular (PMA) free layer versus critical field. PMA film has a much higher Hc and lower Jc. Reprinted with permission from [1].

possible [1]. In 2013, a research group tested single-bit CIDM cells with $(Co/Ni)_n$/ CoFeB free layer, the estimated domain wall velocity is 15 m/S, switching current is 160 µA for a 130 nm wide free layer @20nS pulse width [2].

To understand scaling properties, especially the write error rate and data retention properties, Hall structure was used to simulate the free layer behavior of the MTJ of the DW memory cell [3]. The learnings from the Hall structure study are as follows: (i) Write current and time scale with device size. (ii) Data retention time correlates with the domain de-pinning energy, which correlates with magnetic defects. Sufficient thermal stability (>100 k_BT) is found in devices at 20 nm. (iii) Write error rates can be as small as STT-MRAM. (iv) The domain velocity is on the order of 50 m/s. The switching time can be as short as 2 ns, and switching energy is estimated to be $\varepsilon_{SW} = 1\,V \times 70$ uA $\times$ 1.8 ns ~ 120 *fJ*.

A CIDM MRAM cell can potentially outperform STT-MRAM cell. However, its large cell size and complex cell structure requires an industrial-like pilot facility to build; academic research so far has delivered little success, and the progress is slow.

8.2.2 Multibit Cell: Racetrack

When the ferromagnetic free layer of a single-bit DW cell is further extended into a long stripe, it becomes a multibit Racetrack. Many bits can be stored on the free layer stripe, and each bit is separated by a pair of domain walls from neighboring bits. Figure 8.3 shows the Racetrack concept [4]. An inductive "head" (field coil) writes the bit pattern onto the data stripe. A current in the data stripe shifts the domain walls. In effect, the current shifts the bit string on the data stripe. An MTJ read "sensor" detects the bits. When the bit-string reaches the MTJ, the data on the bit string is detected, or read out. Thus, it is a sequential-access, not a random-access memory.

A more advanced Racetrack device incorporates a SHE material (see next section), such as Pt and Ta in a Pt/FM or a Ta/FM (ferromagnetic metal) bi-layer structure [59, 60, 102, 103]. The magnetic moment of the FM is perpendicular to the film plane. Note that the FM layer can be a composite magnetic heterostructure, such as a synthetic antiferromagnetic layer (SAF). The domain walls in such FM layer with PMA are believed to be homochiral Néel walls (Figure 8.4), which is stabilized by the Dzyaloshinskii-Moriya exchange interaction (DMI; see Chapter 2). The current in the spin Hall material provides laterally polarized spin current, which diffuses from the spin Hall material into the FM. The magnetization of the domain wall moment on the track then rotates in a chiral fashion, and the wall moves along the applied current direction.

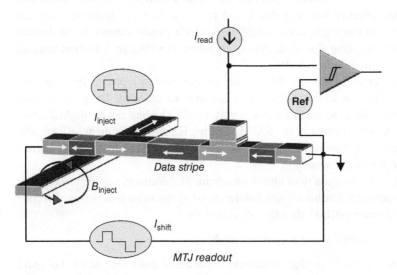

Figure 8.3 A schematic of a Racetrack memory and the operation concept.

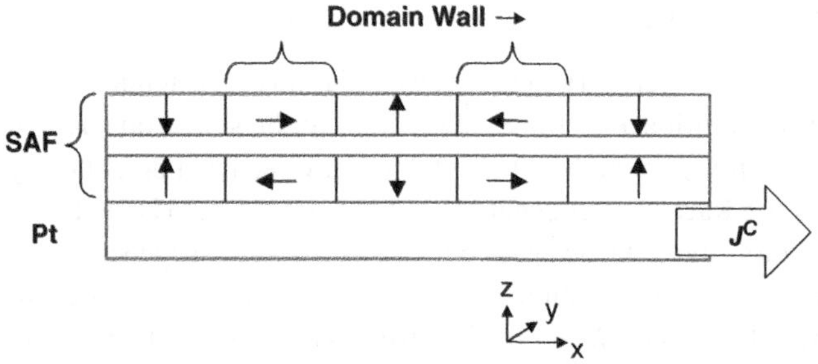

Figure 8.4 A cross-sectional view of synthetic antiferromagnetic (SAF) Racetrack data track over a spin Hall material (Pt) [59].

Let's examine an example of two domain walls between three domain bits on a track in x-direction, "up" (+z-direction), "down" (−z-direction), and "up" again. The magnetization at the middle of the domain wall points to −x-direction in the first wall and the opposite direction in the next wall, and so on. The variation of the magnetization orientation along the FM stripe direction is chiral (always counter-clockwise), due to the existence of interfacial DMI originated from the spin-orbit coupling of the Pt/FM interface. As the current is applied along the x-direction, the SHE from Pt generates a spin current flowing along the +z-direction into the FM (SAF layer for this case) with spin polarization s in the y-direction. The injected or diffused spins then enforce spin-transfer torques onto the magnetic moments, where the effective field is proportional to $s \times m$. This will make all magnetic moments tend to rotate counterclockwise, which can be viewed as the domain walls moving along the +x-direction, therefore resulting in a current-induced motion of chiral domain walls.

The bit density of the data stripe is further improved by replacing the single FM by a SAF. The net moment of each bit can then be canceled. The magnetostatic interference from neighboring bits is very small. The bits can be packed closer as well. Another benefit of the SAF data stripe is that high domain wall velocity can be achieved. Experiments [59] show that velocity can reach as high as 750 m/s in a SAF Racetrack system.

To estimate the density of bits of an advanced Racetrack, one may consider a planar Racetrack. The bit cell size is determined by the stripe pitch in one direction and the domain pitch in the other direction. Or

$$A_{bit} = (stripe\ pitch) \times (domain\ pitch)$$

The bit pitch is the product of domain width and domain wall width. Let's take Fe stripe as an example; the domain wall width is ~150 Fe lattice constant, or

42 nm. In a SAF configuration, the wall is narrower. At 40-nm metal pitch, the estimated bit size = 168 nm^2. The minimum size of a domain is determined by the super-paramagnetic thermal stability limit. Since each bit cell is not gated by the transistor and the domain pitch is not defined by lithography, potentially the bits can be packed very close to each other. Thus, the bit density should be in same order of bit density of hard disk. And layers of this type of Racetrack may be stacked vertically on a wafer as part of the CMOS back-end process. Thus, it opens a path to the extremely dense 3D V – MRAM.

8.3 Spin-Orbit Torque (SOT) Memory

MTJ in STT-MRAM cell generates spin current by filtering the particle current passing through its tunnel junction. There is a different way to generate spin current. SHE splits the conduction electron path based on its spin polarization and results in spin accumulation on opposite sides of a conductor. This effect is strong in heavy metals (HM), such as Pt, W, Ta. In an HM/FM bilayer film stack, effectively a transversal spin current flows from the HM to FM when a longitudinal carrier current flows in the bilayer. Thus, this structure provides an alternative way to inject spin current into an FM and switch its magnetization.

8.3.1 Spin Orbit Torque (SOT) MRAM Cells

In principle, a simple bilayer HM/FM Hall structure can perform as a memory cell. It is shown in Figure 8.5a. When a longitudinal current is applied to the structure (J^C), a spin current from HM to FM carries spin torque into the FM. When the torque exceeds the switching threshold, it switches the FM magnetization. The cell state is read out by applying a small current to read the anomalous Hall voltage.

However, this bilayer structure is not a practical memory cell due to (i) the resistance of the structure is too small to match the impedance of CMOS circuitry, (ii) and the anomalous Hall effect voltage is in the range of nV to μV. Such signal amplitude is too small for practical CMOS-based sense amplifier on memory chips. The sense amplifier works well for signal greater than 50 mV. The high MTJ impedance matches well with CMOS circuitry, and its large tunneling magnetoresistance (TMR) resistance provides sufficient signal to the CMOS sense amplifier.

Thus, a practical SOT MRAM cell must incorporate an MTJ. The MTJ is in a top-pin configuration such that its free layer is in contact with the SOT channel, the heavy metal. The free layer is written by the spin current diffuses up from the SOT

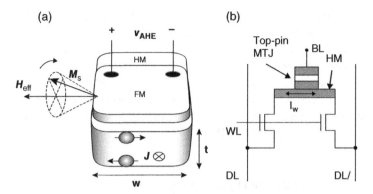

Figure 8.5 (a) Heavy metal/ferromagnet (HM/FM) Hall structure. A longitudinal carrier current in the HM induces a transversal spin current, which diffuses into FM, and induces anomalous Hall effect voltage. (b) Spin-orbital torque (SOT) MRAM cell consists of an MTJ on a SOT channel (heavy metal stripe).

channel. Figure 8.5b shows the cell implementation, an MTJ on a heavy metal stripe. From here on, the heavy metal stripe of a SOT cell is called "SOT channel." It is a 2T-1M cell with four electrical terminals (WL, BL, DL, and DL/).

To form a SOT cell, there are three ways to align the easy axis of an MTJ to the charge current direction in the SOT channel. Assuming charge current in the SOT channel flows along the x-direction, the easy axis of the MTJ can be (i) in the x-direction, the same as the current in the SOT channel, thus, the MTJ is an in-plane MTJ; (ii) in the y-direction, orthogonal is the current direction; thus, the MTJ is an in-plane MTJ; and (iii) in the z-direction, the MTJ is a perpendicular MTJ.

Both anti-damping torque and field-like torque appear to contribute to the switching. External field is required to assist the switching of (i) and (iii) cell configuration, since SOT current alone cannot. Research on this subject is ongoing; many field free cell engineering solutions were proposed [65]. We will focus on only the (ii) configuration, having in-plane MTJ with easy axis orthogonal to the SOT charge current direction, in Section 8.3.1.1. Then, briefly, we will discuss the perpendicular MTJ on SOT channel in Section 8.3.1.2.

8.3.1.1 In-Plane SOT Cell

Here, we discuss the cell operation having in-plane MTJ with easy axis orthogonal to the current direction in the SOT channel. When a current flows in the SOT channel, spin current diffuses from the channel into the free layer of MTJ. The polarization of the spin current is along the easy axis of the MTJ. Thus, the anti-damping torque carried by the spin current can efficiently switch the MTJ, the same way as the switching of STT cell. Based on the STT switching given in

Section 7.3, the switching critical (threshold) current density is a function of switching energy barrier and can be rewritten as

$$J_{c0} \sim \left(\frac{2\,e\,\alpha}{\hbar}\right)\left(\frac{2E_{b,\text{STT}}}{A_{\text{MTJ}}}\right), \tag{8.1}$$

where A_{MTJ} is the MTJ area. Here we assume the polarization of the spin current from the SOT is 100%, or $P = 1$ polarized. This is the spin current density that the SOT channel should supply. Thus, the threshold switching current in the SOT channel is

$$I_{sw} = J_{c0}\left(\frac{W_{\text{SOT}}t_{\text{SOT}}}{\Theta_{\text{SH,eff}}}\right), \tag{8.2}$$

where W_{SOT}, t_{SOT} are the width and thickness of SOT channel, respectively, and $\Theta_{\text{SH, eff}}$ is the effective spin Hall angle, which includes the spin current transmission efficiency through the interface between the SOT channel into the free layer of the MTJ. On a wafer of a given SOT channel thickness, to first order, the switch current I_{sw} is linearly proportional to the product of the SOT channel width and thickness.

Let the MTJ area be $\sim 0.75 \times (l_{\text{MTJ}}w_{\text{MTJ}})$, where l_{MTJ} and w_{MTJ} are the length and width of an in-plane MTJ, respectively. Let l_{MTJ} approximately be W_{SOT}; then Eqs. (8.1) and (8.2) can combined with Eq. (7.7) as

$$I_{sw} \geq 5.3\left(\frac{e\,\alpha\,E_{b,\text{therm}}}{\Theta_{\text{SH,eff}}\hbar}\right)\left(\frac{t_{\text{SOT}}}{w_{\text{MTJ}}}\right). \tag{8.3}$$

And for a given SOT channel width, I_{sw} increases with the STT switching energy barrier of the MTJ. Notice that (Section 7.1) the thermal stability $E_{b,\text{therm}}$ and the switching energy $E_{b,\text{STT}}$ barrier of in-plane MTJs differ only by an added moment term and, thus, the $\geq$ sign. Bits with a larger area (while same H_K) exhibits a longer data retention time. The bits with a longer retention time require a slightly larger SOT switch current. This point will become clear in the next section.

8.3.1.1.1 Cell Engineering and Device Properties The device behavior is affected by the device detailed structure. Here, we will show two SOT device structures. The two SOT cell structures differ by the MTJ etch, and they are: (i) "step" structure, in which MTJ etching stops on MgO tunnel barrier, using a selective etch that etches MgO very slowly [62, 63], (ii) "etch-through" structure, in which etching stops on the (Ta) SOT channel. Each has its pros and cons. Figure 8.6 shows the cell structures.

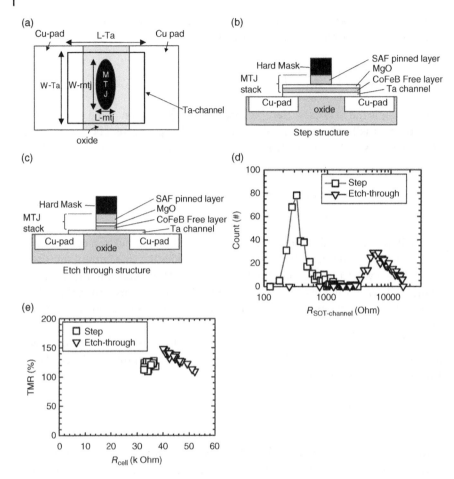

Figure 8.6 (a) SOT memory bit cell top view, (b) cross-section of step structure, (c) etch-through structure. Cu-pad reduces Ta channel resistance, (d) $R_{SOT\text{-}channel}$ distribution, (e) wafer level TMR-Rcell distribution.

From the process point of view, total resistance of the step structure SOT bit cell is in better control, as illustrated in Figure 8.6d. The cell resistance of step structure is ($=R_{MTJ} + R_{channel}/2$, and $R_{channel} = R_{Ta}//R_{Free\text{-}layer}$). The parasitic $R_{channel}$ resistance acts as a series resistance to MTJ and degrades the apparent TMR ratio (Fig. 8.6e). The etch-through structure degrades more, and that may be resulted from etching rate non-uniformity across the 8-inch wafer. The MTJ of the step structure SOT cell is found less likely to be shorted. Both these two arguments are in favor of step structure over etch-through structure.

Comprehensive report of cell read/write behavior has been reported [63, 64, 70]. SOT cell switching behaviors follow STT-MRAM cell. The long-term SOT cell stability has been studied with thermal baking. Thermal baking accelerates failure and shortens the time required to study the data retention failure. In step structure, the free layer extends being MTJ and covers the entire SOT channel (Ta). That gives rise two effects:

a) Larger thermal stability factor, thus better data retention time. It is revealed in the population of bake-induced flipped (retention fail) bit under a series of thermal bakes. This is shown in Figure 8.7a. The step structure shows no flipped bits while the etch-through structure SOT cells show a lot. The number of flipped bits is MTJ size dependence. This retention bake data suggests that for the step structure SOT cell, although the magnetization of the un-etched free layer outside of MTJ is partially damaged by RIE [63], they do contribute to the thermal stability factor, thus, the data retention time. The difference in thermal stability factor between the two structures is estimated to be ~7.5 k_BT at room temperature.

b) The large free layer can turn the MTJ from binary state into multi-state (MS) (Fig. 8.7b). The R-H loop of MS bit is illustrated in Figure 8.8b as compared to the normal bits in Figure 8.8a. The R-H loop of MS bit exhibits a "kink" or "kinks." The SOT wafer is processed with a 250C 0.5 Hr field anneal at the end of the wafer processing and prior to the test. After each baking at

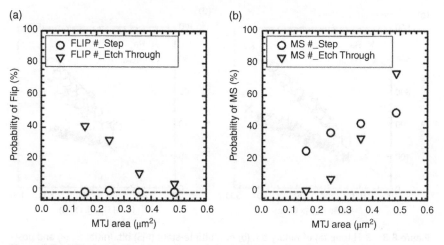

Figure 8.7 MTJ size dependence of post-bake (150C, 4 Hr) (a) flipped bit population, (b) multi-state bit population.

150C without field, additional fresh "kink" cells are observed, in both the step and the etch-through structure. The population of bake-induced MS bit is larger in step structure, indicating the extended free layer promotes the formation of MS states. Figure 8.8c and d are the post-bake R_{ap} of kink-free cells and "kink" cells, respectively. The apparent TMR at zero field of the "kink" cell drops. The MS cell may result from bake-induced domain formation, and the domain wall may pass through MTJ. The population of MS bits increases with bake time. Figure 8.9 shows the post-bake MS cell population of step structure. The population is heavily dependent on the SOT (Ta) channel area and also the MTJ

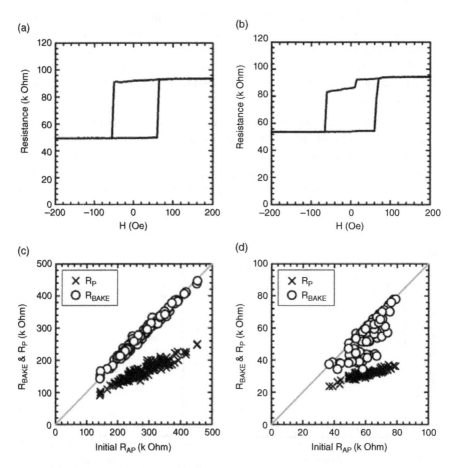

Figure 8.8 R-H loop (a) of binary bit, (b) of multiple-state (MS) bit. Initial R_P (×) and post-bake R_{AP} (O) are plotted against initial R_{AP} of binary state bits c and post-bake multiple-state bits. *Source:* Data are from Step structure.

Figure 8.9 Populations of post-bake multi-state bits vs. MTJ size and SOT channel of step structure SOT cells.

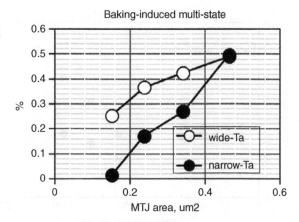

size. Both retention fail and MS fail contribute to the total corruption of stored data in SOT cell. Based on Figures 8.7b and 8.9, clearly, one needs to make MTJ and SOT channel as small as possible to eliminate this kind of latent cell instability (Figure 8.7b).

8.3.1.1.2 Cell Scaling The write characteristics of the in-plane SOT-MRAM resemble those of STT-MRAM such as write current dependence on the write pulse width, temperature effects, and write-error rate (WER) vs. WRITE current [7, 38]. The switching threshold of a typical STT-MTJ is in the order of mid-10^6 A/cm^2. For SOT, the measured switching threshold in the HM is J^C is about 1 order larger, in the mid-10^7 A/cm^2 [5–7].

Experimental data show that the magnitude of write current correlation with the width of SOT channel is strong, and with the area of the MTJ or the fill factor $A_{MTJ}/A_{SOT-channel}$ is weak. Figure 8.10 shows the correlation of write current vs. SOT channel width.

From spin current efficiency point of the view, high resistivity HM is desirable for high Θ_{SH} and the HM thickness should be as thin as possible, few nanometers. The net result is that the resistance of SOT channel, $R_{channel}$, of SOT cell is very high. When integrated into a scaled CMOS circuit with very low power supply voltage, the high $R_{channel}$ limits the write current. High $R_{channel}$ also reduces the effective TMR of the cell and thus, the read margin. One way to overcome this engineering issue is to stitch the SOT channel with low resistance Cu pads, as shown in Figure 8.5.

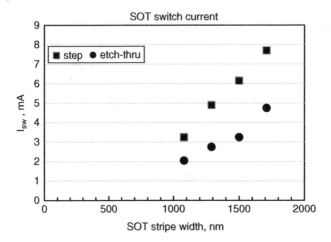

Figure 8.10 Correlation of write current vs. width of SOT channel (heavy metal stripe). The squares are from step structure and dots are taken from etch-through structure. The SOT channel is 10 nm thick Ta, and on which the MTJ length is slightly shorter than the channel width [7, 61].

The SOT cell size is primarily determined by the size of the two gating transistors, which in turn by the switching current. Scaling down SOT cell means the narrowing down of the SOT channel width and reducing MTJ size. To first order, the write current scales with lithographic feature size, $\sim F$. To reach sub-mA write current, SOT channel width needs to be scaled down to 150–250 nm range [38].

Nonetheless, SOT cell is larger than STT cell, since SOT requires two transistors and five wires as compared to one transistor and three wires for a single-bit STT cell. Besides, the transistor of SOT cell is much larger so that it can carry more current. To make a dense SOT cell, the most important task is to lower its cell current.

So far, we have discussed the properties of single-bit SOT cell. To overcome the cell size and power dissipation disadvantages, a multi-bit word SOT has been proposed. We will discuss multi-bit word SOT cell in Section 8.4.2.2.

8.3.1.2 Perpendicular SOT Cell

In a film stack, such as Pt/Co/AlOx with broken inversion symmetry only along the direction perpendicular to the film plane, the anti-damping SOT torque can only be in the direction parallel to the film plane. This means the SOT from this stack can only deterministically switch the free layer of in-plane MTJ, not a perpendicular MTJ. However, perpendicular MTJs, with better scaling properties, have become the mainstream, at least for STT-MRAM. To switch perpendicular MTJs, one is required to break the in-plane inversion symmetry, e.g. with an in-plane field as in experiment carried out by Miron [8]. However, the external field is very undesirable for practical semiconductor memory device applications. Recently, several approaches have been developed to realize a field-free SOT switching of PMA materials by introducing films with lateral inversion asymmetry such as: tilted magnetic anisotropy [9]; interlayer exchange coupling [10]; interlayer dipole coupling [11]; in-plane exchange bias [12–17]. These methods create materials with strong spin orbit interaction and without in-plane symmetry at surface, so that the material can offer perpendicular anti-damping torque [18, 19], or current-driven exchange-bias [20]. MTJ with a built-in field from hard magnet was also reported [57]. A Co hard magnet is incorporated into the hard mask of each MTJ pillar. The Co provides an in-plane bias to the MTJ. Success of "field-free" switching has been reported [56, 58]. Time- and spatially-resolved observation of the switching dynamics of Pt/Co/AlOx pillar has been studied by Baumgartner [21]. Switching starts from domain wall nucleation at the edge of the pillar.

8.3.2 Materials Choice for SOT-MRAM Cell

What will be the optimal materials system for generating SOT for SOT-MRAM cells? It is important to note that, there are more than 6000 papers (as of Jan. 2020) reporting on the SOT properties of various materials systems. Although we have mentioned briefly that the effective spin Hall angle $\Theta_{SH,eff}$ can be large in some common heavy transition metals, such as Pt (~0.10), Ta (~−0.12), and W (~−0.30), an examination on their device performance efficacies should also be addressed. Besides conventional transition metals, emergent materials systems such as topological insulators (TIs), Weyl semimetals, and transition metal dichalcogenides (TMDs) are also potential SOT source materials for SOT-MRAM applications. Some of these materials are claimed to have giant effective spin Hall angle greater than 1. We will first review some classical transition metals with large spin-orbit interactions for such applications, and then examine those with exotic spin-charge transport properties later. A benchmarking based on the published SOT switching data will be provided.

8.3.2.1 Transition Metals and their Alloys

The $5d$ transition metals are the most promising SOT source due to their strong built-in spin-orbit interactions and the resulting strong SHEs. These materials are not uncommon in CMOS-related processing technology as well, which makes them more accessible than other exotic materials systems. Pt is one of the most studied materials, since late 2000s, several seminal works have pointed out that Pt has a spin Hall angle of the order of ~0.01 [66, 67, 68]. However, later experiments have shown that the intrinsic spin Hall angle of Pt can be even higher, ~0.20 or greater, depending on the characterization techniques, such as spin-torque ferromagnetic resonance (ST-FMR) [69], harmonic voltage measurement [70] (for more details of this method, see [71] and [72]), direct SOT switching measurement [73], hysteresis loop shift measurement [74], and current-driven domain wall motion measurement [75] ...etc. Through the refinement of characterization techniques, Pt is now believed to have an effective spin Hall angle of ~0.10–0.20 (See Figure 8.11). Note that this number will be affected by the thickness of the Pt layer, due to the spin diffusion effect [76].

Ta, on the other hand, was discovered to have a spin Hall angle larger than Pt in 2012, in which the very first demonstration of a prototypic SOT-MRAM cell was reported [6]. By utilizing the giant SHE from Ta, which has a spin Hall angle of $\Theta_{SH,eff} \sim -0.15$ (note the negative sign), Liu and colleagues at Cornell University showed that the SOT generated is strong enough to drive magnetization switching in the free layer of an adjacent MTJ. Later on, Pai et al. showed that W in the resistive (β phase) form can produce even large effect, with $\Theta_{SH,eff} \sim -0.30$ [5]. The resistive form of W, either β phase [78] or amorphous phase [79], therefore is an attractive candidate for above-mentioned SOT-MRAM applications, either in-plane or out-of-plane.

It has also been reported that $5d$ transition metal alloys, nitrides, oxides, and borides can generate sizable SHE and even possess greater spin Hall angles than pure transition metals. For example, Hf(Al)-doped Pt [80], Au-doped Pt [81], antiferromagnetic PtMn [82], W-doped Au [83], TaN [84], WO_x [85], and TaB [86] all

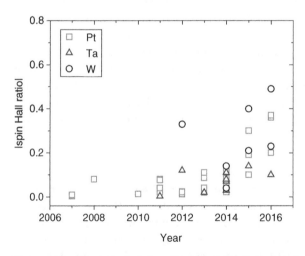

Figure 8.11 "Evolution" of the effective spin Hall angle (spin Hall ratio) for the most important heavy metals, namely Pt, Ta, and W. Note that the increasing trend should not be considered as the enhancement of the SHE, rather, it should be viewed as getting closer to the real values due to the refinement of characterization techniques and the elimination of measurement artifacts. *Source:* Data summarized from Hoffmann [77].

show larger spin Hall angles while compared to their pure metallic counterparts. These results are important since they indicate that the resulting SOT efficiency can be tuned by common materials engineering process during the deposition step.

8.3.2.2 Emergent Materials Systems

A new class of condensed matter, called "topological insulator" has been developed and proposed to be an even better candidate for charge-to-spin conversion while compare to its transition metal counterpart. These materials are typically single crystalline chalcogenide compounds with exotic band structures. Growth of such materials system is typically done by molecular beam epitaxy (MBE) to obtain high quality single crystal texture. The most important transport property of such materials system is the topologically protected surface state, which has spin-momentum locking for the transport electrons flowing only on the surfaces of those materials, as illustrated in Figure 8.12. It shows anti-damping torque when current flows in low crystalline symmetry direction. It is claimed that a much lower in-plane current is required to switch the FM in the TI/FM film stack. A large charge-to-spin conversion efficiency (spin Hall angle) of ~1–1.75 in the thin epitaxial Bi_2Se_3 films, where the topological surface states are dominant. The current density required for the magnetization switching has been reported to be extremely low, ~6×10^5 A/cm^2, which is one to two orders of magnitude smaller than that with heavy metals.

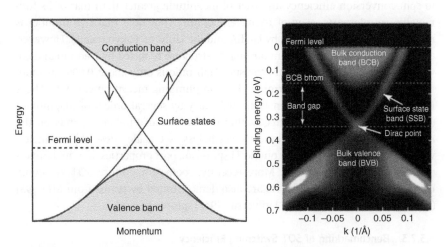

Figure 8.12 (a) Illustration of the band structure of a topological insulator (*Source:* Adapted from Wiki). (b) Angle resolved photoemission spectra (ARPES) of a topological insulator Bi_2Se_3. Note that ideally the Fermi level should lie in the gap between bulk conduction band (BCB) and bulk valence band (BVB). *Source:* Reproduced from [87].

One of the earliest studies on the SOTs in a TI/FM bilayer system was reported by K.L. Wang group from UCLA, United States, in 2014 [88]. In this seminal work, the team showed that an epitaxial $(Bi_{0.5}Sb_{0.5})_2Te_3/(Cr_{0.08}Bi_{0.54}Sb_{0.38})_2Te_3$ bilayer structure, which is grown by MBE, can be used to study the SOTs originates from the $(Bi_{0.5}Sb_{0.5})_2Te_3$ (TI) layer. The effective spin Hall angle of $(Bi_{0.5}Sb_{0.5})_2Te_3$, at least at a cryogenic temperature of ~2K (in order to keep the Cr-doped layer magnetic), was determined by a harmonic voltage technique to be in between 140 and 425. This number is almost three orders of magnitude larger than that of heavy transition metals. However, using a different approach (unidirectional spin Hall magnetoresistance [89]), researchers from University of Tokyo recently reported a very different number in a similar bilayer system (~five orders of magnitude larger than that of transition metals) [90]. Furthermore, the experiments mentioned above were both performed at a very low temperature (~2K), which might not be desirable for any realistic device applications. The most intuitive approach, then, will be using normal ferromagnetic metals as the FM layer, such that the Curie temperature is way above room temperature and measure the strength of the SOT in such TI/(metallic)FM heterostructures.

A simple bilayer structure, Bi_2Se_3/Py(permalloy), was adopted by Mellnik and co-workers from Cornell and Penn State University. As they reported on Nature in 2014 [91], the Bi_2Se_3 TI layer can generate spin-torque acting upon the adjacent metallic Py layer with a spin Hall angle about 2–3.5 at room temperature, which was determined by ST-FMR technique. This is of course a very exciting result, which indicates that a TI/(metallic)FM bilayer structure could possess charge-to-spin conversion efficiency an order of magnitude greater than that of its full-metallic counterparts, even at room temperature (but note that this number is much less than those reported by UCLA and the Univ. of Tokyo teams). However, a followed-up study using a similar bilayer structure reported by Deorani and co-workers from NUS showed that the spin Hall ratio is only about 0.0093 at room temperature, which was determined by spin-pumping measurement [92]. These inconsistent results of the spin Hall angle vary by several orders of magnitudes (0.0093–3.5 for Bi_2Se_3) and suggest that the estimated number depends on the adopted measurement technique. Nevertheless, set aside these discrepancies, we should expect more studies on the TI spin-transport properties at room temperature in the upcoming years. More recently, room-temperature SOT switching using Bi-based chalcogenide TIs are also demonstrated by teams from MIT [93] (MBE-grown) and Minnesota University [94] (sputter-grown).

8.3.2.3 Benchmarking of SOT Switching Efficiency

To benchmark SOT switching efficiency from various materials systems, both conventional and emergent, can be a difficult task. This is mainly due to two reasons: (i) Too many types characterization methods have been employed to characterize and estimate the effective spin Hall angle, as can be seen in Hoffmann's nice review [77] and (ii) the lack of control experiments and standard samples for SOT characterization. For simplicity and for the purpose of this book, we will focus on those reports that have current-induced SOT switching data. The works that only provide characterization (of the spin Hall angle) results will be left out. The figure of merit that we will use to compare the performance of SOT switching from different materials systems is the power consumption (in terms of) per switching event. This can be calculated as $P = \rho J_c^2$, where the ρ and J_c are resistivity and critical switching current density in the spin Hall metals or emergent materials channel. Summaries as organized from some selected materials systems are shown in Table 8.1 and Figure 8.13. It is interesting to note that although Bi-based

Table 8.1 Current-induced switching data from various materials systems.

Materials system	J_c (MA/cm^2)	ρ ($\mu\Omega$-cm)	P (10^9 W/cm^3)	Reference	Comment
Pt	23	26	13.8	[73]	Pt/Co, PMA
Ta	5.5	190	5.75	[6]	Ta/CoFeB, IMA
W	2.5	186	1.16	[95]	W/CoFeB, PMA
$Au_{0.25}Pt_{0.75}$	12	83	12	[96]	$Au_{0.25}Pt_{0.75}$/Co, PMA
$Pt_{0.25}Pd_{0.75}$	22	57.5	27.8	[97]	$Pt_{0.25}Pd_{0.75}$/Co, PMA
PtMn	9	180	14.6	[82]	PtMn/Hf/CoFeB, PMA
$Pt_{0.7}(MgO)_{0.3}$	11.5	58	7.67	[98]	$Pt_{0.7}(MgO)_{0.3}$/Co, PMA
$[Pt/Hf]_n$	17	144	41.6	[99]	$[Pt/Hf]_n$/Co, PMA
$Pt_{0.57}Cu_{0.43}$	2.4	82.5	0.5	[100]	$Pt_{0.57}Cu_{0.43}$/Co, PMA
Bi_2Se_3	2.8	1060	8.31	[93]	Bi_2Se_3(MBE)/CoTb, PMA
Bi_xSe_{1-x}	0.43	7143	1.32	[94]	Bi_xSe_{1-x}/Ta/CoFeB/Gd/CoFeB, PMA
BiSb	1.1	400	0.48	[101]	BiSb(MBE)/MnGa, PMA

Note that only room-temperature switching data are included. Adjacent FM layer might also affect the switching performance, therefore the FM layers and their anisotropies are listed in the comment column.

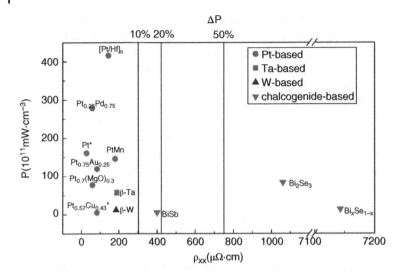

Figure 8.13 Power consumption (P) for SOT switching vs. SOT source materials resistivity (ρ_{xx}) for Pt, Ta, W, Pt-based alloys, and TI (chalcofenide)-based materials systems. ΔP represents the percentage of energy dissipation due to shunting effect by assuming using a CoFeB (1 nm) as FM layer. It can be seen that more resistive channel material will lead to more energy dissipation.

chalcogenides (TIs) are characterized to have humongous effective spin Hall angle, the power consumption results are not outperforming those from heavy transition metals by too much. The current shunting effect from resistive emergent materials, such as Bi_2Se_3, will also cause significant energy dissipation (>50%) in the FM layer. Ideally, a materials system with low resistivity ρ and low J_c (therefore high $\Theta_{SH,eff}$ based on Eq. (8.3)) are desirable for low power consumption SOT-MRAM applications.

8.4 Magneto-Electric Effect and Voltage-Control Magnetic Anisotropy (VCMA) MRAM

8.4.1 Magneto-Electric Effects

When an electric field is applied to a metal film, electrons accumulate on one surface and deplete on the opposite surface. The surface electrons terminate the electric field, such that the electric field does not penetrate into the bulk of metal film. This phenomenon is called screening. The screening electrons distribute over a finite thickness from the metal surface, which is called screening length, typically a depth of the order of atomic dimensions from the surface.

In ferromagnetic metal films, the screening electrons are spin-dependent due to exchange interactions. The spin dependence of the screening electrons leads to spin imbalance of the excess charge, which modifies the chemical potential at the film surface. That results in notable changes in the surface magnetization and the surface magneto-crystalline anisotropy energy (MAE).

Through the screening electrons, a voltage alters the magnetic properties of a magnetic film. It is called magneto-electric (ME) effect. The ME effect is solely a surface effect, and has little dependence on film thickness, since the electric field does not reach into the interior of the film. The magnitude of the ME effect is, to first order, linear with the density of screening electrons. The sensitivity of the ME effect on [95] Fe/MgO interface is larger than Fe/vacuum surface. Because of the dielectric constant of MgO, more screening electron charge is on the Fe/MgO surface than on Fe/vacuum surface [22].

Recall (Chapter 2) that in 3d metals, there are five 3d orbitals, each with different orbital magnetic quantum number m_L and the Crystalline anisotropy results from the interplay between orbital valence and spin-orbit interaction [23]. At the Fe/MgO interface, the Fe 3d orbital and the oxygen 2p orbital hybridize. Screening electrons induced by an E-field perpendicular to the [95] Fe/MgO film plane re-distributes the electron occupancy among orbitals. For an E-field pointing from MgO into Fe, the occupation decreases in the orbitals involving d_{xz} and d_{yz} and the electron occupation in orbital involving d_{xy} enhances, hence a decrease in the perpendicular orbital magnetic moment, but barely changes the in-plane orbital moment. The re-distribution changes the orbital contribution to MAE [24–28].

Early experimental studies were focused on FM/Noble metal, on Fe/Pt and Fe/Pd film [29, 30]. One notable experiment was performed [31, 32]. Figure 8.14 shows the sample (left) and the M-H loop (right) with applied H-field in direction to the film plane. An electric field is generated by applying a voltage through a thick polyimide insulator over the top electrode of the Au/ CoFe/MgO film stack. Thus, there is no current flow in the sample, only electric field across the stack. The M-H loop is measured with an *H* field perpendicular to the film surface under various dc electric fields. A positive voltage enhances the perpendicular anisotropy, while a negative voltage decreases the perpendicular anisotropy. The magneto-electric effect is evident.

As CoFeB/MgO based magnetic tunnel junction becomes technologically important, the study of the ME effect is focused on its interface. [28, 31, 33, 34]. The VCMA sensitivity is $\beta = d\varepsilon_{aniso.}/dE$, where ε_{aniso} is the anisotropy energy per unit area, E is the electric field. The unit is $(Joule/m^2)/(V/m) = Joule/Vm$.

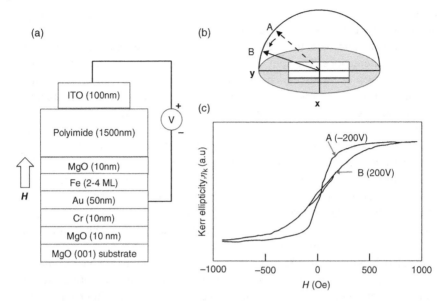

Figure 8.14 Voltage-induced anisotropy modulation in Au/CoFe/MgO stack. (a) test setup. (b) change of M_S due to change of stack anisotropy. (c) Kerr image under magnetic field H sweep. The external magnetic field sweeps in the direction normal to the film plane. The positive voltage on MgO weakens the out-of-plane anisotropy [31].

Measured data of the ME effect on the MgO/FM interface ranges widely [31, 33, 35–37] and is highly dependent on sample preparation and details of film stack. In a SOT-MRAM structure, -30 fJ/Vm was derived [38].

The effective anisotropy of an FM film is the balance of two components: one is perpendicular anisotropy from the film surface and the other in-plane from the film bulk. The effective in-plane anisotropy is

$$K_{\text{eff}} = E_\| - E_\perp = K_b - 2\pi M_s^2 + \sum K_{\text{interf}}/t, \tag{8.4}$$

where K_b is volume anisotropy constant, $\sum K_{\text{interf}}$ is the sum of interfacial or surface anisotropies, and t is the iron film thickness. Altering the surface component (the last term of Eq. (8.4)) changes the net anisotropy of the film K_{eff}. For example, applying an electric field to a ferromagnetic film can turn the film effective anisotropy from in-plane to perpendicular, and vice versa. The ME effect lasts only while the electric field is applied. Thus, the ME effect can be used to tentatively adjust the film anisotropy for memory applications. MRAM that makes use of this control is called Voltage–Control Magnetic Anisotropy (VCMA) assisted MRAM. An example is shown in Fig. 8.15.

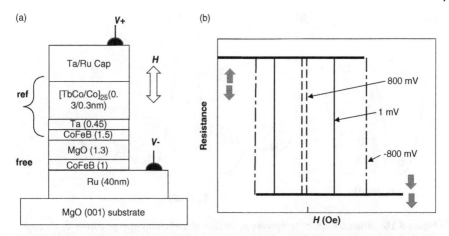

Figure 8.15 Demonstration of VCMA effect in CoFeAl/MgO p-MTJs. (a) Schematic illustration of the film stack of p-MTJs. (b) R-H loops for the p-MTJ under the voltages of −800, 1, and 800 mV. Notice that at V = +800 mV, the H_C collapses to zero, while H_C increases at V = −800 mV [34].

Magnetic anisotropy and mechanical stress are correlated through magneto-striction [39–41]. Changing the film strain could induce anisotropy change, for example the changing the Co content in a CoFe film on specific substrate induces change in film strain, and thus, film anisotropy [42].

8.4.2 VCMA-Assisted MRAMs

For MRAM of any operation mode, the switching energy and the data retention performance has been a trade-off relation. Longer retention MRAM cells require larger energy to switch. VCMA offers an effective way to reduce the switching energy without compromising retention performance. A class of magnetic memories employing such a mechanism is called VCMA MRAM. VCMA provides a means to temporarily reduce the MTJ free layer anisotropy and thus, the switching energy barrier. A voltage pulse across an MTJ temporarily reduces the switching barrier, thus, the switching field of field-MRAM [43], and the switching current of SOT-MRAM [38]. Figure 8.16 illustrates the concept of VCMA-assisted switching.

8.4.2.1 VCMA-Assisted Field-MRAM

As shown in Figure 8.16b, VCMA can assist the field-MRAM switching by temporarily lowering the interfacial anisotropy, thus, the coercivity H_C of the free layer of MTJ during WRITE, thus the cell can be written with smaller write field, therefore

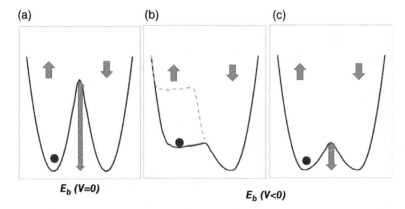

Figure 8.16 The concept of temporary reduction of switching energy barrier E_b through VCMA during switching. (a) Energy barrier when cell is in stand-by. (b) During field-MRAM switching, VCMA is equivalent to adding a temporary aiding field (dashed line), (c) During SOT mode MRAM switching, VCMA is applied, which temporarily reduces the switching energy barrier, thus, the switching current.

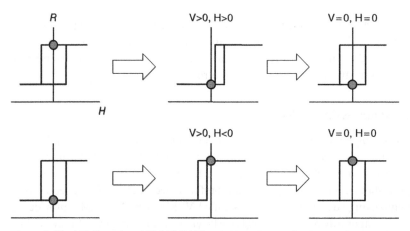

Figure 8.17 VCMA-assisted Field MRAM switching. During field switching, a voltage is applied to the MTJ to temporarily reduce its coercivity, thus, the switching threshold. During standby, the voltage is absent, the cell maintains its anisotropy for thermal stability. Note that, neither the voltage of the field alone with same value can switch the bit. The bits switch only when both are applied.

a smaller the current to generate the WRITE field. After WRITE, the VCMA voltage pulse is terminated, the energy barrier recovers, as shown in Figure 8.16a, the cell is back to stand-by condition with large energy barrier. Figure 8.17 shows the R-H loop before, during, and after field switching.

8.4.2.2 VCMA-Assisted Multi-bit-Word SOT-MRAM

The 1M-2T single-bit SOT MRAM cell described in Section 8.3.2 is neither dense nor energy efficient. The VCMA-assisted multi-bit word SOT MRAM cell is (Figure 8.18a). Multiple MTjs are place on a on a common SOT channel. VCMA provides a bit select mechanism to write a particular bit (or bits). In this case, negative voltage lowers the energy barrier and a positive voltage raises that of bit cells on a word line (SOT channel). And Figure 8.18c shows the VCMA effect on the bit cell write error rate (WER). At -19 MA/cm^2 write current density, about 10 orders of WER can be achieved between $+0.8$ and -0.8 V. The VCMA sensitivity is estimated to be 30 fJ/Vm [38].

Note that, unlike the single bit SOT cell, the WRITE operation of multi-bit word SOT MRAM is conducted in two steps: bits in the Word to be written "1" will be selected and written in one step and those in "0" in another step. Independent of the number of bits on a word, each word is written only twice, once for "0" and once for "1." That saves the number write pulses, thus the write energy.

Potentially, the switching energy and access performance of the VCMA-assisted SOT MRAM can be lower than in the case of STT-MRAM and should be of the same order of magnitude as in DRAM.

8.4.2.3 VCMA-Assisted Precession-Toggle MRAM

In STT-mode switching, the free layer magnetization precesses over many periods to achieve magnetization reversal. Precession-toggle mode is a mode that magnetization reversal is achieved in only half a precession period. So, it is a much faster way to achieve magnetic reversal. This switching mode is found in an MTJ with a

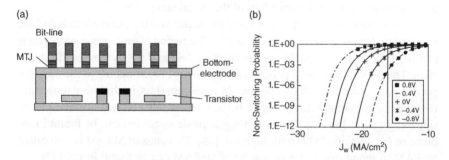

Figure 8.18 (a) Multi-bit word SOT cell. A common write current flow through the bottom electrode. MTJs on the bottom electrode is biased to select the write bits. (b) Write error rate of cells with MTJ bias voltage at -0.8, -0.4, 0, $+0.4$, $+0.8$ V. MTJ sample: The MTJ stack (top-to-bottom) Ta (5 nm)/IrMn (8 nm)/CoFe (1.9–2.0 nm)/Ru (0.8 nm)/CoFeB (1.9–2.0 nm)/ MgO (1.4 nm)/CoFeB (1.2 nm)/Ta (10 nm), E-field pointing from MgO into CoFeB free layer increases anisotropy and disable the write [38].

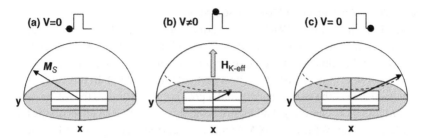

Figure 8.19 Switching under a voltage short pulse. (a) initial state, in which the voltage across MTJ is off, (b) VCMA strengthens out-of-plane H_{eff} magnetization precess around H_{eff} (c) The applied voltage pulse length is terminated when precessing half a period. Magnetic reversal is achieved [44].

high *RA* product, such that the current density is too low to achieve STT-mode switching before junction breakdown.

As illustrated in Figure 8.19a, prior to a voltage is applied to the MTJ, the initial anisotropy H_{K_eff} and magnetization are both nearly in plane and is along the easy axis (y-axis). Once a voltage is applied (Figure 8.19b), the out-of-plane anisotropy is raised by the VCMA mechanism, the net film anisotropy H_{K_eff} turns out of plane. Under this new anisotropy direction, the film magnetization begins to precess around H_{K_eff}. When the voltage pulse is terminated in half a precession period, the film anisotropy returns to initial state, pointing nearly in plane. The magnetization precess around the original (Figure 8.19c). The magnetization stays in the reverse direction. [44] The write voltage in this mode is always unipolar. The voltage must be strong enough to alter the direction of H_{K_eff}. Each voltage pulse toggles the magnetization, independent of the initial state.

The switching current is in the order of 10^5 A/cm^2 and the pulse width is 0.65 ns, the current density is too low for STT-mode. The period of each precession cycle is $(1 + \alpha^2)/(\gamma H_{eff})$ [45], where H_{eff} is the effective field and γ is gyromagnetic coefficient. The typical precession period is smaller than 1 ns, switching to the opposite polarity take places in half of a period of precession, thus, in the sub-nanosecond range. Thus, precession-toggle mode is the most energy efficient, in the order of single-digit fJ/bit [46, 47]. Precession-toggle mode operation can be found in in-plane or out-of-plane MTJs with high *RA*. [48]. This class of MRAM is also called MeRAM. A detail discussion of scaling of MeRAM can be found in ref. [37].

Although this mode of switching is the fastest among all MRAM switching modes, there are several drawbacks. Among those, the most severe one is that it needs precision control of the write pulse width to ensure low WER [45]. Bit-to-bit VCMA strength variation leads to a distribution of H_{K_eff}, thus the precession period. That in turn puts a limit on the WRITE pulse width operation window. Circuit innovations for such applications have been reported [49, 50].

Similar to the toggle mode field MRAM, precession-toggle mode is required to operate in READ-BEFORE-WRITE to know the initial state. The write cycle time is not short, in practice.

8.5 Relative Merit of Advanced Switching Mode MRAMs

Here, we compare the properties of advanced switching mode MRAMs described in this chapter against field-MRAM and STT-MRAM. The comparison is shown in Table 8.2. The product-like STT-MRAM data is taken from [51, 52], the most mature STT-MRAM published data. From left to right, each are the primary consideration for cell size, write access time, endurance, requirement of MgO thickness and data retention. The right most column states the maturity of a technology. From academic research, industrial research to production in a time sequence.

The MRAM cell size has been predominantly determined by the size of the cell transistor and the number of connecting wire per bit cell and wire pitch, not the MTJ. Thus, the number of transistors in a cell is important. Both the single-bit CIDM and single-bit SOT modes are 1M-2T cell, thus the cell size is inherently larger. Although the operating current density of these two modes is higher, on

Table 8.2 The cell size and performance of other switching-modes compared to Field- and STT-MRAM.

Switching mode	WRITE pulse (ns)	Endurance, cycles 10^n	RA Ω-μm^2	Data retention 10^n S	Degree of maturity (2019)
Field MRAM	25	$n = 15$	1 k	$n = 9$	Production 2006
STT-MRAM	5–10	$n = 7$–11	<5	$n = 7$–8	Production 2019
CIDM MRAM	1–10	$n > 15$	10–20	–	Academic research
SOT MRAM	1–10	$n = 15$	~10–20	=Field MRAM	Early Industrial research
Precession toggling MRAM	<1	$n = 15$	–	=Field MRAM	Academic research

the order of mid-10^7A/cm^2 for SOT MRAM and 10^7–10^8 A/cm^2 for CIDM MRAM, the cell current may not necessarily be larger, depending on the cross-section area of the SOT-channel of the SOT cell, and the free layer of the CIDM cell, respectively. The write current, not write current density, determines the size of the cell transistor [53]. Comparing SOT and STT-MRAM, SOT is more suitable for higher access speed, at higher power consumption. One attractive feature of the Racetrack CIDM MRAM is that the total number of connective wires per bit is very small, since a whole string of bits are serially read out by 1 MTJ, or 1 BL. Thus, potentially, its bit density is not limited by wire pitch. Table 8.3 compares the potential density of these three cells.

Despite of the rich academic research in CIDM MRAM magnetics, its industrial research has not been up to pace, leaving many un-answered questions, such as cell size, domain de-pining threshold, CMOS compatibility, reliable operating range, thermal stability and switching error rate, etc. More studies are required to turn a CIDM memory cell into a practical device. The research of Racetrack has been dormant for a while and emerges with new promises [60].

One potential application of a large (long stripe) CIDM cell is to let it act as an analog memory, not a binary (digital) memory cell property. For example, the domain wall can be driven to intermediate positions between "1" and "0" position, under the pinned layer. Thus, the MTJ resistance is in value between R_{AP} and R_p. Thus, it acts as a Memrister [54, 55], a non-volatile tunable resistor. Thus, it can be

Table 8.3 Relative write current density and cell size of STT, SOT, and CIDM.

Switching mode	I_W density (uA/nm^2)	I_W flow through	# of wire/ bit	Cell (1 M-nT)	Relative cell transistor size/bit
STT	0.06–0.1	MTJ	1WL-2BL	n = 1	X
1-bit SOT cell	0.2 – 0.3	SOT channel	1WL-1BL-2DL	n = 2	Y1
multi-bit SOT cell			(1WL-nBL-2DL)/n	n = 2 + n	Y2
1-bit CIDM cell	2–3	FM stripe	1WL-1Din-1Dout	n = 1	Z1
multi-bit Racetrack			non-random access	–	–
					X < Y1, Y2 < Y1, Y2 < Z1

used for storing analog information, which is desirable in neural network of artificial intelligence application.

While STT-MRAM is competing for last-level cache of processor chip, single-bit SOT MRAM may find its way to higher-level cache, because it can offer subnanosecond switching speed and much better endurance properties. These properties are required for L1 and L2 cache. SOT MRAM has gained sufficient attention such that industrial level research has begun. Multi-bit word SOT memory receives more attention from industry than CIDM MRAM.

The VCMA-assisted Precession-toggle mode MRAM is the least mature, and is in its infancy stage. It is a 1M-1T cell. So the density is comparable to the STT cell. Potentially, it is the fastest switching device, completing the switch in half of a precession cycle. However, the window of the write pulse must be precisely controlled to ensure good write error rate. Innovative write circuit design is required.

Homework

Q8.1 Estimate and compare the bit-cell size of four-bit-word SOT cell and an STT cell Reference to cell layout in Figure 8.18a and assume:

1) A 4-bit-word SOT cell contains a Ta write stripe, 2 transistors and 4 MTJs (2T-4M). The MTJ area is 80×240 nm^2. The word write stripe is 5 nm thick, 250 nm wide, and the write current density is 20 MA/cm^2.
2) STT cell: 80 nm diameter, the write current density is 6 MA/cm^2, 1T-1M cell.
3) Gating transistor current density is 1 mA/um (gate width) The transistor length, along the gate length direction is 320 nm.

A8.1 The write current of an STT cell and an SOT cell are 288 and 250 μA, respectively. The transistor gate width of an STT cell and an SOT cell are 288 and 250 nm, respectively. The two transistors of an SOT cell total gate width is 500 nm plus transistor isolation. All 4 MTJs can be placed in-line over 500 nm, thus, the MTJ area can be ignored. Each SOT bit size is 125 nm $\times P_{tx}$, where P_{tx} is the transistor pitch in the gate length direction. Similarly, an MTJ in STT cell equals to the gating transistor area. It is 288 nm $\times P_{tx}$. Thus, to first order, bit cell size of a 2T-4MTJ cell can be smaller than a 1T-1MTJ STT cell.

References

1 S. Fukami, T. Suzuki, K. Nakahara, N. Ohashima, Y. Ozaki (2009) Low-current Perpendicular domain wall motion cell for scalable high-speed MRAM. *VLSI Symp. Tech.*, 230–1.

2 Suzuki, T., Tanigawa, H.; Kobayashi, Y.; Mori, K.; Ito, Y.; Ozaki, Y.; Suemitsu, K.; Kitamura, T.; Nagahara, K.; Kariyada, E.; Ohshima, N.; Fukami, S.; Yamanouchi, M.; Ikeda, S.; Hayashi, M.; Sakao, M.; Ohno, H. (2013) Low-current domain wall motion MRAM with perpendicularly magnetized CoFeB/MgO magnetic tunnel junction and underlying hard magnets, *Digest of VLSI Technology (VLSIT)*, 2013 Symposium on, T138–T139.

3 Fukami, S., Yamanouchi, M., Kim, K.-J., Suzuki, T., Sakimura, N., Chiba, D., Ikeda, S., Sugibayashi, T., Kasai, N., Ono, T., Ohno, H. (2013) 20-nm magnetic domain wall motion memory with ultralow-power operation. *Digest of Electron Devices Meeting (IEDM)*, 2013 IEEE International, DOI: 10.1109/IEDM.2013.6724553, pp. 3.5.1–3.5.4.

4 Parkin, S.S.P., Hayashi, M., and Thomas, L. (2008) Magnetic Domain-Wall racetrack memory. *Science* 320: 190. https://doi.org/10.1126/science.1145799.

5 Pai, C.-F., Liu, L.Q., Li, Y. et al. (2012) Spin transfer torque devices utilizing the giant spin Hall effect of tungsten. *Appl. Phys. Lett.* 101: 122404.

6 Liu, L., Pai, C.-F., Li, Y. et al. (2012) Spin torque switching with the giant spin Hall effect of tantalum. *Science* 336: 555. http://arXiv.org > cond-mat > arXiv:1203.2875.

7 Rahaman Sk A., et al., *IEEE IEDM MRAM Poster Abstract*, paper 24 (2018)

8 Miron, I.M., Gaudin, G., Auffret, S. et al. (2010) Current-driven spin torque induced by the Rashba effect in a ferromagnetic metal layer. *Nat. Mater.* 9: 230. https://doi.org/10.1038/NMAT2613.

9 You, L., Lee, O., Bhowmik, D. et al. (2015) Switching of perpendicularly polarized nanomagnets with spin orbit torque without an external magnetic field by engineering a tilted anisotropy. *Proc. Natl. Acad. Sci. U. S. A.* 112: 10310.

10 Lau, Y.C., Betto, D., Rode, K. et al. (2016) Reduction of in-plane field required for spin-orbit torque magnetization reversal by insertion of Au spacer in Pt/Au/Co/Ni/Co/Ta. *Nat. Nanotechnol.* 11: 758.

11 A. K. Smith, M. Jamali, Z. Zhao, and J. P. Wang, e-print arXiv:1603.09624.

12 Fukami, S., Zhang, C., Dutta Gupta, S. et al. (2016) Magnetization switching by spin-orbit torque in an antiferromagnet-ferromagnet bilayer system. *Nat. Mater.* 15; 535.

13 Oh, Y.W., Baek, S.C., Kim, Y.M. et al. (2016) Field-free switching of perpendicular magnetization through spin–orbit torque in antiferromagnet/ferromagnet/oxide structures. *Nat. Nanotechnol.* 11: 878.

14 Chen, J.-Y., Mahendra, D.C., Zhang, D. et al. (2017) Field-free spin-orbit torque switching of composite perpendicular CoFeB/Gd/CoFeB layers utilized for three-terminal magnetic tunnel junctions. *Appl. Phys. Lett.* 111: 012402.

15 van den Brink, A., Vermijs, G., Solignac, A. et al. (2016) Field-free magnetization reversal by spin-Hall effect and exchange bias. *Nat. Commun.* 7: 10854.

16 Kong, W.J., Ji, Y.R., Zhang, X. et al. (2016) Spin-orbit torque induced magnetization anisotropy modulation in Pt/(Co/Ni)4/Co/IrMn heterostructure. *Appl. Phys. Lett.* 109: 132402.

17 Wu, D., Yu, G., Chen, C.-T. et al. (2016) Spin-orbit torques in perpendicularly magnetized Ir22Mn78/Co20Fe60B20/MgO multilayer. *Appl. Phys. Lett.* 108: 212406.

18 Garello, K., Miron, I.M., Avci, C.O. et al. Symmetry and magnitude of spin–orbit torques in ferromagnetic heterostructures. *Nat. Nanotechnol.* https://doi.org/ 10.1038/NNANO.2013.145.

19 D. MacNeill, G.M. Stiehl, M.H.D. Guimaraes, R.A. Buhrman, J. Park, D.C. Ralph, Control of spin-orbit torque through crystal symmetry in WTe$_2$/ferromagnet bilayer, arXiv:1605.02712

20 Ming-Han Tsai, Po-Hung Lin, Kuo-Feng Huang, Hsiu-Hau Lin, and Chih-Huang Lai, Spin-orbit-torque MRAM: from uniaxial to unidirectional switching, http:// arXiv.org > physics > *arXiv:1706.01639*

21 M. Baumgartner, K. Garello, J. Mendil, C. O. Avci, E. Grimaldi, C. Murer, J. Feng, M. Gabureac, C. Stamm, Y. Acremann, S. Finizio, S. Wintz, J. Raabe, P. Gambardella, Time- and spatially-resolved magnetization dynamics driven by spin-orbit torques, arXiv:1704.06402v1 [cond-mat.mtrl-sci] 21 Apr 2017

22 Niranjan, M.K., Duan, C.-G., Jaswal, S.S., and Tsymbal, E.Y. (2010) Electric field effect on magnetization and magnetocrystalline anisotropy at the Fe/MgO(001) interface. *Appl. Phys. Lett.* 96: 222504.

23 Van Vleck, J.H. (1937) On the anisotropy of cubic ferromagnetic crystals. *Phys. Rev. B* 52: 1178.

24 Hjortstam, J., Trygg, J., Wills, J.M. et al. (1996) Calculated spin and orbital moments in the surfaces of the 3d metals Fe, Co, and Ni and their overlayers on Cu(001). *Phys. Rev. B* 53: 9204.

25 Kyuno, K. et al. (1996) First-principles calculation of the magnetic anisotropy energies of Ag/Fe(001) and Au/Fe(001) multilayers. *J. Phys. Soc. Jpn.* 65: 1334–1339.

26 Duan, C.-G., Velev, J.P., Sabirianov, R.F. et al. Surface magnetoelectric effect in ferromagnetic metal films. *Phys. Rev. Lett.* 101: 137201.

27 Zhang, S. (1999) Spin-dependent surface screening in ferromagnet and magnetic tunnel junction. *Phys. Rev. Lett.* 83: 640.

28 Shimabukuro, R., KohjiNakamura, T.A., and Ito, T. (2010) Electric field effects on magnetocrystalline anisotropy in ferromagnetic Fe monolayers. *Phys. Ther.* E42: 1014–1017.

29 Weisheit, M. et al. (2007) Electric field-induced modification of magnetism in thin-film ferromagnets. *Science* 315: 349–351.

30 Bonell, F. et al. (2011) Large change in perpendicular magnetic anisotropy induced by an electric field in FePd ultrathin films. *Appl. Phys. Lett.* 98: 232510.

31 Maruyama, T. et al. (2009) Large voltage-induced magnetic anisotropy change in a few atomic layers of iron. *Nat. Nanotechnol.* 4: 158–161.

32 Suzuki, Y., Kubota, H., Tulapurkar, A., and Nozaki, T. (2011) Spin control by application of electric current and voltage in FeCo-MgO junctions. *Phil. Trans. R. Soc. A* 2011: 369. https://doi.org/10.1098/rsta.2011.0190369.

33 Endo, M., Kanai, S., Ikeda, S. et al. (2010) Electric-field effects on thickness dependent magnetic anisotropy of sputtered $MgO/Co_{40}Fe_{40}B_{20}/Ta$ structures. *Appl. Phys. Lett.* 96: 212503.

34 Zhenchao Wen, Hiroaki Sukegawa, Takeshi Seki, Takahide Kubota, Koki Takanashi, and Seiji Mitani, Voltage control of magnetic anisotropy in epitaxial Ru/Co2FeAl/MgO heterostructures, https://arxiv.org/pdf/1611.02827

35 Rajanikanth, A., Hauet, T., Montaigne, F. et al. (2013) Magnetic anisotropy modified by electric field in V/Fe/ MgO(001)/Fe epitaxial magnetic tunnel junction. *Appl. Phys. Lett.* 103: 062402.

36 Nozaki, T., Shiota, Y., Shiraishi, M. et al. (2010) Voltage-induced perpendicular magnetic anisotropy change in magnetic tunnel junctions. *Appl. Phys. Lett.* 96: 022506; Nozaki, T., et al. (2013). Voltage-induced magnetic anisotropy changes in an ultrathin FeB Layer sandwiched between two MgO layers. *Appl. Phys. Express* 6: 073005.

37 Amiri, P.K., Alzate, J.G., Cai, X.Q. et al. Electric-field-controlled magnetoelectric RAM: progress, challenges, and scaling. *IEEE Trans. Magn.* 51 (11): 3401507.

38 H. Yoda, N. Shimomura, Y. Ohsawa, S. Shirotori, Y. Kato, T. Inokuchi, Y. Kamiguchi, B. Altansargai, Y. Saito, K. Koi, H. Sugiyama, S. Oikawa, M. Shimizu, M. Ishikawa, K. Ikegami, and A. Kurobe, Voltage-Control Spintronics Memory (VoCSM) Having Potentials of Ultra-Low Energy-consumption and High-Density, *IEEE Dig. Of IEDM 2018*, paper 27.6

39 Sander, D. (1999) The correlation between mechanical stress and magnetic anisotropy in ultrathin films. *Rep. Prog. Phys.* 62: 809–858. http://www-old.mpi-halle.mpg.de/mpi/publi/pdf/1339_99.pdf.

40 Clark, A.E., Restorff, J.B., Wun Fogle, M. et al. (2008) Temperature dependence of the magnetostriction and magnetoelastic coupling in $Fe_{100-x}Al_x$ (x=14.1, 16.6, 21.5, 26.3) and Fe50Co50. *J. Appl. Phys.* 103: 07B310.

41 Yu, G. et al. (2015) Strain-induced modulation of perpendicular magnetic anisotropy in Ta/CoFeB/MgO structures investigated by ferromagnetic resonance. *Appl. Phys. Lett.* 106: 072402; 106: 169902 (2015) (Erratum).

42 Ong, P.V., Kioussis, N., Amiri, P.K., and Wang, K.L. Electric-field-driven magnetization switching and nonlinear magnetoelasticity in Au/FeCo/MgO heterostructures. *Sci. Rep.* 6: 29815. https://doi.org/10.1038/srep29815.

43 Wang, W.G. and Chien, C.L. (2013) Voltage-induced switching in magnetic tunnel junctions with perpendicular magnetic anisotropy. *J. Phys. D Appl. Phys.* 46: 074004. (12pp) doi: https://doi.org/10.1088/0022-3727/46/7/074004.

44 Shiota, Y. et al. (2012) Induction of coherent magnetization switching in a few atomic layers of FeCo using voltage pulses. *Nat. Mater.* 11: 39–43.

45 Yoichi Shiota, Takayuki Nozaki, Shingo Tamaru, Kay Yakushiji, Hitoshi Kubota, Akio Fukushima, Shinji Yuasa, and Yoshishige Suzuki, Evaluation of write error rate for voltage-driven dynamic magnetization switching in magnetic tunnel junctions with perpendicular magnetization, http://iopscience.iop.org/article/10.7567/APEX.9.013001

46 Kanai, S., Yamanouchi, M., Ikeda, S. et al. (2012) Electric field-induced magnetization reversal in a perpendicular-anisotropy CoFeB-MgO magnetic tunnel junction. *Appl. Phys. Lett.* 101: 122403.

47 Grezes, C., Ebrahimi, F., Alzate, J.G. et al. (2016) Ultra-low switching energy and scaling in electric-field-controlled nanoscale magnetic tunnel junctions with high resistance-area product. *Appl. Phys. Lett.* 108: 012403.

48 Wei-GangWang, M.L., Hageman, S., and Chien, C.L. Electric-field-assisted switching in magnetic tunnel junctions. *Nat. Mater.* https://doi.org/10.1038/NMAT3171.

49 Hiroki Noguchi, Kazutaka Ikegami, Keiko Abe, Shinobu Fujita, Yoichi Shiota, Takayuki Nozaki, Shinji Yuasa, and Yoshishige Suzuki, Novel Voltage Controlled MRAM (VCM) with Fast Read/Write Circuits for Ultra Large Last Level Cache, *IEEE Dig. Of IEDM 2016*, paper 27.5

50 Shin, K.-S., Im, S., and Park, S.-G. (2016) Low-power write-circuit with status-detection for STT-MRAM. *J. Semicond. Technol. Sci.* 16 (1).

51 G. Jan, et al., Fully functional 8Mb test chip, *IEEE Symp VLSI Tech*, 2014. pp 50–51.

52 Thomas, L., Jan, G., Le, S., and Wang, P.-K. (2015) Quantifying data retention of perpendicular spin-transfer-torque magnetic random access memory chips using an effective thermal stability factor method. *Appl. Phys. Lett.* 106: 162402. https://doi.org/10.1063/1.4918682.

53 Jabeur, K., Di Pendina, G., and Prenat, G. (2017) Study of spin transfer torque (STT) and spin orbit torque (SOT) magnetic tunnel junctions (MTJS) at advanced cmos technology nodes. *Electr. Electron. Eng. Int. J. (ELELIJ)* 6 (1) https://doi.org/10.14810/elelij.2017.6101.

54 J. Grollier, A. Chanthbouala, R. Matsumoto, A. Anane, V. Cros, F. Nguyen van Dau, and A. Fert, Magnetic domain wall motion by spin transfer, arXiv:1207.3489v1 [cond-mat.mtrl-sci] 15 Jul 2012

55 Chua, L.O. (1971) Memristor - the missing circuit element. *IEEE Trans. Circuit Theory* 18: 507.

56 Cubukcu, M. et al. (2018) Ultra-fast perpendicular spin–orbit torque MRAM. *IEEE Trans. Magn.* 54 (4): 9300204.

57 Kevin Garello, et al., Spin-Orbit Torque MRAM for ultrafast embedded memories: from fundamentals to large scale technology integration, 2019 IEEE 11th *International Memory Workshop* (IMW), 10.1109/IMW.2019.8739466

58 K. Garello, et al., Manufacturable 300mm platform solution for Field-Free Switching SOT-MRAM, *Digest of IEEE Symp. On VLSI Technology*, FJS4–5, T194

59 Yang, S.-h., Ryu, K.-S., and Parkin, S. Domain-wall velocities of up to 750 m s−1 driven by exchange-coupling torque in synthetic antiferromagnets. *Nat. Nanotechnol.* https://doi.org/10.1038/NNANO.2014.324.

60 Stuart Parkin, https://www.youtube.com/watch?v=kB0ixO5lrzQ

61 D.Y. Wang, et al., A statistical study of the reliability of SOT MRAM cell structures by thermal baking, *Abstract of IEDM 2019*, MRAM Poster paper.

62 S. Z. Rahaman, et.al., IEEE, EDL, v.39, No.9, p.1306, (2018)

63 Yao-Jen Chang, et al., IEDM, MRAM Poster, No.7 (2018)

64 K. Honjo, et al., First demonstration of field-free SOT-MRAM with 0.35 ns write speed and 70 thermal stability under 400°C thermal tolerance by canted SOT structure and its advanced patterning/SOT channel technology, *IEEE Digest of IEDM*, paper 28.5 (2019)

65 Fukami, S., Anekewa, T., Zhang, C., and Ohno, H. A spin–orbit torque switching scheme with collinear magnetic easy axis and current configuration. *Nat. Nanotechnol.* https://doi.org/10.1038/NNANO.2016.29.

66 Kimura, T., Otani, Y., Sato, T. et al. (2007) Room-temperature reversible spin Hall effect. *Phys. Rev. Lett.* 98: 156601.

67 Guo, G.Y., Murakami, S., Chen, T.W., and Nagaosa, N. (2008) Intrinsic spin Hall effect in platinum: first-principles calculations. *Phys. Rev. Lett.* 100: 096401.

68 Morota, M., Niimi, Y., Ohnishi, K. et al. (2011) Indication of intrinsic spin Hall effect in 4d and 5d transition metals. *Phys. Rev. B* 83: 174405.

69 Liu, L.Q., Moriyama, T., Ralph, D.C., and Buhrman, R.A. (2011) Spin-torque ferromagnetic resonance induced by the spin Hall effect. *Phys. Rev. Lett.* 106: 036601.

70 Garello, K., Miron, I.M., Avci, C.O. et al. (2013) Symmetry and magnitude of spin-orbit torques in ferromagnetic heterostructures. *Nat. Nanotechnol.* 8: 587.

71 Kim, J., Sinha, J., Hayashi, M. et al. (2013) Layer thickness dependence of the current-induced effective field vector in Ta vertical bar CoFeB vertical bar MgO. *Nat. Mater.* 12: 240.

72 Hayashi, M,, Kim, J., Yamanouchi, M., and Ohno, H. (2014) Quantitative characterization of the spin-orbit torque using harmonic hall voltage measurements. *Phys. Rev. B* 89: 144425.

73 Liu, L.Q., Lee, O.J., Gudmundsen, T.J. et al. (2012) Current-induced switching of perpendicularly magnetized magnetic layers using spin torque from the spin Hall effect. *Phys. Rev. Lett.* 109: 096602.

74 Pai, C.F., Mann, M., Tan, A.J., and Beach, G.S.D. (2016) Determination of spin torque efficiencies in heterostructures with perpendicular magnetic anisotropy. *Phys. Rev. B* 93: 144409.

75 Emori, S., Martinez, E., Lee, K.J. et al. (2014) Spin hall torque magnetometry of Dzyaloshinskii domain walls. *Phys. Rev. B* 90: 184427.

76 Nguyen, M.H., Ralph, D.C., and Buhrman, R.A. (2016) Spin torque study of the spin Hall conductivity and spin diffusion length in platinum thin films with varying resistivity. *Phys. Rev. Lett.* 116: 126601.

77 Hoffmann, A. (2013) Spin Hall effects in metals. *IEEE Trans. Magn.* 49: 5172.

78 Hao, Q. and Xiao, G. (2015) Giant spin Hall effect and switching induced by spin-transfer torque in a $W/Co_{40}Fe_{40}B_{20}/MgO$ structure with perpendicular magnetic anisotropy. *Phys. Rev. Appl.* 3: 034009.

79 Liu, J., Ohkubo, T., Mitani, S. et al. (2015) Correlation between the spin Hall angle and the structural phases of early 5d transition metals. *Appl. Phys. Lett.* 107: 232408.

80 Nguyen, M.-H., Zhao, M., Ralph, D.C., and Buhrman, R.A. (2016) Enhanced spin Hall torque efficiency in $Pt_{100-x}Al_x$ and $Pt_{100-x}Hf_x$ alloys arising from the intrinsic spin Hall effect. *Appl. Phys. Lett.* 108: 242407.

81 Obstbaum, M., Decker, M., Greitner, A.K. et al. (2016) Tuning spin Hall angles by alloying. *Phys. Rev. Lett.* 117: 167204.

82 Ou, Y.X., Shi, S.J., Ralph, D.C., and Buhrman, R.A. (2016) Strong spin Hall effect in the antiferromagnet PtMn. *Phys. Rev. B* 93: 220405.

83 Laczkowski, P., Rojas-Sanchez, J.C., Savero-Torres, W. et al. (2014) Experimental evidences of a large extrinsic spin Hall effect in AuW alloy. *Appl. Phys. Lett.* 104: 142403.

84 Chen, T.Y., Wu, C.T., Yen, H.W., and Pai, C.F. (2017) Tunable spin-orbit torque in Cu-Ta binary alloy heterostructures. *Phys. Rev. B* 96: 104434.

85 Demasius, K.U., Phung, T., Zhang, W.F. et al. (2016) Enhanced spin-orbit torques by oxygen incorporation in tungsten films. *Nat. Commun.* 7: 10644.

86 Kato, Y., Saito, Y., Yoda, H. et al. (2018) Improvement of write efficiency in voltage-controlled spintronic memory by development of a **Ta−B** spin Hall electrode. *Phys. Rev. Appl.* 10: 044011.

87 Chen, Y.L., Analytis, J.G., Chu, J.H. et al. (2009) Experimental realization of a three-dimensional topological insulator, Bi_2Te_3. *Science* 325: 178.

88 Fan, Y.B., Upadhyaya, P., Kou, X.F. et al. (2014) Magnetization switching through giant spin-orbit torque in a magnetically doped topological insulator heterostructure. *Nat. Mater.* 13: 699.

89 Avci, C.O., Garello, K., Ghosh, A. et al. (2015) Unidirectional spin Hall magnetoresistance in ferromagnet/normal metal bilayers. *Nat. Phys.* 11: 570.

90 Yasuda, K., Tsukazaki, A., Yoshimi, R. et al. (2016) Large unidirectional magnetoresistance in a magnetic topological insulator. *Phys. Rev. Lett.* 117: 127202.

91 Mellnik, A.R., Lee, J.S., Richardella, A. et al. (2014) Spin-transfer torque generated by a topological insulator. *Nature* 511: 449.

92 Deorani, P., Son, J., Banerjee, K. et al. (2014) Observation of inverse spin Hall effect in bismuth selenide. *Phys. Rev. B* 90: 094403.

93 Han, J.H., Richardella, A., Siddiqui, S.A. et al. (2017) Room-temperature spin-orbit torque switching induced by a topological insulator. *Phys. Rev. Lett.* 119: 077702.

94 Mahendra, D.C., Grassi, R., Chen, J.-Y. et al. (2018) Room temperature high spin-orbit torque due to quantum confinement in sputtered BixSe(1-x) films. *Nat. Mater.* 17: 800.

95 Wang, T.C., Chen, T.Y., Yen, H.W., and Pai, C.F. (2018) Comparative study on spin-orbit torque efficiencies from W/ferromagnetic and W/ferrimagnetic heterostructures. *Phys. Rev. Mater.* 2: 014403.

96 Zhu, L., Ralph, D.C., and Buhrman, R.A. (2018) Highly efficient spin-current generation by the spin Hall effect in $Au_{1-x}Pt_x$. *Phys. Rev. Appl.* 10: 031001.

97 Zhu, L.J., Sobotkiewich, K., Ma, X. et al. (2019) Strong damping-like spin-orbit torque and tunable Dzyaloshinskii-Moriya interaction generated by low-resistivity Pd1-xPtx alloys. *Adv. Funct. Mater.* 29: 1805822.

98 Zhu, L.J., Zhu, L.J., Sui, M.L. et al. (2019) Variation of the giant intrinsic spin Hall conductivity of Pt with carrier lifetime. *Sci. Adv.* 5.

99 Zhu, L.J., Zhu, L.J., Shi, S.J. et al. (2019) Enhancing spin-orbit torque by strong interfacial scattering from ultrathin insertion layers. *Phys. Rev. Appl.* 11.

100 C.-Y. Hu and C. F. Pai, In preparation (2020).

101 Khang, N.H.D., Ueda, Y., and Hai, P.N. (2018) A conductive topological insulator with large spin Hall effect for ultra-low power spin-orbit-torque switching. *Nat. Mater.* 17: 808.

102 Emori, S., Bauer, U., Ahn, S.-M. et al. (2013) Current-driven dynamics of chiral ferromagnetic domain walls. *Nat. Mater.* 12: 611–616.

103 Ryu, K.-S., Thomas, L., Yang, S.-H., and Parkin, S. (2013) Chiral spin torque at magnetic domain walls. *Nat. Nanotechnol.* 8: 527–533.

9

MRAM Applications and Production

9.1 Introduction

Like metal-oxide-semiconductor field-effect transistors (MOSFETs) in the silicon industry, magnetic tunnel junctions (MTJs) have become the workhorse of the magnetic recording drive industry since the mid-2000s. A decade afterward, MTJ devices are successfully integrated into silicon chip and have become the highest speed nonvolatile memory (nvM) the magnetic random-access memory (MRAM) memory. At the time of this manuscript preparation, MRAM has stepped out of the laboratory and is being manufactured by silicon chip foundries. Although magnetic memory is now under the limelight, a variety of MTJ-based magnetic sensors, such as current sensors and hard-disk drive (HDD) recording head sensor, remain as very large business, larger than the MRAM product. It is believed that with the participation of silicon foundries and memory producers, MRAM technology will proliferate and become part of the memory landscape.

Section 9.2 will describe the MTJ intrinsic properties, which manifest as the MRAM unique attributes; some appear as MRAM strength and others as weakness. Those MRAM attributes determine its position, among many emerging and incumbent nonvolatile memories, in the computer memory hierarchy.

Section 9.3 describes the opportunities of MRAM in the memory landscape. Section 9.3.1 describes the MRAM position in the embedded memory, and Section 9.3.2 describes that in the discrete memory. MRAM has been proven to cost less to manufacture and better perform than the embedded flash in the logic chips of advanced CMOS nodes. It is also a contender for replacing the embedded SRAM/DRAM of low-level cache of CPU chips, due to its much lower standby power. Some MRAM developers are eyeing specialty discrete DRAM mass market where nonvolatility is beneficial, for example, MRAM

Magnetic Memory Technology: Spin-Transfer-Torque MRAM and Beyond,
First Edition. Denny D. Tang and Chi-Feng Pai.
© 2021 The Institute of Electrical and Electronics Engineers, Inc.
Published 2021 by John Wiley & Sons, Inc.

as persistent memory, or the buffer memory of solid-state drive (SSD). Using discrete DRAM as a benchmark, we will point out the required further improvement to the density and chip architecture design of the current MRAM for this application.

In Section 9.3.3.1, we will show how the data nonvolatility property of MRAM improves the overall performance and reliability of the nonvolatile dual-in-line memory module (nvDIMM) applications. We will also show that it is an ideal device in certain new applications, such as Internet of Things (IoT) (Section 9.3.3.2) and the memory of artificial intelligence (AI) application (Section 9.3.3.3). Many battery-operated IoT Edge devices need working memory and local storage that consumes very low power, as well as cybersecurity to protect the frequent data transmission to host, such as the cloud. Deep learning AI function and data encryption of cybersecurity are compute-heavy and require frequent storing of many intermediate data. MRAM plays a dual role of working memory and storage device; it also has been shown to dissipate less power than those made of conventional SRAM and flash. The stochastic switching property of MRAM loans itself naturally to cybersecurity, such as physically unclonable function (PUF) and random number generator, for secret device ID and for counterfeit protection. Its device properties allow PUF to be manufactured with a high degree of immunity to side attack in the manufacturing process flow.

Section 9.4 describes the status of MRAM manufacturing, which began in 2006. MRAM is a relatively new entry to the memory market, compared to DRAM (began in the 1970s, 40 some years ago) and NAND flash (30 some years ago). MRAM manufacturing starts at lower-density product in larger CMOS nodes, learning to perfect the art at an affordable development cost, and gradually migrates to a higher-density product in advanced CMOS nodes and gains acceptance. It succeeds to penetrate the lower-density embedded memory and poises to contend the higher-density specialty discrete memory market.

9.2 Intrinsic Characteristics and Product Attributes of Emerging Nonvolatile Memories

In the 2000s, phase-change RAM (PCM), ferroelectric RAM (FeRAM), MRAM, and resistive RAM (ReRAM) were all considered as contenders of the "universal memory." The universal memory is expected to perform like SRAM and store data like NAND. In mid-2010s, after 10 years of research and development, it becomes clear that due to its unique intrinsic properties, MRAM stands out as working memory and the rest as "storage class memory" (SCM). In today's computer memory architecture, SRAM and DRAM are working memories. NAND is SCM.

9.2.1 Intrinsic Properties

These emerging nonvolatile memories are operated in very different physics principles. Table 9.1 lists the difference between MRAM and PCM.

It is interesting to point out that among all nonvolatile memories, only MRAM is a natural binary resistor device. The bi-axial film magnetic anisotropy drives magnetization toward the two ends of easy axis; thus, MRAM has only two stable states, and no other stable states in between. This biaxial switching property does not change with temperature or manufacturing procedure.

The resistance states of both PCM and ReRAM are not binary in nature. And their resistance value is strongly affected by the programming condition. Programming the device with correct resistance value may require multiple attempts of try-and-fail steps.

The next interesting distinction between MRAM and others is fundamental. The MRAM switching mechanism is based on the spin current exchange coupling to reverse the magnetization and thus the change of resistance state. The MRAM switching does not involve the movement of atoms, while all others do. In PCM, the material is melted and then quickly quenched in a short time to keep atoms in amorphous state or slowly cooled down to poly-grain state. The resistance of the former is high, and the latter is low. The resistance value depends on the quenching (or cooling) condition. Thus, it is not a binary resistor by nature. Similarly, in ReRAM switching, the metal-oxide dielectric breaks down and forms oxygen atom and vacancy pair. The atom and vacancy displacement from original site

Table 9.1 Intrinsic properties of MRAM vs. PCM.

Device physics	MRAM	PCM (3D X'pt)*
Binary device	Yes, binary resistance state	No, analog resistance state
Resistance switching mechanism	Atoms *do not move*, spin current switches magnetization	Atom moves, material melts and changes between poly-crystalline & amorphous
Device operating temp	T_{MTJ} 100C or below	625C to melt, 400C to re-crystallize
Energy consumption	Inherently lower $(J_{SW} \sim 4 \ 10^6 \ A/cm^2)$, 0.05mA (65nm)	Inherently higher $(J_{SW} \sim 2 \ 10^7 \ A/cm^2)$, 0.5mA (20nm, Optane)

* The consumer trade name "Optane" = PCM + selector "Ovanic threshold switch" (OTS).

changes the oxide resistance. Again, that involves atom movement. It is interesting to point out that the failure mechanism of MRAM is the switching mechanism of ReRAM: movement of oxygen atom and vacancy.

As a result, MRAM intrinsically requires the least amount of energy to switch, operated at the lowest voltage and the least current density among all emerging memories. Since there is no atom movement in operation, MRAM does not suffer from the consequence of atom movement in other emerging nonvolatile memory. Since the operating power density is lower and, thus, less self-heating, the device temperature rises up the least amount. It enjoys the highest life expectancy and product write endurance reliability among all [44].

In the 1R-1 T cell configuration, an MRAM gating transistor needs to pass the least amount of write current; thus, the transistor gate width is smaller, and therefore the cell is denser. The PCM product had developed a type of threshold switching diode to replace the gating transistor of the 1R-1 T cells. The threshold switch is a two-terminal diode. As a result of that, the cell size is no longer gauged by the transistor size. The metallic threshold switch can be made the same size as the PCM diode. The cell is defined as the cross-point of word line and bit line. Very high-density cell array had been developed and manufactured. In principle, this switch can be implemented to MRAM to achieve the high-density MRAM. That will require the development of tight control of threshold.

The weakness of MRAM is that the resistance ratio of the two states is not as large as ReRAM or PCM. The current production state-of-the-art TMR ratio is ~200%, meaning the resistance ratio is 3 : 1. When the MTJ is in series with a transistor, the transistor resistance dilutes the bit cell high- and low-current ratio to typically 2 : 1. When MTJ TMR ratio is raised to ~600–1000%, very high density MRAM X-point array will become practical.

9.2.2 Product Attributes

Here, we compare the emerging nonvolatile memories against the two dominant memories in the market: the volatile DRAM and the nonvolatile NAND flash.

First: access time. As mentioned previously, MRAM access performance is the only nonvolatile comparable to DRAM, in both read and write (See Figure 9.1). Due to its short access latency, it can command a higher bit price. Since it is nonvolatile, it could enjoy a price better than DRAM. Thus, MRAM is being considered for the specialty DRAM market.

Second: MRAM is the least battery-drain nonvolatile memory. Although its write current is higher than incumbents, its write voltage is around 0.5 V, an order smaller. Thus, write operation energy consumption, (V_w x I_w x t_{wp}), is the lowest, where V_w, I_w, and t_{wp} are write voltage, write current, and write pulse width, respectively (Figure 9.2). The combination of short write time and low energy

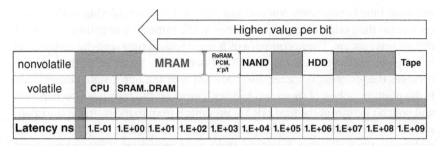

Figure 9.1 Access latency of MRAM versus other volatile and nonvolatile memory/storage devices. Only MRAM can reach DRAM access latency. All others are hindered by the long write cycle time.

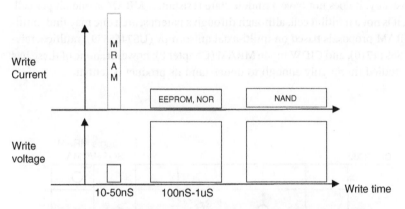

Figure 9.2 Although MRAM requires higher current to write, its write energy consumption is the lowest among all, due to its very low operating voltage and the very short write pulse. Compared to SRAM and DRAM, it requires no standby power. Thus, it is the lowest energy drain nonvolatile memory.

consumption, both an asynchronous and synchronous addressing scheme, works well for MRAM.

Third: MRAM can serve broad market: short data retention/high endurance and long data retention/median endurance applications. The MRAM write current density increases monotonically with the thermal stability factor Δ [1]. For shorter retention time memory applications, the required write current is lower, and thus the MgO degradation is milder; therefore, the MgO junction can tolerate more write cycles. For example, to compete in the high-density cloud journal applications, MRAM can be tuned toward low stability factor to reduce the write current and smaller cell size. The retention time is short, say, a day. For embedded flash replacement market, MTJ is tuned toward higher thermal stability. The data retention time is, say, years. Nonetheless, MRAM has been enjoying much higher endurance than other emerging memories at the same data retention. Figure 9.3 depicts this concept. The horizontal axis is the retention time, and the vertical axis is the endurance cycle. DRAM and SRAM occupy the upper-left corner. NAND occupies the lower-right corner. Field MRAM is at the upper-right corner. That is the position of best of these two metrics. However, the high power dissipation and poor bit density of the field MRAM prevents it from participating in the mainstream market. STT-MRAM is expected to cover the middle range, with better endurance than all other emerging nonvolatile memories. The multibit word SOT/VcMA MRAM is poised to be the next MRAM entry to where today's field MRAM occupies in Figure 9.3, since it offers much better endurance.

Fourth: Operated based on uniaxial anisotropy, MRAM is a binary resistor. Intrinsically, it does not have a middle state resistance. MRAM is one bit per cell. Thus, it is not a multibit cell, although through a patent search one may find multibit MRAM proposals based on multi-axial anisotropy (US7465589), multicoercivity (US6911710), and CIDW mode MRAM (Chapter 8); however, none of these has been studied thoroughly enough to understand its product potential.

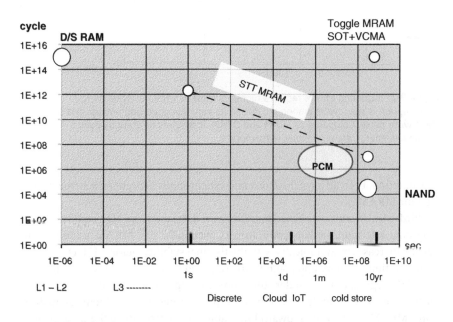

Figure 9.3 Write endurance versus data retention of MRAM versus volatile SRAM or DRAM and storage (NAND) applications.

9.3 Memory Landscape and MRAM Opportunity

The memory production landscape is dominated by NAND and DRAM. They are mainly used as memories in smart phones, tablets, and PC. Figure 9.4 shows the silicon consumption of DRAM and NAND products. Although each PC appears to consume more silicon per unit, the quantity of PC shipment is far below the smartphone. In net, each category consumes same order of memory bits, in the order of 10^{18} bits/year [2]. The huge market prompts the semiconductor industry to scale down memory and to reduce cost, thus, to sustain profit and maintain the Moore's law. Although NAND appears to be scalable further from 2D array to 3D multiple-bit per cell array, DRAM scaling hits a brick wall because the DRAM bit cell capacitor does not scale and is kept constant. At 20-nm node, the aspect ratio of the capacitor tube is more than 100. DRAM cannot be stacked like 3D NAND. This point become clear in Section 9.3.2.

The total volume of PC (include desktop, notebook, tablet) peaked around 400 M unit/year in mid-2010s. The volume has dropped since, displaced by the growth of cell phone volume. In 2017 the volume of phones was 1.4 billion units. The projected IoT production volume is higher than for phones.

Cellphones and IoTs are small form-factor devices, and they are operated by battery. There is less physical room to contain battery. Table 9.2 shows the typical energy stored in the battery of various devices. Based on the information from

	Si area, mm2		
Product	CPU	DRAM	NAND
Smart Phone	102	90	200 – 800
Tablet	102	90	200 – 1600
PC	90 – 160	800–1600	1600
30% annual cost reduction		Brick Wall	

Figure 9.4 Silicon consumption (in mid-2010s) in unit of mm², technology scaling (annual cost reduction) and opportunity of emerging nonvolatile memories in mobile products.

Table 9.2 The battery energy storage capacity in watt-hour (WH) of mobile devices: notebook, tablet, cell phone, and IoT (watch). Based on APPLE product.

Battery	Notebook	Tablet	Phone	Watch
WH	55.00	43.93	5.47	0.78

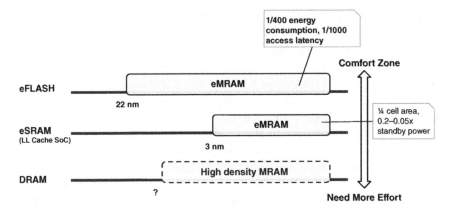

Figure 9.5 STT-MRAM replaces eflash and is being considered to solve the scaling and power issues of last-level cache of processor chip. More effort is needed for a competitive, discrete MRAM chip.

Apple products, the energy stored in the battery of small form factor appliance is far less. Thus, one can conclude that future electronics must be much more energy efficient.

MRAM is beginning to be accepted into the new world of small form factor, battery-operated mobile electronic appliances, outside of the traditional computer market. IoT is a good example. It is preferred to have single-layer nonvolatile, reasonably fast memory to work as a working memory and as a storage device. These two functions are currently covered separately by DRAM and NAND. Such a single-layer memory system requires no data transfer between the working memory and the storage device, eliminating the energy consumption of data transfer between the two and thus conserving the battery life.

9.3.1 MRAM as Embedded Memory in Logic Chips

Embedded MRAM has established a foothold. It comfortably replaces embedded flash and is being considered as embedded SRAM/DRAM cache memory in processor chips. Figure 9.5 shows the footsteps in the market.

9.3.1.1 Integration Issues of Embedded MRAM

Embedded MRAM is the first MRAM technology offering by silicon foundries. From the process integration point of view, integrating the MTJ into a SoC chip is easier than integrating a flash (called eflash) into the SoC chip. MTJ is physically sandwiched between layers of back-end metals and is part of the back-end-of-line

(a)   (b)

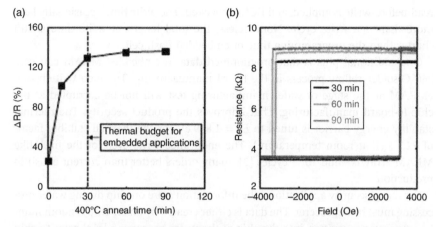

Figure 9.6 (a) TMR of 80 nm diameter MTJ as a function of annealing time at 400 C, and (b) R-H loop of MTJ versus post MTJ heat cycle of integration process [3].

(BEOL) process. The integration of the MTJ into the back-end process imposes no high temperature process cycles and therefore does not affect the front-end CMOS processes or the CMOS properties. On the other hand, eflash is a front-end FET device, and integrating a flash transistor into logic transistor process affects the properties of the logic transistor.

One of the challenges in integrating the MTJ into the advanced node of CMOS technology was to pass the 400 C heat cycle of the low-k copper back-end process. Soon, the MTJ film stack was improved, and the challenge was overcome. The latest MTJ film stack has been proven to work well in the low-k copper back-end (Figure 9.6).

Embedded MRAM performs two major categories of function: (i) to store code as an embedded flash, and (ii) to work as a last-level cache. The first replaces eflash, and the second replaces eSRAM/eDRAM of the processor chip. For these two applications, the MTJ properties are tuned differently – one for long retention time and the other for short retention time but high cell density.

9.3.1.2 MRAM as Embedded Flash in Microcontroller

Flash memory is frequently embedded into microcontrollers to store code and data. The embedded flash is called eflash. In such case, embedded MRAM (sometimes called eMRAM) performs the function of eflash. Compared to eflash, the advantages of eMRAM are comparably dense cell, faster access, and far better endurance cycle. One of the weaknesses of MRAM is its uncertain write success rate – a soft failure due to its stochastic write properties. For eflash applications, read-before-write is applied, and ECC is checked. The write time occasionally lasts more than one write cycle. Nonetheless, the total length of write time is still shorter than the programming time of embedded flash cell.

One additional plus is that programmed data in embedded MRAM can pass 260°C solder reflow process at the board manufacturing. The data written into eMRAM at the stage of wafer manufacturing test will not be corrupted at the chip-to-board manufacturing. That improves the product security. The thermal stability energy barrier is tuned to $E_b = 1.889$ eV (or the thermal stability factor of 72.7 k_BT at room temperature). The endurance write cycle of the flash-like MRAM array is ~mid 10^{7-9} cycles [3], many orders better than current eflash in production.

For many security applications, the information store on chip during wafer processing must be kept secret. The data is either created by an on-chip random number generator or written into the chip at the wafer processing level prior to chip packaging. Thus, there is one less chance of leaking the secret to security attacker. Again, the MTJ can do many tasks that flash cannot. For example, the MTJ array is a good random number generator. This point will be discussed in Section 9.3.3.2.

The combination of easiness in process integration (fewer masking layers), smaller cell size, faster access, and the passing of solder reflow makes embedded MRAM a better choice than embedded flash. Foundries have ceased offering embedded flash in logic chip at 28-nm node and beyond due to the availability of eMRAM technology.

9.3.1.3 Embedded MRAM Cell Size
In 2016, an 8-Mb embedded MRAM paper was published [4]. MTJs are integrated into 28-nm logic CMOS node. The MTJ diameter is 45 nm. The cell size is 0.0364 μm (or effectively $46F^2$, F = 28 nm). It represents the state of the art. The chip operates at 40 MHz. The chip provides 10-year data retention at 85°C. Endurance exceeds 10^7 write cycles. Scaling continues, and 22-nm eMRAM becomes available from foundries. There is no roadblock to further scaling beyond 16-nm. At such advanced nodes, logic transistors are mostly finFET. A finFET transistor is capable of supplying much more current than planar transistors and occupies less surface area. That improves the cell size scaling of eMRAM.

Figure 9.7 compares the embedded memory cell size between SRAM, eDRAM, and MRAM. eDRAM does not scale, because one cannot easily integrate DRAM into a logic chip as in discrete DRAM. Clearly, from the density point of view, MRAM is attractive. This point will be further discussed in Section 9.3.2.

9.3.1.4 MRAM as Cache Memory in Processor
There are several layers of cache memory in CPU. SRAM of different sizes are used for L1 and L2 caches, each with data access latency. An Intel i7 processor can run up to a 3.3 GHz clock. An i7 cache consists of three levels. L1 and L2 are associated

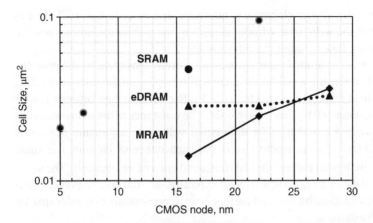

Figure 9.7 Comparison of embedded SRAM, eDRAM, and eMRAM cell size from 28- to 5-nm CMOS technology [6]. The SRAM size data is from different manufacturers.

with each processor core, and L3 is shared among cores. The size of L1 consists 32 kBytes data and 32 kByte instruction; that of L2 is 256 kbytes and L3 is 2 Mbyte per core [6, 7].

MRAM is the only one emerging nonvolatile memory that is being considered for CPU last level cache (LLC) application, since the access time and cell size of MRAM are comparable to those of eDRAM.

For desktop/server applications, speed is more important than power dissipation; STT-MRAM access latency is longer than SRAM L1, L2 caches and, thus, cannot meet the speed challenge. Nonetheless, it can fit into L3 (or last) cache (LLC) [50, 51]. For battery-operated mobile devices, such as phone and IoT, the access latency and number of access are relaxed. MRAM can move up to higher-level cache.

For CPU cache application, it has been pointed out that the data retention time requirement for the new generations of CPU may be shorter one second, not years [8]. The content in the LLC is totally revived (replaced by new content) within one second, the corresponding thermal stability factor is 21, and the write current density is expected to be greatly reduced [9]. Similarly, endurance requirement has been studied [10], and the last level cache requires less than 10^{10} cycles, depending on details of cache configuration. Such an endurance property may be within the reach, judging from the present pace of progress in industrial research labs [44].

9.3.1.5 Improvement of Access Latency

To push the MRAM into the higher-level cache applications, one needs to shorten the data access latency, or the read/write access time, so that it can compete against the SRAM L2 cache. The write latency of MRAM is limited by the write current density a tunnel barrier can endure and, thus, the endurance of the

MTJ tunnel barrier. The development of new switching-mode MRAMs (Chapter 8) should shorten the write time without degrading the endurance, since the write current does not flow through the tunnel barrier.

The read latency is limited again by the read current density set by the read disturb rate. When sensing a cell bit data, current is applied to the selected bit line, which is drained through the selected cell. A differential voltage between the bit line and a reference cell bit line is developed. The resistance of the reference cell is halfway between R_{ap} and R_p. The differential voltage as a function of time is $\Delta V_{BL}(t) = \frac{t}{C_{BL}}(I_{bit} - I_{REF})$, where C_{BL} is the capacitance of bit line, t is time, and I_{bit}, I_{REF} are bit-cell current and reference cell current, respectively. The sense amp is latched when ΔV_{BL} is large enough, typically ~80–100 mV. To prevent accidental write (read disturb), the read current is set approximately one order smaller than the write current.

Clearly, increasing difference in $(I_{bit} - I_{REF})$ is the solution to the sensing delay. The device solution is to raise the TMR of the MTJ. The circuit solution is to change I_{REF}. Here, we describe two solutions. First is called covalent reference sensing (Figure 9.8) [11]. The concept is depicted in Figure 9.8b. The sense amplifier is

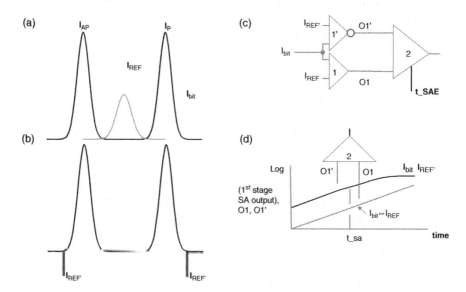

Figure 9.8 (a) Conventional sensing scheme with single reference value "I_{REF}," which is the average of P-state and AP-state cell current. The distribution of the AP- and P-state resistance, depending on sigma(R), may cause I_{REF} to overlap with the distribution of I_{bit}, and that causes sensing failure. (b) Two-reference sensing scheme (covalent references) doubles the sensing margin. The cell current I_{bit} is compared to I_P and I_{AP} in two first-stage amplifiers, and their output is the input of a second-stage amplifier (c). (d) Sensing timing diagram: The output signal amplitude of the two first-stage amplifiers develops at a different rate: the one with $I_{bit} - I_{REF}$ is slower, while the one with $I_{bit} \neq I_{REF}$ is faster. The second-stage sense amplifier is activated when one of the differential signal output reaches a threshold.

made of three amplifiers configured into two stages, two in the first stage and one in the second stage (see Figure 9.8c). Rather than comparing the cell current I_{bit} to that of a reference I_{REF} of $I_{REF} = 0.5(I_P + I_{AP})$, I_{bit} is compared to two references in the first-stage amplifier: one reference is I_{AP} and the other I_P. The output of the sense amplifier is related to the input signal amplitude as $(V_{out}(t) = V_{in}\ exp^{at})$, where a is a circuit-related constant. It takes a shorter time for V_{out} to reach the sensing threshold level when V_{in} is large, Figure 9.8d. For example, when a cell in R_{AP} state is compared to a reference I_P, the input to the sense amp is larger than that with a reference I_{AP}, which is ~0. The bottom first stage sense amp output reaches the second stage sensing threshold sooner than the top one. The differential signal from the two first-stage amplifiers is the input of the second-stage amplifier, which is enabled when one of the first-stage amplifier output reaches the threshold. The covalent sensing scheme is faster than the single-stage sense amplifier with $I_{REF} = 0.5(I_P + I_{AP})$.

The second circuit method is to use a 2 M-2 T twin-cell and self-reference sensing scheme (Figure 9.9). The differential signal on the two bit lines of a twin cell is twice of conventional cell. Thus, the sense amp out is faster. A 3.3 ns read access time with 2 M-2 T cell has been reported [12]. The penalty of this method is the doubling of the cell size.

Many innovations in write circuit have been reported. The innovations clearly address the weakness of the MTJ basic write properties. A self-timed write pulse width control in the write circuit detects the occurrence of switching and terminates the write current [13, 17]. This circuit tailors write pulse width to individual bit, reduces the write power, and also improves the MTJ reliability.

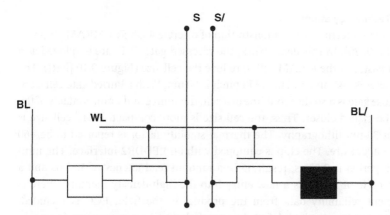

Figure 9.9 A 2 M-2 T (twin-cell), the two MTJs store complementary data. The write current is 50 µA at 3.0 ns write pulse width. The transient read current pulse is large, and the read pulse is short (~1 ns) to eliminate read disturb [12].

System-level studies [14, 15] find that by having a hybrid MRAM/SRAM cache, the performance SoC processor improves, and its power dissipation reduces. This is achieved by exploiting the relative strength and weakness of SRAM (fast write, large cell size, large standby power) and MRAM (slow and high energy write, smaller cell size, small standby power), and a hybrid cache may use MRAM for the cache for intensive read. Note that in practice, most of the cache blocks are read-intensive in nature, which enables us to design a large portion of the cache using MRAM, thereby exploiting its high density and low leakage power.

The sense signal of MRAM bit cell is limited by the TMR and the spread of the MTJ resistance. The two previous schemes improve the margin by effectively doubling the signal amplitude. Another way to improve the sensing is to compensate the offset the imbalance of the sense amplifier [16].

In 2019, three 1 Gb level STT-MRAM demo chips based on eMRAM technology are published [47–49] for industrial graded applications, MCU and IoT applications, and one for 2 MB L4 cache [50]. These demonstrations illustrate the broad applications of MRAM; each is tuned and trades off between data retention, endurance cycle, and access performance, as described in the previous sections.

9.3.2 High-Density Discrete MRAM

In general, for high-capacity memory applications, the market benchmarks are DRAM and NAND. MRAM is neither, as shown in Figure 9.3. Other memory attributes can be used as benchmark. Nonetheless, memory density and performance are still the most important attributes for MRAM to compete in the discrete market. Since the access performance of MRAM is close to that of DRAM, we will benchmark discrete MRAM versus discrete DRAM.

9.3.2.1 Technology Status

In 2016, the first technology demonstration of discrete 4 Gb STT MRAM chip was published [19, 20]. In this demo work, the recessed-gate FETs are employed as a gating transistor of the MRAM cell to reduce the cell size (Figure 9.10 [left]). The MTJ diameter is ~45 nm, and the MTJ pitch is 90 nm. With a buried gate cell transistor, whose gate is also the word line of a cell, the source and drain contacts of the FET can be placed closer. Thus, the cell size is more compact. A $9F^2$ cell size is realized at 30-nm lithography. The thermal stability factor is reported to be ~60 at room temperature. The chip is equipped with an LPDDR2 interface. The minimal clock period is 2.55 ns, resulting in a random read latency of 50.5 ns and a page-read cycle of 5 nA. As a new entry into the high-density memory, there is not sufficient reliability data from the product in the field; ECC is included. The demo work has reached defect density of about 1 part-per-million (ppm).

The following are a few interesting observations from this demo work:

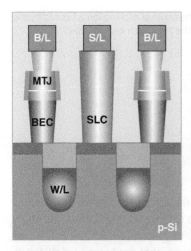 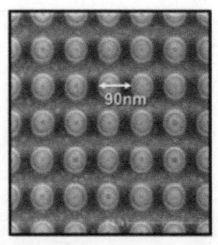

Figure 9.10 Schematics of 4 Gb MRAM cell cross section. (left) Top view of MTJ array. The MTJ pitch is 90 nm. (right) [19].

1) The MRAM in this demo work is placed on the first metal, while the embedded MRAM is placed between the upper-level metal layers. Metal vias are used to connect the gating transistor to the MTJ. Placing the MTJ on the first-level metal drastically reduces the cell size due to the unfriendly via landing pad size layout rules that introduce penalty in cell size. It shows under the proper layout design, MRAM can be as dense as DRAM.

2) The physical height of the MTJ pillar is orders smaller than the DRAM capacitor. The DRAM capacitor does not scale in value, ~30 fF. As the DRAM cell shrinks, the DRAM capacitor is made into a tube shape. The smaller the cell, the taller the tube. At ~20 nm node, the tube is as tall as 1 μm, and the capacitor height/width aspect ratio is ~100 [21]. Shrinking capacitor size is difficult and is considered as a roadblock of DRAM scaling. MRAM is placed on first metal in the high-density MRAM chip, and the MTJ thickness hardly changes with scaling. Such a difference is in favor of MRAM scaling (see Figure 9.10) over DRAM scaling.

One of the current challenges in high-dense MRAM technology is MTJ patterning. Etching the MTJ array in close pitch is a hurdle to overcome. The MTJ metallic film stack is etched with reactive-ion etch followed by a (nonreactive) low-angle ion beam etch (IBE). The by-product of RIE etching is nonvolatile and cannot be pumped out of the etching chamber. The by-product re-deposits on the sidewall of MTJ, shorting the thin MgO tunnel barrier. The low-angle IBE serves to remove the RIE-induced damage on the MTJ sidewall and to clean up the re-deposition.

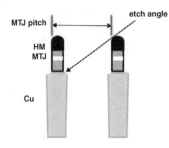

Figure 9.11 The angle of etching ion is limited by the aspect ratio of the MTJ pitch and the stack height.

But, the angle of IBE etching ion is limited by the aspect ratio of the MTJ height (sum of MTJ and hard mask) and the MTJ pitch (Figure 9.11) and, thus, the effectiveness of the IBE.

Continuing effort in finding an etching chemistry that produces a volatile etch by-product has led to the issuing of US patent to an equipment company (US 9806252 B2, "Dry plasma etch method to pattern MRAM stack"). Readers may find it novel.

To build a gigabit-level capacity MRAM chip with comparable die size as DRAM, the MTJ pitch should be in the order of 100 nm (Table 9.3). For reference, the 4 Gb MRAM demo chip was built with 90 nm MTJ pitch, the die size is 107.5 mm^2, and the die size of 8Gb DRAM @19 nm in production is between 52 and 55 mm^2.

9.3.2.2 Ideal CMOS Technology for High-Density MRAM

Currently, the MRAM cell size is dominated by the size of the gating field-effect transistor (FET). FET structure today has been developed into two major

Table 9.3 Estimated MTJ pitch for MRAM die size.

die (mm2)	50	100	150
Gb		pitch, nm	
1	140	197	241
4	70	98	121
16	35	49	60

categories: one for logic application and the other specifically for high-density DRAM. Traditional logic FET provides higher current per gate width, while the DRAM transistor structure provides dense cell layout and very low leakage current.

Contrarily to DRAM, which is a voltage device, MRAM is a current device. To realize high-density MRAM cell layout and fast switching, the ideal FET should provide both high current and dense layout. Figure 9.12 shows the higher current logic transistor (a) and lower current DRAM gating transistor (b). The recess gate transistor suffers from longer effective channel length and larger channel resistance. On the other hand, the subthreshold leakage of the recess gate transistor is lower. Without a gate structure protruding over the silicon surface, the source and drain contacts can be placed closer, and the wiring pitch is more compact. At same lithographic node, the MRAM cell size of recessed-gate transistor is about a factor of four smaller than that of the logic transistor (Figure 9.12c).

Compared to DRAM, one of the fundamental MRAM cell size limitations is from the cell wiring. Figure 9.13 illustrates the difference. DRAM is a 1C-1 T cell, and MRAM is 1 M-1 T cell, Although from cell circuit topology point of view, both cells require two metal wires (y) and one gate wire (x), they seem equivalent, and DRAM is denser due to the fact that one of DRAM cell wires is commonly shared among cells, called the plate. The common plate does not occupy the cell area. The MRAM cell needs both wires to perform WRITE operations. In addition, read and write current flow in MRAM bit lines, and thus their line resistance, affects access performance. The metal pitch issue can limit MRAM cells to at least 2× DRAM size.

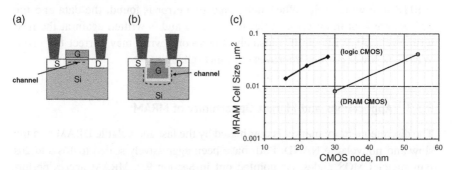

Figure 9.12 Cross section of (a) planar logic transistor, (b) recessed-gate DRAM transistor. The contact of DRAM transistor can be placed closer for dense memory layout. (c) A comparison of MRAM cell size in logic process a DRAM process.

(a)

(b)

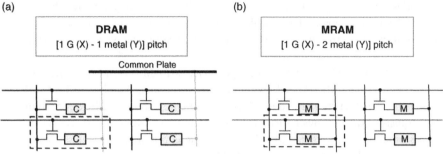

Figure 9.13 Cell schematics of (a) DRAM and (b) MRAM.

9.3.2.3 Improvement to Endurance and Write Error Rate with Error Buffer in Chip Architecture

For MRAM, the write error rate (WER) and reliability of MgO tunnel barrier impose conflicting requirements to the MTJ write current. To achieve low WER, the write should be large; however, that degrades the MgO tunnel barrier reliability. One way to relieve this conflict is to use a strong error correction code (ECC). The penalty of strong ECC is the long parity bits that degrade effective bit density and the long error correction computation time that degrades the access performance.

An error buffer solution has been proposed to relieve this requirement [52]. A weak ECC code is used. MRAM bits are written at a smaller current, at which the error rate is in the order of $\sim 10^{-3}$ instead of 10^{-9}. A VERIFY step follows each WRITE in a write cycle. When noncorrectable error is found, the data and the address are kept in an error buffer temporarily and is written again at the next write cycle. At WER of 10^{-3}, usually, the second write may correct the error. WRITE at lower current improves the MgO reliability.

9.3.3 Applications and Market Opportunity of MRAM

The discrete memory market is dominated by the fast and volatile DRAM and the slow and nonvolatile NAND. Both have been aggressively scaled to 10s and 20s nanometer CMOS nodes. As pointed out in Section 9.2, MRAM access performance overlaps that of DRAM. Its nonvolatility is what DRAM lacks. Section 9.3.2 had shown how the choice of CMOS impacts the density of MRAM. We will take a close look of possible market opportunity. To displace DRAM, one needs to invest in a production facility to scale MRAM as aggressive as DRAM so that the two can compete on the same footing.

Figure 9.14 Use of discrete DRAM.

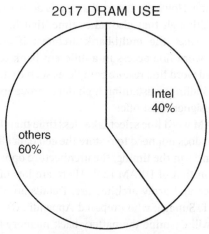

2017 DRAM USE

Intel 40%

others 60%

Let's examine the worldwide market of discrete DRAM. In 2017, the market size is $70B. About the annual production, computers consume 40%, the rest by other applications (Figure 9.14). For computer applications, density (cost) and data throughput (performance) are the two primary value metrics. In the "others" applications, the DRAM value metric includes many other considerations, such as power consumption during standby, data security, service quality. One of "others" is "specialty DRAM."

"Specialty DRAM" is a kind of DRAM that does not fit well for a particular application. DRAM is used simply because the lack of more suitable candidate; for example, the battery-back nonvolatility RAM, the high quality of service RAM (no access interruption, or no time to refresh), etc. It also means a particular DRAM specification that is not offered or no longer offered by major suppliers. One example of this category of DRAM is the DRAM for hard disk drives. On the other hand, we do not expect MRAM can replace the PC main memory in the "Intel" category. It is a comfort zone of DRAM.

To enter the discrete market and to gain acceptance, discrete MRAM should "look" like the DRAM in mass production, such that the cost overhead of switching from DRAM to MRAM is minimized. It means the MRAM input/output (I/O) interface should be DRAM compatible, and the commands are transparent to the DRAM controller. Today, the majority of DRAM productions is with double-data rate in data transfer (DDR) interface, and more and more are in low-power DDR (LPDDR) interfaces for mobile devices [22].

The page-mode access is the strength of DRAM. Once the word line is precharged and ready for access, the data in cells of the selected word line is capacitively coupled to the bit lines and stored into the sense amp buffer. The existence or absence of charge in the cell capacitor determines the cell state. This step takes little time. However, the read access is destructive; one must write back after each read. Although that also takes time, that hardly affects the read access performance, since the multi-bank architect is created to shield the restore time. A new word line access in a different bank can be issued before the previously accessed word line restore completes such that, outputs of two pages can be seamless (no idling), sustaining high data throughput. This is the feature that MRAM chip designer must offer.

MRAM word line select takes less time than DRAM. The read is nondestructive; MRAM does not need to restore the cell data after read. Nonetheless, due to many differences in the timing, the architecture optimization of MRAM could be different from that of DRAM [20]. There are lots of rooms for innovation in terms of interface and array architecture. Details are discussed in Appendix C (courtesy of Dr. T. Sunaga, who prepared Appendix C).

We will examine the battery-back memory applications in Section 9.3.3.1, the IoT and cybersecurity applications in Section 9.3.3.2, in-memory computing, AI in Section 9.3.3.3, and MRAM-based memory-driven computers in Section 9.3.3.4.

9.3.3.1 Battery-Backed DRAM Applications

One of the DRAM applications is battery-backed memory. It is built in the form of nvDIMM (see Figure 9.15). The module is made up of DRAM, NAND flash, a power interrupt sensor, and a controller that activates a backup power source

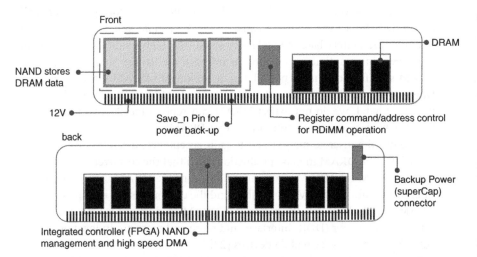

Figure 9.15 A schematic of nonvolatile DIMM, made of up DRAM chips, NAND chips, and a controller that receives power interruption signal and activates the transfer of data in volatile DRAM into nonvolatile NAND for storage. The backup power source (on the motherboard) is not shown.

on the module once power interruption is detected. The backup power source can be either a pack of batteries or capacitors, called "SuperCap," located in the system. When power is on, data is stored in volatile DRAM, since its access latency is much faster than the nonvolatile NAND. When power interrupt occurs, the controller switches into the backup power to copy the contents in DRAM to the NAND. Thus, the module is nonvolatile.

MRAM is idea for this application, since MRAM can match DRAM cell in access performance. The IEEE standard committee has defined type-N module (nvDIMM-N). If one replaces DRAM in the nvDIMM-N with MRAM, then NAND chips, batteries, or caps are no longer needed, and the memory controller becomes much simpler. The cost of MRAM is preferably comparable to the combined cost of NAND, DRAM, controller, and backup power source; among these the dominant cost is DRAM. Thus, the primary challenge is to shrink the MRAM cell such that the MRAM wafer cost is within few times of the DRAM wafers.

One major nvDIMM application is cache for SSDs. For this application, DRAM can be replaced by MRAM. It has been shown that the I/O data transaction rate (IOPS) of SSD improves nearly 2× [23] because once data is stored in the fast MRAM, SSD can acknowledge the data reception and is ready for another data transaction. Data is safe in MRAM even when power interruption takes place afterward. If DRAM is the buffer, the SSD cannot acknowledge data reception until data is written from the DRAM into the NAND. And that takes time.

To win this application, the interface of discrete MRAM chip should follow the spec of the DDR-DRAM used in nv-DIMM-N. The overall reliability of the MRAM should also match that of the DRAM. From a reliability point of view, it is hard to compete against the 50-year maturity of DRAM technology. Nonetheless, the MRAM must address this point: the reliability of thin MgO tunnel barrier. The time-dependent-dielectric-breakdown (TDDB) study methodology of the DRAM capacitor is well-established. When implemented in the MgO tunnel barrier, the large current flowing through the tunnel barrier heats up the junction. The accumulated product reliability data is still coming out of the manufacturing line. It is believed that the MTJ reliability will continue to improve.

9.3.3.2 Internet of Things (IoT) and Cybersecurity Applications

As the Internet is becoming a major medium of mobile communication, more and more wireless electronic appliances (things) are connected to the Internet and communicate through the Internet. "IoT" stands for "Internet of things" and includes things around us in our daily life. IoTs are usually battery-operated mobile electronic devices equipped with sensors, processors, and radios. They periodically collect personal or environmental data and communicate with other mobile devices or data centers ("the cloud").

Due to the limited form factor and thus battery charge, the local processor in an IoT works at a very low duty cycle. The MRAM technology has been recognized as the best candidate for IoT applications due to (i) the programming energy is the lowest among all nonvolatile memories, and (ii) although the endurance of today's STT-MRAM is not infinite, it is more than sufficient to work in the IoT environment as storage and as SRAM of the local processor. It has been demonstrated that MRAM with power-gating peripheral circuit design techniques can lower the standby power of the MRAM chip to the nano-watt level [24].

In additional to the conservation of battery life, MRAM technology is also important to the security of IoT device communication. Internet connection authentication requires user identification. Internet messages are encrypted to protect privacy.

"Hardware security" is regarded as indispensable to IoT. Modern Internet appliances demand compact and low-power security devices that provide PUF for authentication/counterfeit protection and for the generation and secure storage of secretive data encryption keys [25]. For the authentication applications, the device should provide a unique and unclonable device "fingerprint" (or secret ID). For the encryption key application, it provides an unpredictable and nonrepeatable random number. A good PUF is a device not likely to be modeled succinctly, nor be predicted or replicated, even using identical hardware.

The die-to-die variation of silicon circuits has been the working principle of the silicon PUF devices. The popular Si PUF types, including the arbiter, ring oscillator, SRAM, flip-flop, and latch, have been evaluated for PUF [26], and SRAM PUF shows the best performance [27] and has been implemented in commercial microcontrollers [28–31].

An SRAM cell consists of two inverters with a feedback loop. The transfer curves, or butterfly curve, of them is designed to be identical, but due to manufacturing mismatch, are not entirely the same on a wafer. When the cell is powered up, the initial state of the cell is either "1" or "0." Thus, the SRAM power-up bit pattern is random, not predictable, and varies from die to die.

As CMOS manufacturing process matures, the device variation diminishes, and the power-up state of some bits does not settle to the same state at every power-up. Thus, the bit pattern of an SRAM PUF varies from time to time. The power-up bit pattern becomes not repeatable. Thus, using the power-up SRAM bit pattern for ID applications becomes problematic. Other factors, such as temperature, power supply variations, circuit board noise, etc., can also lead to inconsistence in power-up bit patterns. This bit-error issue usually requires costly overhead such as "helper data" [29] to identify the consistent bits and save its address for later use. In addition, SRAM devices are volatile and cannot store the secret key and therefore must be paired up with a nonvolatile memory to securely store the security key.

Many MRAM PUF schemes are proposed based on random distribution of MTJ R_{ap} [31–34]. Some address the read-back error problem. They can reduce error bit down to a small number but cannot completely eliminate error. We will describe a different MRAM PUF scheme that eliminates the read-back error entirely. Rather than based on the distribution in R_{ap} state resistance, it is based on the stochastic switching properties variation of switching threshold of individual bits of MRAM [35, 36] (see Figure 9.16).

The random pattern in the MRAM can be created with two methods. One is H-method (with H-field write), and the other V-method (with spin-transfer-torque write). The H-method is to apply an external hard-axis field to all bits of the MRAM array and then release. Upon the release of the external field, the magnetization of an MTJ free layer returns to the easy axis direction and is in either of the two binary resistor states, ideally 50% each. In actuality, the bit pattern is dictated by the random position-dependence of device parameters, such as a small deviation of the easy axis from normal direction of the film in a perpendicular MTJ [36]. Thus, the bit pattern in each MRAM die is different.

The V method is based on both the stochastic switching property of STT-write of MRAM and the process-induced random distribution of MTJ device parameters. This randomization procedure is executed with writing MTJ with a current density at high WER, say, near 50%. The former makes bit pattern nonrepeatable on a given die under identical stimulation, and thus *unclonable*; the latter results in different bit pattern on different die. MRAM random number generator is also called "spin dice."

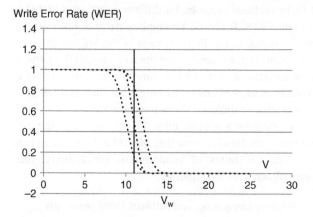

Figure 9.16 MRAM PUF based on random minor deviation in write error rate of individual bits in an array and stochastic switching properties (WER). One can create nonrepeatable random bit pattern by writing the array in the high WER range.

(a) Voltage (V) method

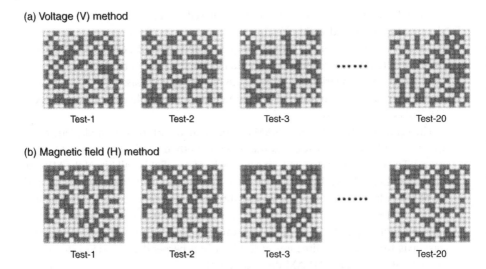

| Test-1 | Test-2 | Test-3 | Test-20 |

(b) Magnetic field (H) method

| Test-1 | Test-2 | Test-3 | Test-20 |

Figure 9.17 Random bit pattern generated by V-method (a), and H-method (b) [35].

Figure 9.17 illustrates the bit patterns of 256-bit MRAM PUF randomized with these two methods, and each method is repeated 20 times with identical stimulation on the same PUF. The bit pattern is different every time and does not repeat.

Although the MRAM PUF bit pattern is highly random, the intra-chip Hamming distance (defined as the percentage of unchanged bit in the array over two consecutive randomization steps) is ~50% more than 20 times. The inter-chip Hamming distance of 4 PUFs (defined as bit-by-bit difference between PUFs) is also ~50%, as shown in Figure 9.18. It should be emphasized that once randomized, the bit pattern can be stored in the PUF segment of the MRAM chip. As long as the address of the PUF bits is inaccessible from the chip pins, the bit pattern is securely stored and stays unchanged until the next randomization procedure is exercised. Since the bit pattern is unpredictable, cloning embedded MRAM PUF bits onto another device is extremely difficult, if not impossible.

Embedded MRAM array is one good way to perform this function since the pattern is not detectible from chip pins. Studies show that MRAM PUF performs better than all other candidates, in terms of randomness, repeatability, and uniqueness, and takes least silicon area to implement.

9.3.3.3 Applications to In-Memory Computing, and Artificial Intelligence (AI)

As MRAM technology becomes mature, many new applications are being considered. One is in the field of AI [18, 45]. AI memory needs are quite different from traditional workloads, requiring faster access to data and nonvolatility to reduce

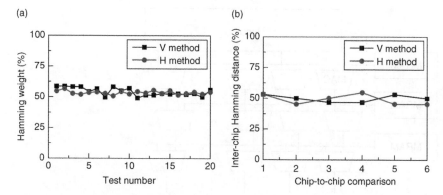

Figure 9.18 (a) Intra-chip Hamming distance of a 256b chip over many randomization attempts; (b) Inter-chip Hamming distance of different chips [35].

energy consumption due to the very high data traffic between the processing unit and the data storage unit. MRAM, as a fast nonvolatile memory, fits the application well. One implementation is shown in Figure 9.19 [37]. In the convolution neural network (CNN), the coefficient memory SRAM is replaced by a denser STT-MRAM. The stand-by power of the CNN function is reduced by 8× at 25C and 20× at high temperature. The chip area of original 9 MB SRAM for coefficient can fit 40 MB MRAM. The energy-consuming data transfer between SRAM and storage, on chip or off chip, is also eliminated. The chip achieves 9.3 tera operation per second per watt (TOPS/W) with peak power less than 300 mW. Thus, a fast nonvolatile MRAM enables energy-efficient computing.

9.3.3.4 MRAM-Based Memory-Driven Computer

Looking further out in the future, MRAM is scaled to the sub-10 nm node, and the MRAM discrete chip is in the tens to hundreds Gb range. Once such high-density MRAM becomes available, it could accelerate the penetration into the traditional DRAM market.

High-density MRAM can serve as persistent memory in memory-centric computing [38, 39]. Today, processor-centric (von Neumann) computers spend a lot of time and power moving data back and forth from storage to working memory. Power dissipation has been a major roadblock that hampers the scaling of processors. When high-density MRAM becomes available as working memory, such data transfer is eliminated. In addition, the software overhead that manages the communication between working memory and storage is simplified.

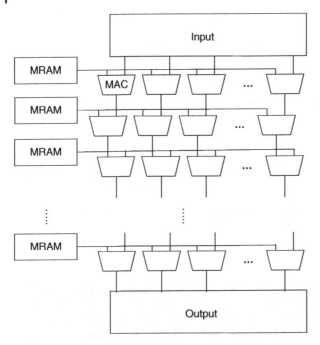

Figure 9.19 Convolutional neural network domain-specific architecture (CNN-DSA) accelerator for extracting features out of an input image. It processes 224 × 224 RGB images at 140 fps with ultra-power-efficiency, a record of 9.3 TOPS/W and peak power less than 300 mW [37]. Such energy-efficient computing is gaining momentum in IoT and edge, mostly are battery operated.

9.4 MRAM Production

This section gives a historical prospective of how and when MRAM turns into real product from academic research.

9.4.1 MRAM Production Ecosystem

STT-MRAM research has moved from academic laboratories into industrial research and development facilities. High-throughput MTJ film sputtering tools with throughput in tens wafer per hour are becoming available. The production cost reduces. The MTJ film stack is becoming more sophisticated, and the number of layers increases way beyond imagination (some MTJ stacks consist of more than 50 layers). Today, the MTJ is far more robust than ever, and the defective MTJ on a chip can be as low as 1 ppm. The electronic design automation system (EDA) tool company has incorporated the MTJ SPICE model for chip design. Silicon IP

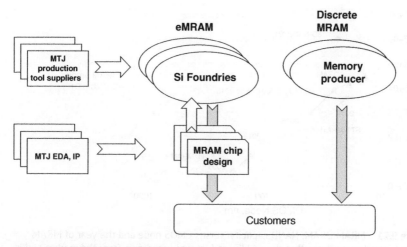

Figure 9.20 MRAM production ecosystem.

companies offer predesigned IP, including MRAM macros for circuit design companies. The technology ecosystem has taken shape since mid-2010. The progress in STT-MRAM technology accelerates.

Figure 9.20 shows the MRAM production ecosystem. Silicon foundries lead the mass production. Two kinds of companies participate in the technology development and production of MRAM. Foundries find that embedded MRAM at 40 nm and more advanced CMOS node performs better than and costs less than embedded flash. Besides, MRAM can be programmed during wafer production. That further reduces the cost and improves the product security from side attacks.

9.4.2 MRAM Product History

The common product practice is to develop an MRAM product on a matured CMOS platform node. For example, one may develop an MRAM product at 28-nm platform node after 28-nm products are in the stage of mass production so that the development effort can be focused on the MTJ integration issues, not the transistor issues. Consequently, MRAM scaling always lags the DRAM in terms of device miniaturization and thus the MRAM density.

Figure 9.21 shows a short history of MRAM product capacity and CMOS nodes in manufacturing. The data are plotted together with DRAM and NAND products. The first generation of the MRAM product is field MRAM. It was introduced in 2006. It was scaled from 4 to 16 Mb and, later in 2019, to 32 Mb. In 2013, the second-generation MRAM, 64 Mb STT-MRAM, was introduced; and 256 Mb in 2016, 1 Gb in 2019. These MRAMs are built with foundry logic CMOS processes, and

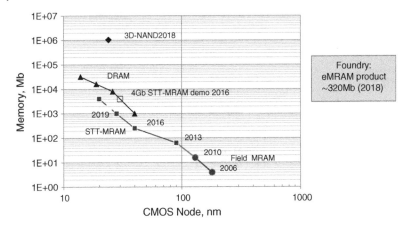

Figure 9.21 MRAM DRAM, NAND capacity, versus CMOS node and the year of MRAM product introduction. Note that the 64 Mb product was withdrawn from the market and is replaced with 256 Mb in 2016.

thus the scaling follows foundry's logic CMOS scaling pace, from 180 nm node for 4 Mb field MRAM to 20 nm for 1 Gb. In this range of dimension, logic scaling is behind DRAM scaling in time.

During the time of this manuscript preparation, the highest-density 4 Gb MRAM was demonstrated at 30 nm node [19]. For the first time, MRAM is shown to be comparable to DRAM density when the MTJ is integrated into a DRAM CMOS. This work demonstrates the importance of the CMOS process on the MRAM density. Another significant point is that MRAM can achieve DRAM density at the same CMOS node.

As it has been described in Chapter 8, potentially the spin-orbit-torque (SOT) MRAM with voltage modulation of anisotropy bit selection mechanism may become mature enough to be introduced as a third-generation MRAM product.

9.4.2.1 First-Generation MRAM – Field MRAM (Also Called Toggle MRAM)

The first-generation MRAM technology, field MRAM, went into production in 2006. It was a 4 Mb field-write mode MRAM product built at 180 nm logic CMOS node. It is also called Toggle MRAM, since it is based on the toggle switching scheme as described in [40]. The storage element is in-plane magnetized MTJ with the AlO_x tunnel barrier. The I/O interface is an SRAM–like asynchronous I/O interface, and read and write access times were both 35 ns at the time of product introduction, subsequently reduced to 25 ns. The product is designed to replace battery-backed "nonvolatile" SRAM. The product was proven rugged: it provides 20 years of data retention time, and it can be operated over infinite R/W cycles.

Later, the temperature spec extends from commercial spec (0 C–70 C) to industrial spec (−40 C to 85 C) and eventually automobile spec −40 C to 115 C). The combination of fast access, infinite write endurance cycles, and 10-year data retention time is a unique set of product attributes and not achievable from all other memory available in the market.

The chip is mounted in a special chip package, which shields the external magnetic field; thus, the chip does not fail under a magnetic field up to 250 Oe. A follow-up 16 Mb AlO_x MTJ-based product at 130 nm CMOS node was introduced in 2010. Since then, the part number proliferates, all based on the same MTJ. The first generation MRAM (field MRAM) proves the reliability of the MTJ.

The product penetrates into the specialty market and offers a unique application that no incumbent memory product can replace: low-latency access, nonvolatile, radiation hard, and high-endurance memory. However, the chip active power is large. Large size transistors are required to drive the write current. Thus, the peripheral circuits occupy most of the chip area, and the cell array utilization (ratio of cell area to total chip area) is poor. These two drawbacks keep field MRAM from penetrating into the mainstream memory market.

It is a high-power memory chip. To generate the necessary write field (Oersted field), the current on the selected write word line (WWL) is in the 10 mA range, and the current on the selected bit line is in the same range (Figure 5.7). Although such current is orders smaller than that of the write head of the hard disk drive, it is high among nonvolatile semiconductor memories. The scaling effort of field MRAM ceased when the spin-transfer-torque mode MRAM technology began to mature in 2010s. The latter has proven to be superior in chip power and bit density. Industry has switched its development focus away from field MRAM and has developed the more energy-efficient spin-torque-transfer MRAM in 90-nm CMOS nodes and beyond.

9.4.2.2 The Second-Generation MRAM – STT-MRAM

The second-generation MRAM (STT-MRAM) proves the scalability of the MRAM technology. The combination of the large device endurance, long data retention time, and high-speed access properties of MRAM is unique, not available in the legacy memory market.

In 2012, the first commercial spin-torque transfer mode (STT) MRAM product chip was announced [41]. The storage element is an in-plane MTJ with MgO as the tunnel barrier. The 64 Mb STT-MRAM chip is designed with a DRAM-like synchronous I/O interface. The chip is built on 90-nm logic CMOS technology.

Electronic system designers began to exploit this unique set of properties for new applications. The conventional von Neumann computer memory architecture is being re-examined in light of this new set of device properties. Many new applications of the MTJ technology began to emerge. Notable new applications are for the battery-powered mobile electronics, such as IoT devices [42] and for solving current most pressing power-dissipation issues of processors. The former addresses low battery life, and the latter addresses the system cooling costs. The AI electronics demands fast, dense, and nonvolatile memory. This set of memory attributes happens to fit STT-MRAM well. STT MRAM built on silicon on insulator (SOI) can be operated up to 150 C [46]. It finds automobile Grade-I controller applications.

9.4.2.3 The Potential Third-Generation MRAM – SOT MRAM

In a laboratory, SOT devices showed interesting properties such as better reliability and its potential of better density than STT-MRAM [5]. There are a few drawbacks:

1) The operating current density is undesirably very high today. The high operating current issue can be solved by a multibit word cell topology (see Figure 9.22). The high write current is shared by multiple bits on a word line; thus, the effective write current of each bit is lower, comparable to that of the STT-MRAM bit. The voltage-control magnetic anisotropy (VCMA) scheme is needed to select bits when WRITE (Chapter 8). Other potential solution of the high SOT efficiency material solution, such as topological insulator, is in the academic research stage.

2) H Field-free operation of perpendicular MTJ in SOT mode is yet mature. In-plane MTJ SOT cell is the only practical solution at the time of this manuscript preparation.

3) The high-yield integration process of SOT device into CMOS is yet reported. The etching process must be developed to etch MTJ and stops exactly on

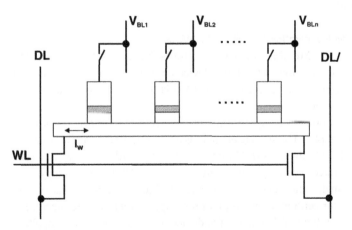

Figure 9.22 Multibit word SOT MRAM.

the nm-thick heavy-metal SOT stripe. One possible solution is to change the device structure as reported in [43], the step-SOT structure easies the etching requirement. Nonetheless, a thorough investigation is required to qualify the structure.

Although there are material solutions to address (i), many solution schemes to address (ii), and the device structure solution to (iii) as stated in Chapter 8, all these solutions are at the academic or early industrial research stage. All require further extensive study to understand the caveat and trade-offs. Until the best solution is identified, we are refraining from making any availability projection.

Homework

Q9.1 The MTJ endurance requirement is highly dependent on the size of MRAM array and how it is used. A CPU of an IoT with a single 512 Mbyte embedded MRAM as memory. No other memory is in this IoT device. The memory is accessed at 10^7 access per second, and 40% of the access is write operation. The memory I/O bus is 32 b wide. Assume that the cells are uniformly written. What is the endurance requirement of this particular MRAM?

A9.1 The frequency of write action is 0.4×10^7 Write/s. The i/o bus = 32b. In average, each bit of the IoT (512 M × 8 bits) is written = $32 \times 0.4e7/(512$ M × 8) = 2.98 10^{-2} Write/(s.bit). 3-yr = 9.46 10^7 s. The average #Write/bit = 2.98 $10^{-2} \times 9.46$ 10^7 (sec) =2.8 10^6 WRITE in three years. Endurance of 10^7 will suffice. For other cache endurance requirements, one may find them in reference [10].

References

1 Guenole Jan, Luc Thomas, Son Le, Yuan-Jen Lee, Huanlong Liu, Jian Zhu, Jodi Iwata-Harms, Sahil Patel, Ru-Ying Tong, Vignesh Sundar, Santiago Serrano-Guisan, Dongna Shen, Renren He, Jesmin Haq, Zhongjian Jeffrey Teng, Vinh Lam, Yi Yang, Yu-Jen Wang, Tom Zhong, Hideaki Fukuzawa, and Po-Kang Wang (2018). Demonstration of Ultra-Low Voltage and Ultra Low Power STT-MRAM designed for compatibility with 0x node embedded LLC applications. *Symposium on VLSI Technology Digest of Technical Papers*. VLSI. p. 65.
2 IEEE ISSCC conference (2015). *Digest of Memory Forum* (February 22). IEEE.

3 Thomas, L., Jan, G., Zhu, J. et al. (2014). Perpendicular spin transfer torque magnetic random access memories with high spin torque efficiency and thermal stability for embedded applications, (invited). *J. Appl. Physiol.* 115: 172615.

4 Y. J. Song, J. H. Lee, H. C. Shin, K. H. Lee, K. Suh, J. R. Kang, S. S. Pyo, H. T. Jung, S. H. Hwang, G. H. Koh, S. C. Oh, S. O. Park, J. K. Kim, J. C. Park, J. Kim, K. H. Hwang, G. T. Jeong, K. P. Lee, and E. S. Jung (2016). Highly Functional and Reliable 8Mb STT-MRAM Embedded in 28nm Logic. *Digest of IEDM 2016, paper 27.2.* IEEE. p. IEDM16–663, 978-1-5090-3902-9/16/$31.00 ©2016 IEEE

5 H. Yoda, N. Shimomura, Y. Ohsawa, S. Shirotori, Y. Kato, T. Inokuchi, Y. Kamiguchi, B. Altansargai, Y. Saito, K. Koi, H. Sugiyama, S. Oikawa, M. Shimizu, M. Ishikawa, K. Ikegami, and A. Kurobe (2016). Voltage-Control Spintronics Memory (VoCSM) Having Potentials of Ultra-Low energy-Consumption and High-Density. *IEEE Digestof IEDM.* IEEE.

6 WikiChip, 5 nm lithography process, https://en.wikichip.org/wiki/5_nm_lithography_process (accessed July 27, 2020).

7 John L. Hennessy, David A. Patterson (October 30, 2012). Memory Hierarchy Design - Part 6. The Intel Core i7, fallacies, and pitfalls. EDN Magazine.

8 Adwait Jog, Asit K. Mishra, Cong Xu, Yuan Xie, N. Vijaykrishnan, Ravishankar Iyer, Chita R. Das (2011). Cache Revive: Architecting Volatile STT-RAM Caches for Enhanced Performance in CMPs. CMU/*Intel Technical Report* CSE-11-010.

9 Clinton W. Smullen, IV, Vidyabhushan Mohan, Anurag Nigam, Sudhanva Gurumurthi, Mircea R. Stan (2011). Relaxing Non-Volatility for Fast and Energy-Efficient STT-RAM Caches. *IEEE 17th International Symposium on High Performance Computer Architecture.* IEEE. p. s: 50–61.

10 J.J. Kan, C. Park, C. Ching, J. Ahn, L. Xue, R. Wang, A. Kontos, S. Liang, M. Bangar, H. Chen, S. Hassan, Kim, M. Pakala, and S. H. Kang (2017). Systematic Validation of 2x nm Diameter Perpendicular MTJ Arrays and MgO Barrier for Sub-10 nm embedded STT-MRAM with practically unlimited endurance. *IEEE Transactions on Electron Devices*, IEEE. Vol. 64, No. 9.

11 Chankyung Kim, Keewon Kwon, Chulwoo Park, Sungjin Jang, Joosun Choi (2015). A covalent-bonded cross-coupled current-mode sense amplifier for STT-MRAM with 1T1MTJ common source-line structure array, IEEE International Solid-State Circuits Conference - (ISSCC) Digest of Technical Papers, paper 7.4, IEEE.

12 H. Noguchi et al., Highly Reliable and Low-Power Nonvolatile Cache Memory with Advanced Perpendicular STT-MRAM for High-Performance CPU, *IEEE Symp. VLSI Circuits Dig. Tech.* Papers, June 2014.

13 R. Bishnoi, F. Oboril, M. Ebrahimi and M. B. Tahoori (2016). Self-Timed Read and Write Operations in STT-MRAM, *IEEE Transactions on Very Large Scale Integration (VLSI) Systems*, vol. 24, no. 5, pp. 1783-1793, May 2016, doi: 10.1109/TVLSI.2015.2496363.

14 Xuanyao Fong, Yusung Kim, Rangharajan Venkatesan, Sri Harsha Choday, Anand Raghunathan, and Kaushik Roy, Spin-Transfer Torque Memories: Devices, Circuits, and Systems *Proceedings of the IEEE* (Volume:PP, Issue: 99), Page(s): 1–40, ISSN: 0018–9219 DOI: 10.1109/JPROC.2016.2521712, Date of Publication:07 April 2016.

15 Seunghan Lee, Hwaseong 445–330, South Korea.; Kyungsu Kang; Jongpil Jung; Chong-Min Kyung Hybrid L2 NUCA Design and Management Considering Data Access Latency, Energy Efficiency, and Storage Lifetime, IEEE TRANSACTIONS ON VERY LARGE SCALE INTEGRATION (VLSI) SYSTEMS, *IEEE Transactions on Very Large Scale Integration (VLSI) Systems* (Volume:PP, Issue: 99) Page(s):1–14, 24 March 2016

16 Chang, M.-F. et al. (March 2013). An offset-tolerant fast-random-read current-sampling based sense amplifier for small-cell-current nonvolatile memory. *IEEE J. Solid State Circuits* 48 (3): 864–877.

17 Q. Dong et al. A 1Mb 28nm STT-MRAM with 2.8ns read access time at 1.2V VDD using single-cap offset-cancelled sense amplifier and in-situ self-write-termination. *IEEE International Solid - State Circuits Conference - (ISSCC)*, San Francisco, CA, 2018, pp. 480-482, doi: 10.1109/ISSCC.2018.8310393.

18 Ranjana Godse, Adam McPadden, Vipin Patel, Jung Yoon, Memory Technology enabling the next Artificial Intelligence revolution, *Digest of 2018 IEEE Nanotechnology Symposium (ANTS)*, pp.1–4

19 S.-W. Chung, T. Kishi, J.W. Park, M. Yoshikawa, K. S. Park, T. Nagase, K. Sunouchi, H. Kanaya, G.C. Kim, K. Noma, M. S. Lee, A. Yamamoto, K. M. Rho, K. Tsuchida, S. J. Chung, J. Y. Yi, H. S. Kim, Y.S. Chun, H. Oyamatsu, and S. J. Hong, 4Gb bit density STT-MRAM using perpendicular MTJ realized with compact cell structure, *IEEE IEDM 2016 Technical Digest*, 27.1, 2016.

20 Kwangmyoung Rho, Kenji Tsuchida, Dongkeun Kim, Yutaka Shirai, Jihyae Bae, Tsuneo Inaba, Hiromi Noro, Hyunin Moon, Sungwoong Chung, Kazumasa Sunouchi, Jinwon Park, Kiseon Park, Akihito Yamamoto, Seoungju Chung, Hyeongon Kim, Hisato Oyamatsu, Jonghoon Oh, A 4Gb LPDDR2 STT-MRAM with Compact 9F2 1T1MTJ Cell and Hierarchical Bitline Architecture, *IEEE Digest of ISSCC*, 2017, paper 23.5

21 Sungjoo Hong, et al. (2010). Memory technology trend and future challenges, *IEEE Digest of IEDM*. IEEE. DOI: https://doi.org/10.1109/IEDM.2010.5703348

22 JESD209-2F Low Power Double Data Rate 2 (LPDDR2) standard, JEDEC Solid State Memories Committee (JC-42.6), June 2013

23 Mertens, R. (2012). Buffalo introduces new SSDs that use MRAM cache, https://www.mram-info.com/buffalo-introduces-new-ssds-use-mram-cache (accessed 2020)

24 Shinobu Fujita, Kumiko Nomura, Hiroki Noguchi, Susumu Takeda, Keiko Abe, Novel Nonvolatile Memory Hierarchies to Realize Normally-Off Mobile Processors,

ASP-DAC 2014, Toshiba Corporation, R&D Center, Advanced LSI technology laboratory

25 Charles Herder et al: Physical Unclonable Functions and Applications: A Tutorial, *Proceedings of the IEEE* | Vol. 102, No. 8 August 2014

26 G. E. Suh and S. Devadas (2007). Physical unclonable functions for device authentication and secret key generation. *IEEE Proc. Design Automation Conf.*, IEEE. pp. 9–14, (2007).

27 Stefan Katzenbeisser, Ünal Kocabaş, Vladimir Rožić, Ahmad-Reza Sadeghi, Ingrid Erbauwhede, and Christian Wachsmann, PUFs: Myth, Fact or Busted? A Security Evaluation of Physically Unclonable Functions (PUFs) Cast in Silicon, http://www.iacr.org/archive/ches2012/74280281/74280281.pdf; https://eprint.iacr.org/2012/557.pdf (2012)

28 NXP product application note: A700x family Secure authentication microcontroller (2013), https://www.mouser.co.cr/datasheet/2/302/A700X_FAM_SDS-119904.pdf (accessed July 27, 2020).

29 Intrinsic ID, WHITE PAPER, Flexible Key Provisioning with SRAM PUF, https://www.intrinsic-id.com/resources/white-papers/white-paper-flexible-key-provisioning-sram-puf/

30 Mikhail Platonov, Josef Hlav'a˘c, R'obert L'orencz, "Using Power-up SRAM State of Atmel ATmega1284P Microcontrollers as Physical Unclonable Function for Key Generation and Chip Identification, COST, European Cooperation in Science and Technology, 33-33, (2014);(accessed 2017)

31 Geert-Jan Schrijen, et. al., Physical Unclonable Functions to the Rescue A New Way to Establish Trust in Silicon, embedded-Word 2018, http://www.intrinsic-id.com/wp-content/uploads/2018/05/Physical-Unclonable-Functions-to-the-Rescue-A-way-to-establish-trust-in-silicon-Schrijen-Garlati-Embedded-World-2018.pdf (accessed 2020)

32 Xian Zhang, Guangyu Sun, Yaojun Zhang3, Yiran Chen, Hai Li, Wujie Wen, Jia Di, A Novel PUF based on Cell Error Rate Distribution of STT-RAM, *IEEE IEDM* 2014,

33 Jayita Das, Kevin Scott, Srinath Rajaram, Drew Burgett, and Sanjukta Bhanja, MRAM PUF: Using Geometric and Resistive Variations in MRAM Cells, IEEE TRANSACTIONS ON NANOTECHNOLOGY, VOL. 14, NO. 3 MAY 2015

34 Xiaochun Zhu, Steven M. Millendorf, Xu Guo, David Merrill Jacobson, Kangho Lee, Seung H. Kang, Matthew Michael Nowak (2015). US Patent Application Pub. No.: US 2015/0071432 A1, PHYSICALLY UNCLONABLE FUNCTION BASED ON RESISTIVITY OFMAGNETORESISTIVE RANDOM-ACCESS MEMORY MAGNETIC TUNNEL JUNCTIONS.

35 Chen, Y.-S., Wang, D.-Y., Hsin, Y.-C. et al. (2017). On the hardware implementation of MRAM physically Unclonable function. *IEEE Trans. Electron Devices* 64 (11).

36 Marukame, T., Tanamoto, T., and Mitani, Y. (2014). Extracting physically Unclonable function from spin transfer switching characteristics in magnetic tunnel junctions. *IEEE Trans. Magn.* 50 (11): 3402004.

37 Baohua Sun, Daniel Liu, Leo Yu, Jay Li, Helen Liu, Wenhan Zhang, Terry Torng, MRAM Co-designed Processing-in-Memory CNN Accelerator for Mobile and IoT Applications, arXiv:1811.12179v1 [eess.SP] 26 Nov 2018

38 S. P. Park, S. Gupta, N. Mojumder, A. Raghunathan, and K. Roy (2012). Future cache design using STT MRAMs for improved energy efficiency: Devices, circuits and architecture. IEEE *Proc. Design Autom. Conf.* (Jun. 2012). IEEE. pp. 492–497.

39 Kimberly Keeton. Memory-driven computing. (2017). https://www.youtube.com/watch?v=eSP9euiV4-M (accessed July 27, 2020).

40 Tang, D. and Lee, Y.J. (2010). *Magnetic Memory, Fundamentals and Technology.* Cambrdige University Press.

41 STT-MRAM: Introduction and market status, https://www.mram-info.com/stt-mram (accessed July 27, 2020).

42 S. Kang (2010). Embedded MRAM for mobile applications. *Non-volatile memories workshop* (April 11–13, 2010). UCSD.

43 S. Z. Rahaman, I. J. Wang, C. F. Pai, J. H. Wei, D. Y. Wang, H. H. Lee, Y. C. Hsin, S. Y. Yang, Y. J. Chang, Y. C. Kuo, Y. H. Su, G. L. Chen, H. Y. Lee, K. C. Huang, C. I. Wu, and D. L. Deng (2018). Device size-dependent Spin-Orbit-Torque switching properties in a stepped MTJ with CMOS-compatible 8-inch fab processes. *Abstract of IEDM MRAM Posters.* IEDM.

44 Jimmy J. Kan, Chando Park, Chi Ching, Jaesoo Ahn, Yuan Xie, Fellow, IEEE, Mahendra Pakala, and Seung H. Kang., A Study on Practically Unlimited Endurance of STT-MRAM, IEEE TRANSACTIONS ON ELECTRON DEVICES, VOL. 64, NO. 9, SEPTEMBER 2017

45 Chunmeng Dou, Wei-Hao Chen, Yi-Ju Chen, Huan-Ting Lin, Wei-Yu Lin, Mon-Shu Ho, Meng-Fan Chang (2017). Challenges of Emerging Memory and Memristor Based Circuits: Nonvolatile Logics, IoT Security, Deep Learning and Neuromorphic Computing. *IEEE 12th International Conference on ASIC* (ASICON). IEEE

46 Lee, et. al. (2018). 22-nm FD-SOI Embedded MRAM Technology for Low-Power Automotive-Grade-l MCU Applications, *IEEE Digest of IEDM*, 18437874

47 Sanjeev Agarwaral, et al., (2019). Demonstration of a reliable 1Gb standalone STT MRAM for industrial applications. *IEEE Digest 2019 IEDM, paper 2.1*. IEEE.

48 K. Lee, et al. (2019). 1Gb high density embedded STT MRAM in 28-nm FDSOI technology. *IEEE Digest of 2019 IEDM, paper 2.2*. IEEE.

49 V.B. Naik, et al. (2019). Manufacturable 22nm FD-SOI embedded MRAM technology for industrial graded MCU and IoT applications. *IEEE Digest of 2019 IEDM, paper 2.3*. IEEE.

50 J.G. Alzate, et al. (2019). 2MB Array-level demonstration of STT-MRAM process and performance towards L4 applications. *IEEE Digest of 2019 IEDM, paper 2.4.* IEEE.

51 K. Lee, et al. (2014). 1Gbit high density embedded STT MRAM in 28nm FDSOI Technology. *IEEE Digest of IEDM, paper 2.2.* IEEE.

52 Louie, B.S. and Berger, N. (2018). Spin Transfer Torque MRAM Device With Error Buffer. US Patent 1011544.

Appendix A

Retention Bake (Including Two-Way Flip)

For long retention bake, weak bits (with smaller thermal stability factor) flip from one state to another, and they may flip back (Figure A.1a). As a result, the apparent flip rate decreases [1]. The analysis involves a thermal stability factor of p2ap Δ_p and ap2p Δ_{ap}. Let $K_p = \exp(-\Delta_p)$ and $K_{ap} = \exp(-\Delta_{ap})$. When flip back is included,

the apparent flip rate is $K(t) = 1 - \dfrac{\left\{ K_p + K_{ap} \exp\left[-\dfrac{t}{\tau_0}\left(K_p + K_{ap}\right)\right]\right\}}{K_p + K_{ap}}$, where τ_0

is 1e-9 seconds and the flip attempt time is constant.

For long bakes, independent of initial bit states, the final equilibrium number of bits in the two states is a function of the two thermal stability factors. For example, with two equal energy barriers, after a long bake, 50% bits settle in p-state, and 50% settle in ap-state. For nonequal energy barriers, there are more bits in a state than in the other. Figure A.1b–d illustrates the distribution. Notice that in actuality the thermal barrier is distributive, and bit distribution smears out. Nonetheless, in a short bake where the flip bits is less than a few percent, one may consider simple analysis without flip back. This point is illustrated in Figure A.1b; the flip rate around time $= 10^7$ seconds is linear in a log–log plot.

Magnetic Memory Technology: Spin-Transfer-Torque MRAM and Beyond,
First Edition. Denny D. Tang and Chi-Feng Pai.
© 2021 The Institute of Electrical and Electronics Engineers, Inc.
Published 2021 by John Wiley & Sons, Inc.

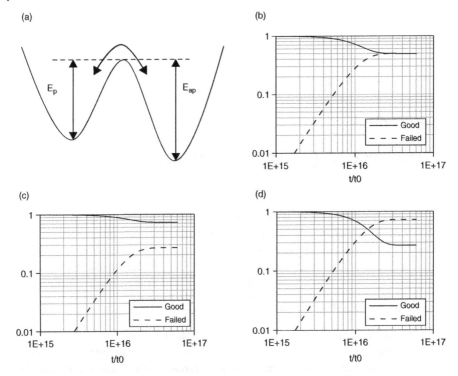

Figure A.1 (a) A schematic of thermal stability energy barrier, where the bit may fail (flip) from one state to another and flip back. Assume all bits are in one state initially. As bake time increases, the fraction of FailBit (dashed line) increases and good (non-flip) bits (solid line) drops, for thermal stability factor 40, 40 (b), 41, 40 (c), 40, 41 (d). Note that this analysis does not include the distribution of Δ, or $\sigma(\Delta) = 0$. $t0 = 1$ ns.

Reference

1 K. Tsunoda, M. Aoki, H. Noshio, S. Fukuda, C. Yoshizaki, A. Takahashi, A. Hatada, M. Nakayashi, Y. Tsuzaki, T. Sugii (2014). *Area Dependence of thermal stability factor in perpendicular STT-MRAM analyzed by Bi-directional data flipping model. IEEE Digest of IEDM. IEEE. p.* 486.

Appendix B
Memory Functionality-Based Scaling

B.1 Introduction

Scalability is one of the major requirements for all viable memory products. Spin-transfer torque magnetic random-access memory (STT-MRAM) cell scaling has been studied extensively, because it reduces the write current substantially. However, it also has various impacts, both advantageous and disadvantageous, on other device parameters and memory circuit operating features. Thus, the scaling method itself is important, and it needs careful considerations to keep the magnetic tunnel junction (MTJ) designs as practical memory cells. Using an example perpendicular spin-transfer torque (p-STT) MTJ design with write and read functionalities as constraint factors, a scaling methodology is provided to demonstrate how it impacts cell characteristics and memory circuit behavior. Besides the write current reduction, the memory functionality-based scaling causes a profound impact on write endurance. Scaling trends of endurance characteristics are shown through analysis of endurance's device parameter dependence.

For STT-MRAM with on-chip error-correcting code (ECC), each memory product tolerates a certain number of error rates. In the endurance case, there are two facets of tolerable error rates; one is a write error rate (WER) ensured from write conditions, and the other is an endurance failure rate (EFR) as criterion after the desirable number of write cycles. This appendix also describes a guideline to assess how scaling affects EFR for given write conditions with particular WERs.

Magnetic Memory Technology: Spin-Transfer-Torque MRAM and Beyond,
First Edition. Denny D. Tang and Chi-Feng Pai.
© 2021 The Institute of Electrical and Electronics Engineers, Inc.
Published 2021 by John Wiley & Sons, Inc.

B.2   Operating Parameters for Write Endurance Failure Analysis

Since scaling causes the most significant impacts on endurance, it is necessary to describe its parameter dependence before discussing scaling. Write endurance failure is a catastrophic low-resistance state regardless of the cell's stored P (parallel) or AP (antiparallel) state. The symptom implies a dielectric breakdown of the MgO isolation layer caused by repeated stresses induced after a number of write cycles. Analysis and evaluation methods for dielectric breakdown of metal-oxide-semiconductor field-effect transistor (MOSFET) gate oxides such as time-dependent dielectric breakdown (TDDB) have been used for STT-MRAM write endurance failures as well. Although the symptom seems to be similar, there are substantial differences between MOSFET and MRAM cases. In the MOSFET case, a few electrons flow in the oxide through Fowler-Nordheim tunneling at a high bias voltage. The small number of electrons injected into the conduction band of the insulator are accelerated by the high electric field and become hot. Thus, the major cause of MOSFET gate oxide breakdown is due to damages in the dielectric layer by hot electrons [1]. The breakdown in the MOSFET case is, therefore, attributed to the voltage or the electric field across the oxide. On the other hand, MgO thickness of MTJ is much thinner than the gate oxide, and a large direct tunneling current flows even at a small write voltage below 1.0 V. The current density is more than several orders of magnitudes higher than the MOSFET gate. Hence, it causes a significant self-heating in the MgO layer [2, 3]. Incorporating a self-heating mechanism in the TDDB breakdown model reveals that the conventional 10-year lifetime has been underestimated the error rate by a factor of 10^7 for 1 ppm fail criterion [4]. Namely, the write endurance of STT-MRAM cells poses complicated stresses on MgO by not only voltage but also current density, associated power area density (which is a product of the current density and the voltage), and its consequential temperature.

As a matter of fact, there are some experimental results to deny voltage-alone dependency and show dependence of write endurance failures on current density, power area density, and MgO temperature. At the same applied voltages across the MTJ, higher ambient temperatures cause more endurance failures [5]. Also, at the same voltages, wider pulse widths raise MgO temperatures more because of longer self-heating periods and result in worse endurance failures [6]. Whichever the heating mechanism is, either ambient or self-heating, endurance failure counts increase as MgO temperatures rise at the same voltage stress.

Among various MTJ designs, process technologies, and test methods, all cases consistently show that AP-state TDDB has higher breakdown voltages and/or longer lifetimes than P-state [7–9]. In addition to TDDB, in actual write endurance tests using functional chips, Figure B.1a shows that AP-state unipolar endurance

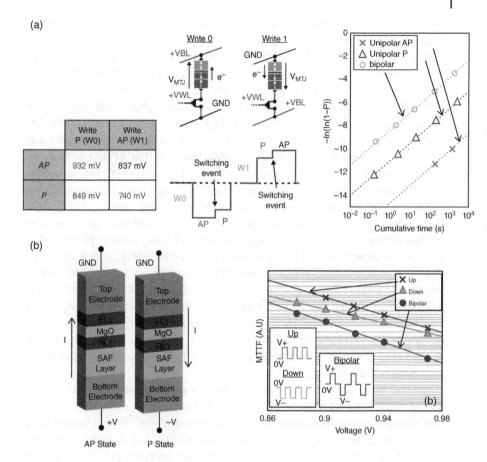

Figure B.1 Bipolar, unipolar AP, and unipolar P endurance test results of two different designs, (a) [9], and (b) [10]. Bipolar repeats write 0 and write 1 alternately. Unipolar AP, "Up" in (b), applies write 1 voltage on AP-state cells, while unipolar P, "Down" in (b), means write 0 voltage stresses on P-state cells.

has a far lower failure rate than P-state, although applied voltages are very close: 837 mV for AP and 849 mV for P [9]. AP-state unipolar endurance means that write 1 voltage pulses, in which the pinned layer at the bottom electrode is biased to positive with respect to the free layer at ground, are applied repeatedly to cells stored AP states. P-state unipolar endurance applies the opposite polarity pulses on P-state cells. Only 12 mV higher voltage cannot explain about 60 times worse failure rate in P-state unipolar endurance. Figure B.1b displays the identical result in a completely different design from Figure B.1a; unipolar endurance tests also show a longer mean-time-to-failure (MTTF) in AP-state than P-state case [10]. The better result of AP-state endurance agrees with higher breakdown voltages and longer lifetimes of AP-state TDDB.

The common reasoning to explain those results of P/AP-state TDDB and unipolar write endurance tests is the difference in MTJ resistances of P and AP states. P-state resistance, R_P, does not depend on a bias voltage, but AP-state resistance, R_{AP}, decreases linearly as the voltage across MTJ rises. Although MR ratio drops significantly at write voltages around 840 mV in Figure B.1a, R_{AP} is still about 30% higher than R_P. This means that AP-state cells have about 30% lower current density and power area density. The significantly better EFR of AP-sate cells, therefore, seems to be due to the substantially reduced amount of stresses in current density, power area density, and thus less self-heating. Figure B.1b more clearly denies voltage dependence of endurance failures. From 0.86 to 0.98 V, both "Up," which is AP-state unipolar, and "Down," P-state unipolar, have the same applied voltages, but MTTFs are always different. It rather evidently proves endurance failure dependence on power area density. At 0.86 V, similar to Figure B.1a, "Up" has much longer MTTF, better EFR. However, as the applied voltage increases, its MTTF becomes shorter, more rapidly than "Down." This is due to a decrease in R_{AP} by the applied voltage, and R_{AP} becomes close to R_P. Thus, the current density of "Up" increases, and so does its power area density, which in turn works to reduce the difference in MTTF. At 0.98 V, R_{AP} is very close to R_P, and MTTFs of "Up" and "Down" become nearly identical because their power area density is almost the same.

Another important observation is that bipolar write endurance has a much worse failure rate [9] or a significantly shorter MTTF than unipolar tests [10]. It suggests that alternating opposite voltage polarities cause momentous impacts, besides stresses by voltage, current density, and power area density. A three-stage breakdown model of bipolar write endurance failures has been proposed [11, 12]. The three stages consist of the following: (i) defect generation by MgO bond breaking at the MgO-metal interface, because injected electrons by a write pulse create Frenkel-typed defects, leaving oxygen vacancies (V^{2+}) and interstitial oxygens (O^{2-}); (ii) defect activation to displace O^{2-} from V^{2+} by a following opposite polarity pulse voltage; and (iii) defect (O^{2-}) diffusion during no write pulse voltage periods to form a breakdown path in MgO. According to the model, current density means colliding electrons per area and per time at the interface, and multiplying the voltage to it, which is power area density, becomes a power source per interface area for defect generations. On top of this, the power area density is a measure for the self-heating of MgO, which increases defect generation rates further. Among bipolar, P- and AP-states unipolar results, both unipolar cases do not have the second stage, defect activation, because of no opposite polarity pulse. Even though defects are generated in the first stage, most of O^{2-}s may recombine with V^{2+}s, and only a small fraction of

O^{2-}s that escape the recombination can diffuse in MgO to make smaller failure counts than bipolar. Between two unipolar endurance tests, even at the similar voltages on AP and P states, AP-state unipolar endurance causes less defect generation due to two aspects: the smaller driving power and the harder bond breaking by lower MgO temperatures. The two aspects stem from the lower power area density by higher R_{AP}. This is the reason for significantly lower failure rates in AP-state unipolar endurance, which the voltage-alone theory cannot explain.

Unlike MOSFET gate oxides, using the voltage across MTJ alone for write endurance characterization of STT-MRAM cells is misleading because of such more complicated failure dependencies on operating parameters and conditions. The power area density is a better-chosen parameter than the voltage to analyze write endurance failures, since it involves the voltage and the current density, and it plays two crucial roles in the endurance fail mechanism: driving power for defect generation and a power source of self-heating. Unipolar endurance stress is a useful means for understanding the nature of endurance failures, but they most likely do not happen in actual write accesses, because, to save write currents, most STT-MRAM chips read multiple data first and write only bits that have to be altered. Thus, it is unlikely to apply write 1 pulse on cells already at AP-state (or write 0 on P-state cells), and bipolar endurance reflects the realistic memory chip operations. Choosing the worst case is also the way for reliability assessments. Therefore, lifetime estimation must rely on bipolar write endurance. Scaling deals with device characteristics that encompass a wide range of MTJ sizes and film resistance area products, R_as, of MTJ cells. Thus, device operating parameters to characterize scaling impacts endurance must have information about MTJ sizes. The power area density automatically has size information in itself, and in this sense, it is also a good parameter for analyzing write endurance failures on scaled devices. Scaling impacts on bipolar endurance needs to be analyzed by observing how the power area density varies by changes in MTJ sizes and R_as.

B.3 Functional Requirements for Scaling

Most of scaling theories and predictions focus on mainly write current reduction. This is a great aspect of STT-MRAM cells, but reducing the write current shrinks a margin between write and read functions, yet as memory cells, they must be written and read correctly. Thus, single-purpose scaling to spoil other functions must be avoided, and meaningful scaling processes to keep memory functionalities valid are necessary. Conventional random access memories such as dynamic random-access memory (DRAM) and static random-access memory (SRAM) have stringent reliability requirements. A huge number of cells must be written and read for trillions of cycles at a few error rates. Although the requirements are extremely relaxed because of on-chip ECC, STT-MRAM cells also must be written and read reliably and safely. In all scaling sizes, therefore, at least MTJs must satisfy those fundamental write/read functions as memory cells. To do so, write and read conditions become constraint factors for scaling. The conditions depend on ECC capabilities such as one- or two-bit correction and codeword length in addition to application requirements. Thus, before discussing scaling, those conditions for a write and a read are defined as prerequisites in the following sections. The reliable write and read functions are fundamental requirements for memory cells, and other constraints that stem from different functions and/or requirements may limit the scaling capability of STT-MRAM.

B.3.1 Write Function – Switching Current Density

The write condition is generally specified by a switching voltage, V_{SW}, with a WER, which is typically 1 ppm for actual memory applications and 50% for technology evaluations or comparisons. Figure B.2a shows such an example of the write condition for various R_as of p-STT-MRAM cells [9].

Using the same magnetic structure but different R_as, from 10.2 to 3.0 $\Omega\mu m^2$, the plot shows WER variations by changing write voltages with a 10 ns write pulse width [9, 13]. The following discussions and analyses exploit this MTJ design as an example of scaling methodology.

Each line to represent R_a is used to estimate a switching voltage for 1 ppm WER, V_{SW_1ppm}, as well as a switching voltage for 50% WER, $V_{SW\_50\%}$. Both V_{SW_1ppm} and $V_{SW\_50\%}$ decrease as R_a becomes smaller. Since scaling down of MTJ size accompanies with R_a reduction, it suggests that the switching voltage drop by scaling, which is a favorable trend for write endurance. However, switching current density, J_{SW}, is a better-suitable parameter rather than V_{SW} to deal with scaling. Since J_{SW} is given by V_{SW}/R_a, the plot that shows WER dependence on V_{SW} and R_a of Figure B.2a is easily converted to J_{SW} versus R_a for two WER lines, 1 ppm and 50%. Figure B.2b shows those two WER lines of switching current density dependence on area resistance, J_{SW_1ppm} and $J_{SW\_50\%}$.

It is interesting to notice that switching current density increases as R_a scales down. As the constant switching current line shows, if the switching current density of R_a at 10.2 $\Omega\mu m^2$ is used to write cells for MTJs with lower R_a, WER increases rapidly, and below 4.5 Ω μm^2 it becomes more than 50%. Thus, a constant switching current scaling method cannot satisfy the 1 ppm WER, and to keep the write function at this error rate with 10 ns write pulse width, scaling has to track the increasing J_{SW}.

(a)

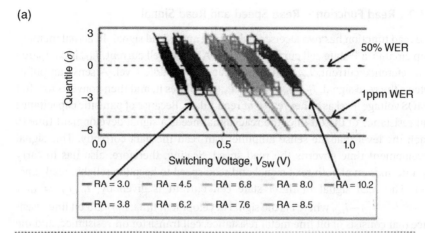

(b)

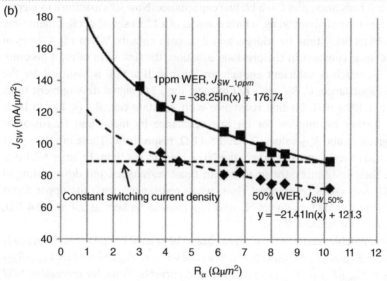

Figure B.2 (a) WER dependence on write voltage levels for different R_as [9]. (b) Dependency of switching current density for 1 ppm and 50% WERs (J_{SW_1ppm} and $J_{SW\_50\%}$) on R_a.

When R_a is reduced while keeping write J_{SW} constant, V_{SW} also drops, thus so does switching power area density, JV_{SW}, which in turn reduces self-heating to raise an effective barrier energy to make switching harder for the given J_{SW}. Lowering JV_{SW} is good for endurance, but ironically this makes it harder to flip cell state, and the cell needs more J_{SW} when R_a is reduced. This is the main reason for the J_{SW} increase by lowering R_a [14].

B.3.2   Read Function – Read Speed and Read Signal

The read function has two facets, a read speed and a read signal. In actual memory chip circuits, a P-state cell current, I_P, or an AP-state cell current, I_{AP}, is compared to a reference current, I_{ref}, which is set at their middle level. A sense amplifier detects the read signal, $I_P - I_{ref}$ or $I_{ref} - I_{AP}$, amplifies it, and then converts to a full CMOS voltage level as either read 0 or read 1 data. Because of parasitic capacitance and resistance of the bit line, the read signal needs a signal development time to reach the level that the sense amplifier can read the data correctly. This signal development time governs the read speed. Scaling, therefore, also has to carry the information of both read signal and read speed to keep the reliable read function. The read signal of the P-state cell current is given by, $I_{read\_P} = I_P *$ $(1 - e^{-t/R_{BLP}*C}) - I_{ref}$, where t is the signal development time; R_{BLP} is a bit line resistance that consists of bit line metal resistance, cell transistor on-resistance, and the P-state cell resistance; and C is a bit line capacitance. Now let's assume a moderate read access time at about 30 ns, which consists of a 12 ns signal development time and an 18 ns delay time by address and data path circuits. When t is 4 times of $R_{BLP} * C$ time constant in the previous equation, the first term of right becomes 98% of I_P, which is sufficient enough for sensing. If a 1 pF is assumed for the bit line capacitance, C, R_{BLP} is 3 KΩ to meet the 12 ns signal development time. To have a 100% dI/I, bit line resistance with AP-state cell, R_{BLAP}, has to be 6 KΩ. A further assumption for bit line resistance by metal and transistor as 600 Ω gives R_P and R_{AP} values, 2.4 and 5.4 KΩ, respectively. These are the upper limits of MTJ resistances to satisfy the 30 ns read access time. In large MTJ diameters, their resistances are smaller than those resistances, but downscaling of the MTJ sizes raises resistances. Thus, when resistances reach the upper limits by reducing the MTJ diameter, R_P and R_{AP} have to be kept at 2.4 and 5.4 KΩ, respectively, for smaller MTJs.

To have reliable reads, the read signal has to be at least |7.5 μA|, which means I_P and I_{AP} are 30 and 15 μA, respectively, at a read voltage, V_{read}. Since $I_P = V_{read}/R_{BLP}$ and $I_{AP} = V_{read}/R_{BLAP}$, V_{read} can control read currents. Thus, for increasing MTJ resistances by scaling, V_{read} has to be increased to maintain I_P and I_{AP} at 30 and 15 μA, respectively. Setting the ceiling for R_P and R_{AP} at 2.4 and 5.4 KΩ, respectively, and keeping I_P and I_{AP} at 30 and 15 μA, respectively, are the requirements to have the read function with the 30 ns access time and the reliable read signal.

B.4   Scaling Procedure

Now, the condition to guarantee the write function to maintain 1 ppm WER with 10 ns write pulse width and the read function with 30 ns access time with the reliable read signal are settled. Namely, the writing has to use the R_a-dependent switching current density shown in Figure B.2b, for the read function, R_P, R_{AP}, are 2.4, 5.4 KΩ, respectively, and V_{read} has to be adjusted to have I_P and I_{AP} at 30 and 15 µA, respectively. The scaling has to be done meeting those constraints. For MTJ diameters from 100 to 10 nm, 10 nm per step, Table B.1 displays such a scaling example using the p-STT MTJ design shown in Figure B.2a.

The scaling starts from the 100 nm diameter by choosing a typical R_P of 1.3 KΩ and increasing R_P 200 Ω per every 10 nm reduction in diameters. When R_P exceeds the prespecified ceiling to satisfy the 30 ns read access time, it remains at 2.4 KΩ from 40 to 10 nm. Dividing each R_P by the area in µm² gives R_a for each MTJ diameter.

Once R_a is known, $J_{SW\_50\%}$ and J_{SW_1ppm} can be calculated by plugging R_a values into fitting equations (not shown in Figure B.2b). The switching voltages are obtained by the equation, $V_{SW} = J_{SW}xR_a$, and using those values, power area densities, JV_{SW}s, are also listed by multiplying J_{SW} by V_{SW} at 50% and 1 ppm WERs. Simply multiplying MTJ areas to J_{SW} results in switching current, I_{SW}, for each MTJ size at two WER cases.

For large memory chips such as more than 256 Mb, 1 ppm WER seems to be too relaxed for even two-bit ECC, and a tighter WER is desirable. WER slope is calculated to find a J_{SW} to meet such a WER. For each MTJ size, it is obtained by taking the differences of $J_{SW\_50\%}$ and J_{SW_1ppm} and dividing them by 5.5 decades of WER from 50% to 1 ppm Thus, it indicates the amount of switching current density to change one decade in WER. The current density to have 0.01 ppm WER, $J_{SW\_0.01ppm}$, for example, can be calculated by adding two times of WER slope to J_{SW_1ppm}. It is used to obtain switching voltages, switching power area densities, and switching currents in the same way as 50% and 1 ppm WER cases.

Regarding the read function, to keep the AP-state read current, I_{Read}, at 15 µA, the read voltage, V_{Read}, is raised gradually from 52 mV at 100 nm MTJ to 88 mV at 50 nm MTJ, and then it is kept at a constant 90 mV when R_P reaches the ceiling of 2.4 KΩ from 40 to 10 nm. Since large MTJ sizes have smaller R_Ps than the ceiling, their signal development times are shorter. Thus, read access times are faster than 30 ns as shown in Table B.1. For example, 100 nm MTJ has an R_P of 1.3 KΩ, and the total bit line RC time constant is 1.9 ns due to 1 pF capacitance and 1.9 KΩ total resistance including 600 Ω metal and transistor resistances of the bit line. Since the signal development time is four times of the RC time constant, the read access time becomes 26 ns with the 18 ns address/data path delays.

The AP-state read current density, J_{Read}, is calculated by $I_{Read}/(\text{MTJ area in }µm²)$ for each MTJ to assess read disturb concerns to be discussed in the next section.

Table B.1 Scaling table of p-MTJ example design from 100 to 10 nm in MTJ diameters.

	MTJ diameter (nm)	p-MTJ scaling table example									
		100	90	80	70	60	50	40	30	20	10
	R_P (Ω)	1300	1500	1700	1900	2100	2300	2400	2400	2400	2400
	R_a $(\Omega\mu m^2)$	10.2	9.5	8.5	7.3	5.9	4.5	3.0	1.7	0.8	0.2
Write function											
WER 50%	$Jsw\_50\%$ $(mA/\mu m^2)$	72	73	75	79	83	89	98	110	127	157
	$Vsw\_50\%$ (mV)	730	696	644	575	494	402	294	187	96	30
	$JVsw\_50\%$ $(mW/\mu m^2)$	52.3	50.8	48.5	45.3	41.1	35.8	28.8	20.5	12.2	4.6
	$Isw\_50\%$ (μA)	562	464	379	303	235	175	123	78	40	12
WER = 1 ppm	Jsw_1ppm $(mA/\mu m^2)$	88	90	95	101	109	119	135	157	188	241
	Vsw_1ppm (mV)	897	863	809	736	645	538	406	265	141	45
	$JVsw\_1ppm$ $(mW/\mu m^2)$	78.8	78.1	76.6	74.1	70.0	64.0	54.6	41.6	26.5	10.9
	Isw_1ppm (μA)	690	575	476	387	307	234	169	111	59	19
	WER slope $(mA/\mu m^2/decade)$	3.0	3.2	3.5	4.0	4.6	5.5	6.7	8.5	10.9	15.2
WER = 0.01 ppm	$Jsw\,0.01\,ppm$ $(mA/\mu m^2)$	94	97	102	109	118	130	148	173	209	271
	$Vsw\_0.01\,ppm$ (mV)	957	924	869	794	700	587	446	294	158	51
	$JVsw\_0.01\,ppm$ $(mW/\mu m^2)$	89.8	89.4	88.4	86.3	82.5	76.3	66.0	51.0	33.1	13.8
	iSW_001ppm (M- I)	737	616	511	418	333	255	186	123	66	21
Read Function	V_{Read} (mV)	52	60	65	72	80	88	90	90	90	90
	I_{Read} (μA), AP state	15	15	15	15	15	15	15	15	15	15
	J_{Read} $(mA/\mu m^2)$, AP state	1.88	2.37	2.92	3.84	5.32	7.76	11.9	21.2	47.8	191
	Read access time (ns)	26	26	27	28	29	30	30	30	30	30
	Read disturb rate	1×10^{-24}	2×10^{-22}	8×10^{-22}	6×10^{-20}	6×10^{-18}	5×10^{-16}	6×10^{-14}	1×10^{-11}	2×10^{-8}	0.9975

B.5 Scaling Impacts

The scaling procedure that takes write and read functions as constraints determine various operating parameters, as shown in Table B.1. Based on this information, it is necessary to discuss how this specific scaling causes impacts cell characteristics.

B.5.1 V_{SW} and JV_{SW}

Figure B.3 shows scaling variations of V_{SW} at write conditions to meet 0.01 and 1 ppm WERs. For comparison, V_{SW} by a constant J_{SW} scaling is also depicted, where the constant J_{SW} scaling takes a switching current density at 100 nm MTJ, 88.0 mA/μm^2 of 1 ppm WER, and applies it to all scaled MTJs. The switching voltages are calculated by $J_{SW}xR_a$ using R_as shown in Table B.1. As Figure B.3a demonstrates, V_{SW}s of both 0.01 and 1 ppm WER cases decrease as MTJ sizes shrink, although their decays are slower than the constant J_{SW} scaling. If write endurance failures depend on only the voltage across MTJ, endurance lifetime seems to improve almost monotonically by scaling down MTJ sizes. However, as discussed earlier, write endurance depends on mainly power area density, and it is necessary to see its variations by scaling.

Figure B.3b displays scaling impacts on switching power area density, JV_{SW}, at write conditions to meet 0.01 and 1 ppm WERs. They also decay, but much slower than V_{SW} and JV_{SW} of the constant J_{SW} scaling. This means that the write endurance lifetime stays almost flat from 100 nm to 70 nm. Gradual improvements appear from 70 to 40 nm, and the endurance lifetime extends significantly below 40 nm. Since scaling reduces R_a as well as MTJ size, V_{SW}s decrease. However, their

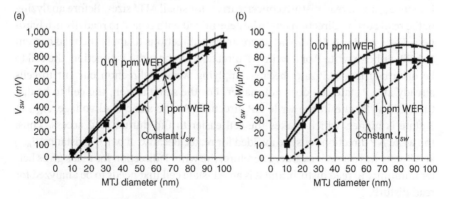

Figure B.3 Scaling impacts of write conditions, 0.01 and 1 ppm WERs and constant J_{SW}. (a) Switching voltage. (b) Power area density.

decays are slower because J_{SW} increases as MTJ sizes get smaller. Thus, JV_{SW}s, which is the product of V_{SW} and J_{SW}, decreases gradually in the whole scaling range.

The J_{SW} dependence on R_a shown in Figure B.2b is based on the same MTJ size. Namely, for a fixed MTJ size, it shows how V_{SW} and thus J_{SW} change as scaling down R_a [9]. However, it is reported that J_{c0} also increases as MTJ size shrinks at the same R_a [15]. Thus, J_{SW} and JV_{SW} are actually smaller at large MTJ sizes than the numbers shown in Table B.1, and the endurance lifetime becomes better in 100–70 nm. On the other hand, smaller MTJ sizes, 30–10 nm, are expected to have larger J_{SW} and JV_{SW}. Particularly from 40 nm to 20 nm, a very rapid increase in J_{c0}, about 25%, is anticipated [15]. Thus, it increases JV_{SW}s of 20 and 10 nm MTJs substantially. Namely, JV_{SW}s of both 0.01 and 1 ppm WERs increase from 100 to 80 nm and stay at a peak around 80–60 nm. The rapid reduction in 40–10 nm degrade and particularly 20–10 nm MTJs have more than 1.5 times larger than the values shown in Figure B.3b. The endurance lifetimes at a given EFR such as 1 ppm and, therefore, are expected to become slightly longer from 100 to 80 nm, then stay at the same level for 80–60 nm, and finally extend below 50 nm. The endurance lifetime seems to exhibit such complicated behavior by the scaling.

B.5.2 Read Disturb

Since the scaling keeps the read current constant, the read current density, J_{Read}, becomes large as MTJ size shrinks. It is directly proportional to $1/(\text{MTJ diameter})^2$. For example, J_{Read} of 10 nm MTJ is 100 times of that of 100 nm. Although J_{SW} also increases, the enlargement ratio between the same dimensions is less than three. Thus, margins between write and read current density become smaller as MTJ size is reduced, and a read disturb concern arises for small MTJ sizes. Before analyzing it, the read current direction has to be determined with regard to read disturb hardness. For a read in STT-MRAM, there are two directions to apply the read current to MTJ, either a write P or a write AP direction. In the write P direction, the cell to be disturbed is at AP sate, because applying write P direction current to cells already at a P state cell does not matter. For the same reason, the cell to be disturbed is at P state in the write AP direction. At the same read voltage, J_{read} at P state is higher, which makes the cell easier to flip during read access. Thus, the write AP direction must be avoided for reads. Because of twice resistance, J_{read} at AP state has a smaller value. Therefore, the write P direction for reads gives better immunity to read disturb, and it is an AP-state cell that needs to be analyzed for read disturb.

As Table B.1 shows, I_{Read} at AP state is kept at 15 μA. Its current density, J_{Read}, is calculated by dividing I_{Read} by each MTJ area in μm^2. Figure B.4a depicts scaling impacts on J_{Read} along with switching current density for 1 ppm and 50% WERs.

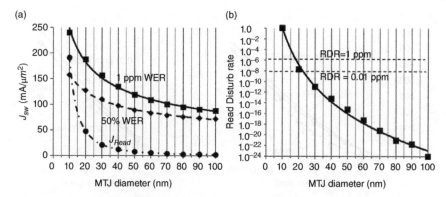

Figure B.4 (a) Scaling impacts on switching current density for 1 ppm and 50% WER cases and read current density. (b) Scaling impacts on read disturb rate.

J_{Read} increases rapidly from 40 nm, becomes closer to J_{SW}s, and eventually exceeds the 50% WER line at 10 nm. Namely, read disturb becomes so severe that the 15 μA I_{Read} at AP state turns to be more effective write current with a WER lower than 50%.

To understand it quantitatively, read disturb rates (RDRs) of all MTJ sizes are calculated in the bottom line of Table B.1. Since 1 − WER is equal to a write success rate, which means the read disturb rate, the WER slope can derive read disturb rate for each MTJ size. For example, at 40 nm MTJ, $J_{SW\_50\%}$ and J_{Read} are 98 and 11.9 mA/μm^2, respectively. Since its WER slope is 6.7 mA/μm^2/*decade*, there are 12.8 decades ([98−11.9]/6.7) from $J_{SW\_50\%}$ to J_{Read}. Because 50% WER is 50% RDR as well, this J_{Read} has 12.8 decades lower than RDR of 50%. For 12.8 decades means 8×10^{12}, the read disturb rate of 40 nm MTJ is given by $0.5/8 \times 10^{12} = 6.25 \times 10^{-14}$. Figure B.4b plots RDRs of all MTJ sizes and how they increase by the scaling. As it is considered that 0.01 ppm is adequate WER for chips with several hundred million cells using two-bit ECC, RDR criterion is also set at 10^{-8}, 0.01 ppm. Thus, MTJ sizes smaller than 20 nm may not be practical as memory cells. As Figure B.3b shows, endurance lifetime improves substantially in MTJ diameters below 50 nm. However, RDR limits the feasible MTJ size to 30 nm. Thus, read disturb is another constraint to the scaling, in addition, to write and read functions.

B.5.3 Switching Current

Since the greatest advantage of STT-MRAM scaling is write current reduction, how this specific scaling reduces switching current is another interesting subject. Figure B.5a shows two write condition cases, 0.01 and 1 ppm in WER, with the constant J_{SW} scaling as a reference.

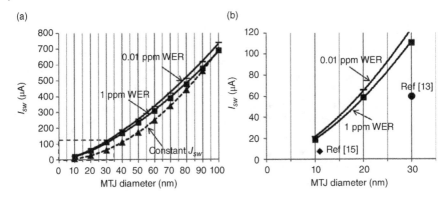

Figure B.5 (a) Scaling impacts on switching current for 0.01 and 1 ppm WER cases and constant J_{SW}. (b) Expanded view of the marked portion of (a) along with published data.

In both 0.01 and 1 ppm write conditions, switching currents decrease, but much slower than the constant J_{SW} scaling as MTJ diameters shrink. Also, compared to published data, the scaling is not so aggressive in write current reductions. For example, 30 nm with 3–4 $\Omega\mu m^2$ in R_a, the write current is about 60 μA [13], while for the same MTJ size, Table B.1 shows I_{SW}, 111, and 123 μA for 1 and 0.01 ppm WERs, respectively. For an 11 nm MTJ, it is reported that the write current is only 7.5 μA with a very high write voltage of 0.75 V, at a very low WER of 7×10^{-10} [16]. Even at 50% WER, the example design in Table B.1 shows 12 μA for 10 nm MTJ. However, this 7.5 μA write current is based on a very high R_P, about 100 KΩ, and R_a is about 9.5 $\Omega\mu m^2$; thus, the designs are completely different. Figure B.5b expands the marked portion of Figure B.5a and adds the published data to indicate the differences. Relatively high write current of the example design is due to increasing J_{SW} by scaling down MTJ sizes and more aggressive reductions in R_as to keep the read function.

B.5.4 Nonvolatile Function – Data Retention

In early STT-MRAM memories with large MTJ sizes around 100–60 nm, a full non-volatile function has been kept, and they intend to replace eFlash memories with limited endurance lifetime around 10^6 write cycles. However, as MTJ scales down, it becomes difficult to preserve nonvolatility, which means a full data retention capability for device life, 10 years, at elevated temperatures, 85–125°C. Scaling extends endurance lifetime, but nonvolatility is incompatible to long endurance lifetimes, and as scaling goes downsize, data retention time shortens. For random access memory applications using smaller MTJ sizes, in which extended write endurance lifetimes are required, STT-MRAMs no longer have nonvolatility, and specify long data retention times. For example, DDR3-compatible 256 Mb STT-MRAM has a data retention time of three months at 70°C [17]. As a memory cell, write and read functions are mandatory necessities, and they are fundamental constraints. Once the cell loses nonvolatility, length of data retention is not important as far as it is more than a day. Hence, although data retention is another constraint for scaling, it is of secondary importance. However, one has to notice that DRAMs and SRAMs keep data as far as power is on, but STT-MRAMs lose data even if power is on when the time comes. If application systems need to keep data during its operation, refresh cycles similar to DRAMs become necessary for STT-MRAMs. For instance, if data retention time is very short such as 10 hours, a typical 256 Mb (312 Mb including check bits of two-bit ECC) STT-MRAM chip has a 4 M address depth with 78-bit simultaneous access, and it needs a refresh cycle, which is exactly the same as normal read with ECC, every $36\,000\,\text{s}/4\,\text{M} = \sim 8$ ms, which causes negligible refresh power. This is much easier than DRAM, which has a refresh cycle every 8 μs with simultaneous access of 32 K bits, having a large peak current and nontrivial refresh power. Any attempt hardly justifies extending data retention time from hours to months by sacrificing endurance lifetime; applying refresh cycles is a far better alternative. Thus, the specific scaling shown in Table B.1 does not take into account data retention and leaves it as it is, assuming refresh cycles if necessary.

B.5.5 Remarks on Temperature

The scaling also does not involve temperature, and the room temperature is assumed to derive all device parameters. However, WER, EFR, and RDR all depend on temperature strongly. As temperature rises, EFR becomes worse, but the switching current density to meet specific WERs can be reduced. Thus, write circuit designs to control bit line voltages and/or write pulse widths by tracking temperature are desirable in actual memory chips. For different temperatures, numbers shown in the scaling table have to be modified by applying such design actions. However, RDR remains as a serious problem at high temperatures, because the decrease in read current density is small, and it cannot be reduced for the sake of stable read, yet cells become flip more easily due to substantial switching current density reduction. Thus, the scaling limit may have to move to larger MTJ sizes by the RDR constraint.

B.6   Write Endurance and its Lifetime Characterization Method

Because endurance failure strongly depends on write conditions, to evaluate endurance hardness, clearly organized criteria become necessary for both write conditions and endurance lifetime. WER is a measure of write capability quality. Typically, 1 ppm is set for WER. Because of ECC, it is good for small density chips such as less than 1 Mb. However, for large density chips, tighter criterion becomes necessary. The level of WER needs to be selected depending on chip density, capability of ECC used, and MTJ characteristics. The selected WER defines the exact write condition for the chip, and endurance must be evaluated by this write condition. Endurance lifetime (ELT) is also typically specified as the number of write cycles that reaches a 1 ppm EFR. For the required lifetime, EFR level also needs to be specified depending on chip density, ECC, endurance robustness of MTJ, and allowability of applications.

Because EFR depends on power area density, ELT versus power area density can be obtained for different EFR levels. Figure B.6 is such a conceptual picture. It has three EFR cases, 1, 0.1, and 0.01 ppm. As power area density decreases, ELT in numbers of write cycles, which are shown on a log scale, of all three EFR criteria become longer. At a given power area density, ELT becomes longer in the number of write cycles when more relaxed EFRs are allowed. To reach the desired ELT at tighter EFRs, MTJ designs must reduce the power area density.

As Table B.1 shows, each MTJ size has its own power area density for different WER such as 0.01 and 1 ppm. Thus, for each WER level, ELT dependence on MTJ size can be obtained for different EFRs in Figure B.7. In both 0.01 and 1 ppm WER, ELTs stay almost flat or become longer very slightly from 100 to 70 nm, because

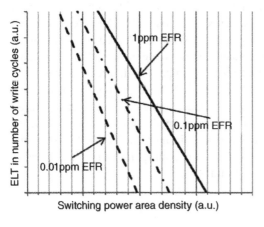

Figure B.6   ELT dependence on switching power area density with EFR criteria, 0.01, 0.1, and 1 ppm WERs.

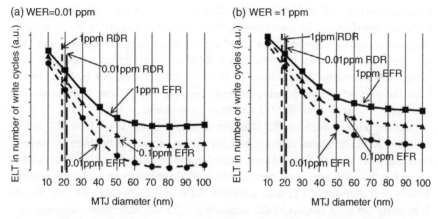

Figure B.7 (a) Scaling impacts on ELT for various combination of WER and EFR. (a) WER = 0.01 ppm, (b) WER = 1 ppm. Vertical lines show 0.01 and 1 ppm RDR.

their power area densities do not reduce much. On the other hand, ELT extends very rapidly from 50 to 10 nm as power area densities drop sharply. Naturally, ELT becomes longer as EFR criteria are relaxed from 0.01, 0.1, and 1 ppm. Similarly, as WER changes from 0.01 to 1 ppm, ELT also extends.

As shown in Section B.5.2, when MTJ scales down, J_{Read} increases to cause read disturb. The vertical two lines in (a) and (b) of Figure B.7 indicate 1 and 0.01 ppm RDRs. Because of rapid increases in J_{Read} of MTJs smaller than 20 nm in diameter, the 1 ppm RDR line is close to 0.01 ppm. For reliable memory operations, read disturb actually limits the use of such sizes, although ELT improves significantly for small MTJ diameters. Thus, RDR poses a practical scaling limit of STT-MRAM cells.

There are two favorite factors that make ELT of smaller MTJs better, although the RDR limit does not change. The first one is MgO temperature. EFR depends on power area density, but it is mainly through MgO temperature due to self-heating. Hence, if the cooling system is better, for the same power area density, the reduced temperature results in a lower EFR. This further means that smaller MTJs have better endurance characteristics, because of their lower MgO temperatures by self-heating [3, 4, 6, 10]. The other aspect is "effective diameter." A finite etching-damaged peripheral zone is more resilient to breakdown, and excluding the damaged zone from the designed size is called the effective diameter. As MTJ size shrinks, the ratio between the effective and designed diameters becomes smaller; thus, it also results in robust endurance [12, 18]. However, the effective diameter is one of the reasons that makes switching current density higher by MTJ size scaling while keeping R_a the same [15]. It increases power area density to reduce ELT. There is another unfavorite factor in small MTJ size. As Table B.1 shows, R_a

becomes less than $2\,\Omega\mu m^2$ in 30 nm MTJ diameter. As R_a is reduced, defect density increases significantly. In read heads of HDD, less than $1\,\Omega\mu m^2$ in R_a is not unusual, where one MTJ is one product, and even if wafer test yield drops to 96%, it is still very good. However, for Mb-class MRAMs, 4% cell defect is unacceptable. Those manufacturing defects must be fixed by redundancy, not by ECC, and it is unlikely to achieve reasonable memory product yield. Thus, realistic scaling limit may appear at a larger MTJ size than RDR poses.

Since the purpose of Figure B.7 is to provide a guideline to know how ELT, WER, EFR, and RDR are related through switching power area density, there are no concrete numbers in both X and Y axes. Actual numbers differ from design to design, various conditions, and constraints. With realistic numbers in switching current density and desirable ELT, both tolerable WER and allowable EFR are chosen depending on chip density, ECC capability, MTJ characteristics, temperature, and application requirements.

B.7 Summary

The scaling method using the p-STT-MRAM cells as an example design was shown. For the whole scaling range from 100 to 10 nm, 10 nm per step, it keeps constant WER and constant I_{Read} as fundamental constraints of practical memory cells. Various impacts how this scaling causes on V_{SW}, JV_{SW}, J_{Read}, RDR, I_{SW}, and data retention were discussed. It was also shown that EFR depends on JV_{SW}, not V_{SW}-alone, and diagrams displayed how ELT is related to given WERs and allowable EFRs. For the desired ELT. the approach can be used to define optimum MTJ designs based on considerations of tolerable WER and allowable EFR depending on chip density, ECC capability, MTJ characteristics, temperature variations, and application system requirements. Although smaller MTJs have better EFR and longer ELT due to decreasing JV_{SW}, RDR limits the MTJ size reduction to around 20 nm, and manufacturing defects may also restrict memory cell usage at an even larger MTJ size.

References

1 Taur, Y. and Ning, T.H. (2009). *Fundamentals of Modern VLSI Devices*, 2e, Chapter 2.5.3 – 2.5.6. Cambridge University Press.
2 Tang, D.D. and Lee, Y.-J. (2010). *Magnetic Memory"*, Chapter 6.6.2.1. Cambridge University Press.

3 Kan, J.J. et al. (2016). *Systematic Validation of 2x nm Diameter Perpendicular MTJ Arrays and MgO Barrier for Sub-10 nm Embedded STT-MRAM with Practically Unlimited Endurance. IEEE Digest of IEDM* 4: 274.1–274.4.

4 Van Beek, S. et al. (2018). *Impact of self-heating on reliability predictions in STT-MRAM. IEEE Digest of IEDM* 4: 252.1–252.4.

5 Lim, J.H. et al. (2018). *Investigating the Statistical-Physical Nature of MgO Dielectric Breakdown in STT-MRAM at Different Operating Conditions. IEEE Digest of IEDM*: 25.3.1–25.3.4.

6 Lim, J.H. et al. (2018). *Area and Pulsewidth Dependence of Bipolar TDDB in MgO Magnetic Tunnel Junction. IEEE Digest of IRPS*: 6D.6-1–6D.6-6.

7 Yoshida, C. and Sugii, T. (2012). *Reliability Study of Magnetic Tunnel Junction with Naturally Oxidized MgO Barrier. IEEE Digest of IRPS* 5: 2A3.1–2A3.5.

8 Kan, J.J. et al. (September, 2017). *A Study on Practically Unlimited Endurance of STT-MRAM. IEEE Tr on Elec. Devices* 64 (9): 3639–3646.

9 Zhu, J. et al. (2019). *Comprehensive Reliability Study of STT-MRAM Devices and Chips for Last Level Cache Applications (LLC) at 0x Nodes. IEEE Digest of VLSI TSA*: T5.4.1–T5.4.2.

10 Pey, K.L. et al. (2019). *New Insights into Dielectric Breakdown of MgO in STT-MRAM Devices. IEEE Digest of EDTM*: 264–266.

11 Carboni, R. et al. (2016). *Understanding cycling endurance in perpendicular spin-transfer torque (p-STT) magnetic memory. IEEE Digest of IEDM* 6 (4): 21.6.1–21.6.4.

12 Carboni, R. et al. (June, 2018). *Modeling of Breakdown-Limited Endurance in Spin-Transfer Torque Magnetic Memory Under Pulsed Cycling Regime. IEEE Tr on Elec. Devices* 65 (6): 2470–2478.

13 Jan, G. et al. (2018). *Demonstration of Ultra-Low Voltage and Ultra Low Power STT-MRAM designed for compatibility with 0x node embedded LLC applications. IEEE Digest of VLSI Tech*: 65–66.

14 Mihajlovic,G. et al., *Origin of the resistance-area product dependence of spin transfer torque switching in perpendicular magnetic random access memory cells*, https://arxiv.org/abs/1905.02673, 2020, July 23rd.

15 Chien, C.W. et al. (2014). *Scaling Properties of Step-Etch Perpendicular Magnetic Tunnel Junction With Dual-CoFeB/MgO Interfaces. IEEE Electron Dev Ltrs*: 738–740.

16 Nowak, J.J., Robertazzi, R.P., Sun, J.Z. et al. (2016). *Dependence of Voltage and Size on Write Error Rates in Spin-Transfer Torque Magnetic Random-Access Memory. IEEE Magnetics Ltrs* 7: 3102604.

17 Everspin datasheet: EMD3D256M08BS1/EMD3D256M16BS1, https://www.everspin.com/spin-transfer-torque-mram-products, 2020, July 23rd.

18 Van Beek, S., Marten, K., Roussel, P.J., and Donaldio, G. (2015). *Four Point Probe Ramped Voltage Stress as an Efficient Method to Understand Breakdown of STT-MRAM MgO Tunnel Junctions. IEEE Digest of IRPS*: MY.4-1–MY.4-6.

Appendix C

High-Bandwidth Design Considerations for STT-MRAM

C.1 Introduction

Cell technologies of spin-transfer torque magnetic random access memory (STT-MRAM) have been matured sufficiently that write current, write pulse width, and read speed became comparable to those of existing high-bandwidth memories such as synchronous DRAM (SDRAM) double data rate (DDR) versions. However, most of the STT-MRAMs aim for embedded applications such as e-flash or e-SRAM replacement, and those memories are still based on primitive SRAM architectures. Discrete high-bandwidth products are limited to Everspin's DDR3/DDR4-compatible 256 Mb/1Gb STT-MRAMs [1, 2], but chips have long latencies and short page sizes because of the difficulties to implement DRAM-like page modes into STT-MRAM arrays.

SDRAM DDR evolutions show a good example of how data rates are enhanced systematically based on basic DRAM cell characteristics. Thus, reviewing it before discussing STT-MRAM high-bandwidth designs is important. Nevertheless, simply copying DDR architecture is not the right way for STT-MRAM high-bandwidth designs because of the inherent difference in operation modes between DRAM and STT-MRAM cells. It needs to analyze STT-MRAM cell first, identify clearly its own strength and weakness, and explore optimum approaches. Based on this strategic method, this appendix provides an example design that exploits a full-bit prefetch architecture to achieve DDR3/DDR4-class data rates, which are based on random row access capabilities rather than DRAM's page-based column access modes. For comparison, 1Gb chip chips are assumed for both SDRAM and STT-MRAM architectures.

Magnetic Memory Technology: Spin-Transfer-Torque MRAM and Beyond,
First Edition. Denny D. Tang and Chi-Feng Pai.
© 2021 The Institute of Electrical and Electronics Engineers, Inc.
Published 2021 by John Wiley & Sons, Inc.

C.2 DRAM Fundamentals

To consider the architecture of any other high-bandwidth memory, revealing how DRAM achieves high-data rates gives important information. First, for better understanding, it is necessary to describe basic read/write operations, DRAM terminologies, operation modes, and the mechanism of DDRs.

C.2.1 Cell and Sense Amplifier – Basic Operations

Figure C.1 shows basic DRAM read/write modes and the circuit diagram of a cell and a sense amplifier. In (c), only one cell and one sense amplifier are shown, but there are a few thousands of cells with different word lines (WLs) vertically and a few thousands of cells connected to the same WL, along with bit line pairs and sense amplifiers horizontally in the actual memory array. During a standby mode, the sense amplifier is disabled by raising SetP to Vdd and pulling SetN down to GND. A bit line (BL) and a complement bit line, BL/, are tied together and precharged to 1/2 Vdd. When a read access begins, BL and BL/ are disconnected from the precharge voltage and become float, and WL rises to turn a cell transistor on. As shown in (a), if the stored data is 1, a cell capacitor is charged to Vdd, and when the cell transistor turns on, a charge sharing between BL and the cell capacitor occurs. Because the cell charge is added to BL, the BL voltage begins to increase. After a signal development time, it becomes slightly higher, about 100 mV, than BL/, and the sense amplifier detects this voltage difference and activates by turning

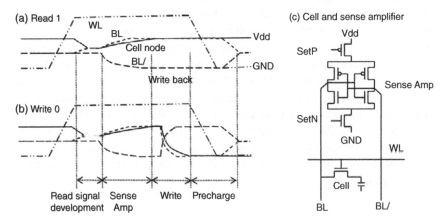

Figure C.1 DRAM operation modes; (a) Read a stored 1 signal and write it back to the cell. (b) Write the data from a 1 to a 0 and store it to the cell. (c) Circuit diagram of a DRAM cell and a sense amplifier.

SetP to GND and SetN to Vdd. It latches the read 1 signal by setting BL at Vdd and BL/at GND, and a data path circuit transfers this signal to a data I/O port.

In this read operation, stored 1 data in the cell is lost by the charge sharing with BL, and it is called a destructive read. Thus, to restore the read data, DRAM needs to have a mechanism to write back the data into the cell capacitor. As shown in (c), the sense amplifier consists of a simple cross-coupled inverter pair, and it is attached to each pair of bit lines. After sensing, it latches the data at the bit line pair, and thus the read data is written back to the cell capacitor automatically while the cell transistor is kept on. After that, WL returns to GND to turn the cell transistor off, and then BL and BL/ are tied together to the 1/2 Vdd during a precharge operation, and the array becomes ready for the next access. If the stored data is 0, there is no charge in the cell capacitor. When WL rises to turn the cell transistor on, the charge in BL flows into the capacitor, and their shared voltage becomes slightly lower than 1/2 Vdd of BL/. Thus, the cell capacitor also loses the stored 0 data resulting in the destructive read. In a write access, the same process as the read operation up to the sense amplifier activation occurs. To write a 0 to the cell that stores 1 data, a write driver switches BL and BL/ to their opposite polarities. The cell capacitor stores the 0 signal and WL goes to GND followed by the precharge operation to complete the write access (b).

C.2.2 Terminologies

RAS, CAS, and page mode: Figure C.2 depicts the timing chart of an old asynchronous DRAM page mode that had been used up to 4 Mb generation before SDRAM was introduced into 16 Mb for the first time.

As shown in Figure C.1, when any access begins, an asserted row address selects one WL, which turns on many cell transistors connected to it. Then the access, in turn, activates many associated sense amplifiers. Among many latched sense

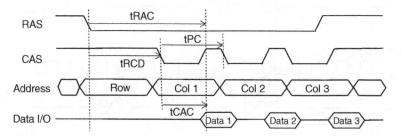

Figure C.2 Asynchronous DRAM page mode; RAS goes low and after tRCD, CAS becomes low, and the chip accepts column addresses at each CAS timing. tRAC, tRCD, tPC, and tCAC are RAS access time, RAS-CAS delay, page cycle time, and CAS access time, respectively.

amplifiers, a column address selects one. This means that row and column addresses do not have to be input simultaneously. Thus, as shown in Figure C.2, at a timing when RAS goes low, a row address is supplied, and a RAS-to-CAS delay time, tRCD, later a column address is asserted when CAS goes low.

Since the single row access can activate many cells, such a large number of data become available from the latched sense amplifiers without the time-consuming sensing process. The page mode uses the latched sense amplifier data by applying column addresses on the same row. In reads, data in sense amplifiers are transferred to data I/Os, and write accesses alter the latched sense amplifier data. As far as accessing addresses are on the same row, such column accesses are done at a faster speed than row accesses. Typical row access time (tRAC) is about 70–80 ns, while the access time from CAS (tCAC) is about 30 ns, and the page cycle time (tPC) is around 40 ns, which means data appears in every 40 ns in the page mode.

Burst mode in SDR and DDR: The burst mode is an advanced version of the column access used in SDRAM. Instead of applying each column address in the page mode, the burst mode asserts the first column address on CAS clock timing only, and the following column address data appear consecutively at clock timings by an internal column address generation circuit. In SDRAM, commands become clock-driven signals from the level-sensitive signals of the old asynchronous DRAM. RAS, CAS, and Pre (precharge command) are asserted at the rising edge of the clock. Figure C.3 shows SDRAM burst read modes. Single data rate (SDR) (a) has data at each rising edge of the clock, while data appear at both rising and

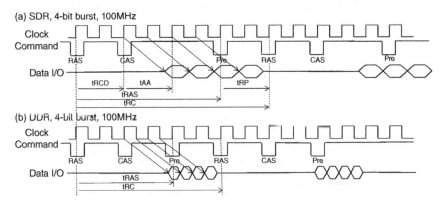

Figure C.3 SDRAM burst read modes; RAS, CAS, and other timing commands are clock-driven. (a) Data appear at each rising edge of the clock in SDR. (b) DDR generates data at both rising and falling edges of each clock.

falling edges of the clock in double data rate (DDR) (b). After 4-bit burst reads, the accessed row is precharged, and the next access goes to a different row in both SDR and DDR timing charts. They are read accesses, but the similar burst modes apply to write accesses as well. Because there is no need for consecutive column address inputs, data appears much faster than the old asynchronous page mode: every 10 ns in SDR 100 MHz and 0.625 ns in DDR4 800 MHz.

Prefetch: In SDR, one data is transferred to data path and data I/O by the clock, but DDR needs to transfer two data at a time in each clock timing because data appear at both rising and falling clock edges. Thus, DDR has a twice-wide data path bus, and two data in the latched sense amplifiers are transferred at a time. This is called a 2-bit prefetch. For an 8-bit burst, in addition to 2-bit, 4-bit, or 8-bit prefetch is used depending on clock speed in DDR2 – DDR4.

Bank and chip organization: SDRAM introduces a bank structure. For example, in the 1Gb, Figure C.4 shows an eight-bank scheme. All banks share address and data path circuits, but each 128 Mb array is independent in operations. Banks are activated by RAS commands with an interval time of RAS-to-RAS delay, tRRD, and they can operate in parallel. The purposes of the bank structure will be discussed in section B.2.5. In a JEDEC standard 1Gb DDR3, the array is organized as eight banks, and each 128 Mb bank has 8 K rows × 1 K columns × 16 data I/Os (or 2 K columns × 8 data I/Os) [3]. Thus, every single row access activates 16 K sense amplifiers in both 8 and 16 I/Os. In all of the data rates shown in this appendix is based on the 16-I/O (2 bytes) case.

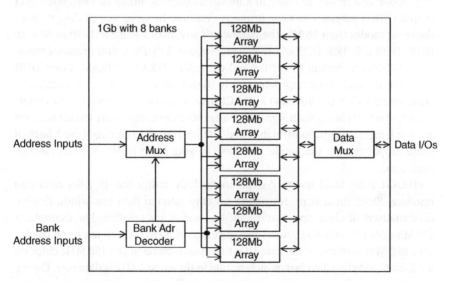

Figure C.4 1 Gb with eight banks. There are eight independent 128 Mb arrays in 1 Gb chip. Banks share address and data I/Os, but accesses can go to banks independently in parallel.

C.2.3 Basic Approach of High-Bandwidth SDRAM

In any memory array, cells are arranged in row and column directions. A selected word line, therefore, turns on many cell transistors. Since DRAM stores charge data in its capacitor cell, turning on cell transistors cause charge sharing with bit lines to result in the destructive read in all cells connected to the selected WL. Fortunately, bit lines are floating during sensing, and thus, the simplest cross-coupled inverter pair can be used without DC current as the sense amplifier, and it fits in the narrow bit line pitch, which also writes back the read data into the cell capacitor. Thus, for example, the 1 Gb chip activates 16 K sense amplifiers at a single row access. This is a significant disadvantage of DRAM, because of large peak current, slow activation time, and thus longer RAS access and long cycle times. On top of this, since the DRAM transistor is designed to focus on tight leak current, its speed is two to three generations behind logic chip transistors for the same CMOS design rule. The cycle time, tRC, improvement from DDR to DDR4 is very small, only 10 ns from 60 ns to 50 ns. However, such a large number of data in latched sense amplifiers as 16 K provides a tremendous opportunity for high-bandwidth design capability. In every row access with 50 ns cycle time, 16 K (2 K bytes) data are latched, which means 40 GB/s bandwidth is available as a data rate resource. It can be accessed by column addresses without any time-consuming sensing operation. Bandwidth evolution of all DRAMs has been relying on this single concept, with many data in the latched sense amplifiers. It stems from the old page mode and moves to faster modifications such as nibble and extended data output (EDO) schemes in asynchronous versions. In synchronous designs with the burst modes from 16 Mb, the bandwidth advances significantly from SDR to DDR, DDR2, DDR3, DDR4, low-power versions, LPDDRx, and graphics memories, GDDRx including RAMBUS and embedded DRAM, eDRAM. From DDR to DDR4, the cycle time improves to only 50 ns from 60 ns, yet data rate expands eight times, 3200 MB/s from 400 MB/s. Clearly, this is not due to scaling, but internal architectures bring such significant advancements, and many circuit features are implemented to support high-speed I/O operations. The page-based latched sense amplifiers are the single and base driving force of the high-bandwidth evolution.

eDRAM is an ideal memory structure to fully utilize the 40 GB/s data rate resource. Since there is no restriction on chip-internal data bus width, flexible combinations of clock speed and data bus width are possible. For example, a 128 Mb eDRAM with 8 M rows × 16 K columns at 50 ns cycle time can be configured to 8 M × 8 column × 2 K data I/Os to run an 8-bit burst at a 160 MHz clock on a 2 K-bit (256 bytes) data bus, which results in the same 40 GB/s data rate. On the other hand, all discrete memory products have a limitation on the number of data I/Os, 16. Thus, even DDR4 utilizes only a small fraction of the resource, 3.2 GB/s out of 40 GB/s. However, this also means that the same concept of many latched data is applicable for further bandwidth enhancement beyond DDR4 by faster I/O schemes.

C.2.4 SDRAM Operation Mechanism

Figure C.5 explains the SDRAM operation mechanism. It shows only one I/O, but there are 16 such identical circuit blocks in each chip. All cases assume to latch 16 K sense amplifiers per chip at a row access; therefore, there are 1 K data in latched sense amplifiers in one I/O.

In (a), SDR processes one data at a time between the sense amplifier and the data path circuit at 100 MHz. Through the 2-byte data I/O, it produces a data rate of 200 MB/s by the same clock rate of 100 MHz as the internal clock. All DDRs, (b), (c), and (d), prefetch data at the sense amplifiers and make a parallel-to-serial transfer between the data path circuit and I/O with wide internal data path bus widths. DDR shown in (b) uses a 2-bit prefetch scheme. In reads, it transfers two data from the sense amplifiers to the data path circuit at a time on 2-bit-wide data path bus to one I/O. It uses the same 100 MHz clock as SDR, but those 2 data appear at rising and falling edges of the clock to make 400 MB/s data rate. Write accesses proceed in a similar way. Two data inserted at rising and falling edges of the clock are transferred to the array on the 2-bit data path bus in parallel, and they are written to the cells simultaneously. For 8-bit burst mode, the chip repeats the 2-bit transfer scheme four times. As shown in (c), DDR2 has a 4-bit data path bus driven by the same 100 MHz clock as SDR and DDR and uses a 4-bit prefetch to realize 800 MB/s data rate. Similarly, DDR3 uses an 8-bit prefetch and an 8-bit wide data path bus with the same 100 MHz data path clock. To move to DDR4, the data path

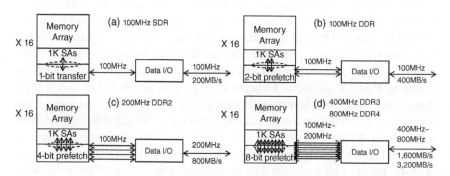

Figure C.5 SDRAM operation mechanism. (a) SDR, (b) DDR, (c) DDR2, and (d) DDR3 and DDR.

Table C.1 Key features of SDRAM family.

	SDRAM family				
	SDR	DDR	DDR2	DDR3	DDR 4
Vdd (V)	3.3	2.5	1.8	1.5	1.2
Main Clock (MHz)	66–133	100–200	200–400	400–600	600–800
Number of I/O	4,8,16	4,8,16,32	4,8,16	8,16	4,8,16
Number of Bank	2	4	4.8(>1Gb)	8	8
Number of Prefetch Bit	None	2	4	8	8
Burst Length	1,2,4,8 page	2,4,8	4.8	4.8	4.8
Max Bandwidth (MB/s)	133–266	400–800	800–1600	1600–2400	2400–3200
Density	16 –256 Mb	64 Mb ~ 1Gb	256 Mb ~ 2Gb	1Gb ~ 8Gb	4Gb ~ 32Gb

clock is increased to 200 MHz to achieve 3200 MB/s data rate at data I/Os using the same scheme as DDR3. Table C.1 shows the key features of the SDRAM family.

C.2.5 SDRAM Performance

Figure C.6 shows read timing chart examples of SDRAM DDR2 and DDR3.

As shown in (a), consecutive column accesses on the same row make data I/O seamless, data appear without any gap, and thus the data rate reaches the maximum, 800 MB/s for DDR2 at 200 MHz using 4-bit burst. As far as addresses stay the same row, this seamless operation continues. However, when access goes to a different row, the array has to be restored by issuing the precharge command, Pre, as shown in Figure C.6b and c. To begin the next new access, the chip has to go through RAS and CAS commands. If such accesses on different rows come every time, which are called random row accesses, the data throughput in SDRAM drop substantially. DDR2 with the same 4-bit burst as (a) has 1/6 of the maximum, as shown in Figure C.6b. This is not shown in Figure C.6, but at a higher clock speed of DDR3 for the 4-bit burst, the data throughput drops to 1/10 of the maximum, 160 MB/s, because a 4-bit burst takes only 5 ns, while cycle time, while tRC, remains 50 ns. Thus, at the same burst length, larger degradation in data

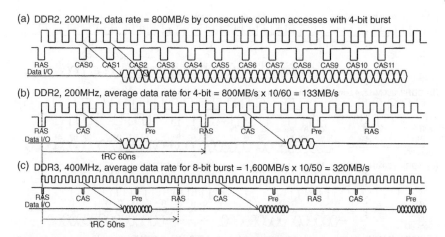

(a) DDR2, 200MHz, data rate = 800MB/s by consecutive column accesses with 4-bit burst

(b) DDR2, 200MHz, average data rate for 4-bit = 800MB/s x 10/60 = 133MB/s

(c) DDR3, 400MHz, average data rate for 8-bit burst = 1,600MB/s x 10/50 = 320MB/s

Figure C.6 SDRAM timing chart examples. (a) Column read access of DDR2. All column addresses are on the same row and data appear on data I/Os without any gap. (b), (c) When an access moves to a different row, the array has to be precharged. The seamless operation stops, and the new burst appears one cycle time later. Thus, the averaged data rates drop substantially.

throughput occurs at higher clock speeds. When the burst length becomes longer, data rates improve. Figure C.6c shows a DDR3 8-bit burst operation, where the averaged data rate is enhanced to twice of the 4-bit burst case, 320 MB/s. However, it is still only 1/5 of the maximum capability, which means SDRAM shows much poorer performance in random row accesses than page-based column accesses.

To make data rates close to the maximum at various accesses, SDRAM implements the multiple-bank structure. There are two purposes in this scheme. The first one is to increase the chances of column accesses so that the chip can keep the maximum data rates as long as possible. For example, as shown in Figure C.7a, the chip activates two banks, B1 and B2, and if accesses stay on any column of the selected two rows, it can continue seamless operations at the maximum data rates by switching two banks, the bank-interleaved mode. Activating multiple banks increases page size from 2 K to 4 K bytes in two banks and 8 K bytes in four banks to enlarge the probability of column accesses.

The other purpose of the bank structure is to improve random row access data rates. For example, Figure C.7b shows DDR2 8-bit burst operations with four banks, in which each asserted row is located in different banks. Activating such four banks serially, 8-bit bursts appear consecutively without gap to realize the maximum data rate, where R, C, and B1-B4 means RAS, CAS, bank1 – 4, respectively. However, there are some other cases that cannot make seamless operations. The first case is short burst length. Figure C.7c shows the same four-bank

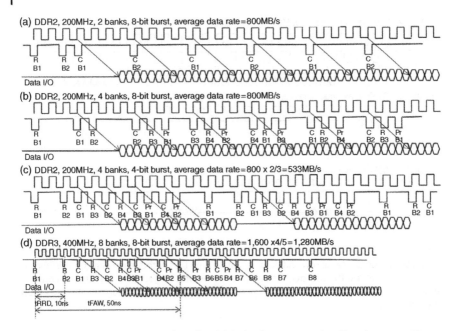

Figure C.7 Timing chart examples of multiple bank access modes. Seamless operations with 8-bit burst, by 2-bank interleaved column accesses (a), and 4-bank random row accesses (b). In the same DDR2, the 4-bit burst with four banks causes gap in data I/O (c). DDR3 has eight banks, but simultaneous activation is limited to four. Thus, even 8-bit burst results in 8-bit-burst long gaps between fourth and fifth burst. The averaged data rate becomes 4/5 of the maximum (d).

operations as (b), but the burst length is four. It cannot continue 4-bit burst after the bank 4 access, and the data throughput decreases to 533 MB/s, although it is better than without bank shown in Figure C.6b, 133 MB/s. The second example is DDR3 shown in Figure C.7d. It has eight banks and an 8-bit burst, but simultaneous bank activation is limited to four because of power reasons. Up to fourth bank activation, RAS commands are asserted every 10 ns interval according to tRRD, RAS-to-RAS delay time, but before activating the fifth bank, the chip has to restore the first accessed bank, B1, in which a precharge is assumed at one clock after C B3 timing. The timing specification of tFAW, four active windows, is required to wait for the completion of the bank 1 precharge. Thus, even with the 8-bit burst, there appears a gap of one 8-bit burst lapsed time, and the data throughput becomes 4/5 of the maximum. The multiple bank structure improves the random row access performance of SDRAMs, but it does not necessarily mean to always provide the maximum data rates. The random row access capability, therefore, still remains a weakness of SDRAM.

C.3 Random Row Access Performance Analysis

The page-based SDRAM architecture delivers the maximum data rates as far as accesses stay on the selected row, or rows, in single or multiple-bank cases. However, random row accesses cause gaps in data I/Os for short burst length, 4-bit, or limited number of banks, and the averaged data rates decrease substantially from the maximum. As clock frequency increases, data throughput degradation becomes worse. There is a fundamental principle behind those symptoms. Basically, if the time that an x-bit burst elapses is equal or longer than the chip cycle time, tRC, all random row accesses become seamless, because the row access completes the precharge and becomes ready for the next row access before the ongoing burst ends. In reality, however, even the 8-bit burst time, 20 ns – 5 ns in DDR2/DDR3/DD4, is shorter than tRC, and random row accesses always result in degraded data rates from the maximum. Thus, the time ratio between burst time and cycle time is an important factor for random row access performance, and it is defined as B/C ratio, BCR. The averaged data rate is calculated by multiplying BCR to the maximum.

Table C.2 shows the BCRs of various SDRAMs. In every DDR generation, the clock frequency becomes twice, and burst time reduces to 1/2, but tRC improves from 60 ns in DDR/DDR2 to 50 ns in DDR3/DDR4. Thus, BCR drops significantly, resulting in a large reduction in data throughput. For example, the averaged data rates of DDR4 is only 5% of the maximum in 4-bit burst, and 10% in even 8-bit burst for random row accesses with single bank operations.

Table C.2 BCR and averaged data rates for random row access in single bank case.

								Random row access average data rate MB/s				
		Chip specification										
		Max data rate	Cycle time	Burst time (ns)		BCR		Single bank			Multiple banks	
Type	Clock (MHz)	(MB/s)	(ns)	4-bit	8-bit	4-bit	8-bit	4-bit burst	8-bit burst	#of Bank	4-bit burst	8-bit burst
SDR	100	200	80–120*	40	80	0.50	0.67	100	133	2	200	200
DDR	100	400	60	20	40	0.33	0.67	133	267	4	400	400
DDR2	200	800	60	10	20	0.17	0.33	133	267	4	533	800
DDR3	400	1,600	50	5	10	0.10	0.20	160	320	8**	640	1,280
DDR4	800	3,200	50	2.5	5	0.05	0.10	160	320	8**	640	1,280

Note: * SDR cycle tunes are 80ns and 120ns for 4-bit and 8-bit burst cases respectively.
**DDR3/DDR4 have 8 banks, but simultaneously activation is limited to 4 banks.

Multiple banks improve random row access data rates. Activating each bank in the tRRD interval, which is the same as the burst time, puts one set of x-bit burst in the same tRC. Thus, in an n-bank chip, total burst time becomes n times of x-bit burst, and BCR is enhanced to n times. When this calculation becomes equal to or greater than one, operations become seamless accesses. Those are SDR, DDR in both 4-bit and 8-bit burst, and DD2 with 8-bit burst cases, as shown in Table C.2. For example, DDR2 8-bit burst has a BCR of (20 ns burst time)/ (60 ns tRC) times four banks, which is larger than one. This, in turn, means that the cycle time of the same bank does not have to be 60 ns, but can be extended to this BCR times 60 ns, which is 80 ns. This is the situation that Figure C.7b shows. To keep the seamless operation to maintain the maximum data rate, 800 Mb/s, bank 1 is activated for the second time 16 clocks later, 80 ns, from the first activation.

However, DDR3 and DDR4 cannot make seamless operations because of the small BCRs and limited simultaneous bank activation to four. Data rates shown in multiple banks column of Table C.2 are the highest possible numbers of random row accesses, which are substantially lower than the maximum.

In summary, page-based column accesses of SDRAM can achieve the maximum possible data rates, but its random row access performance is inferior to the column accesses. The degree of data throughput degradation depends on burst time, cycle time, and number of banks. BCR and number of banks systematically show the averaged data rates for random row accesses, and the principle is applicable to any other high-bandwidth memory.

C.4 STT-MRAM Fundamentals

As understanding DRAM cell characteristics is important to design SDRAM high-bandwidth memory chips, reviewing STT-MRAM fundamentals is imperative to define chip architecture that has high-bandwidth performance similar to or better than SDRAM.

C.4.1 Cell and Basic Operation

Figure C.8 illustrates STT-MRAM cell and basic read and write operations.

In (a), the configuration of an MTJ, a cell transistor, and a bit line pair, W1BL/ W0BL, is shown. During read, a small voltage less than 100 mV is applied to W0BL with respect to W1BL (b). To write 0, W0BL is connected to a bit line voltage supply, and W1BL is set at ground. Thus, electrons flow from the pinned layer to the free layer to make the cell to P-state. The opposite connection makes write 1 to switch the cell to AP state by the electron flow from the free layer to the pinned

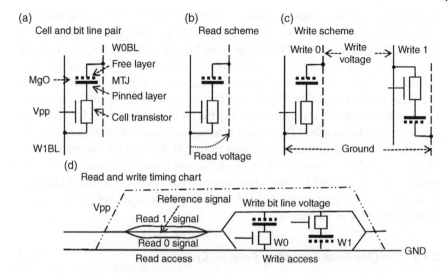

Figure C.8 (a) Configuration of an MTJ, a cell transistor, and bit lines. (b) Read access wiring connections. (c) Cell to bit line pair connections for write 0 and 1. (d) Read and write timing chart.

layer (c). Figure C.8d shows a timing chart or read and write. Before access, W0BL and the reference bit line are tied to an intermediate voltage between Vdd and ground. When a word line voltage, Vpp, rises, the cell transistor turns on. A read signal of either 1 or 0 appears across W0BL and the reference bit line, which generates the middle level of cell currents between data 1 and data 0. A sense amplifier (not shown) detects the read signal as either data 1 or 0. After the read and turning off the sense amplifier, a write pulse voltage is applied to W0BL or W1BL depending on write 0 or 1, as shown in (c).

During both read and write accesses, DC currents flow in the cell. The only difference between read and write is the number of currents, much smaller read currents than those of write. Therefore, STT-MRAM operating schemes are quite different from DRAM. Relatively large DC pulse currents are required to write STT-MRAM cells, while DRAM needs displacement currents to store the data into the cell capacitors. The read signal by DC current is robust, having better immunity to noise compared to floating node voltages of DRAM. The most important fact is that STT-MRAM is a nondestructive read. One selected word line turns on many cell transistors connected to both selected and unselected bit lines. Since STT-MRAM cell stores data as MTJ resistances, R_P and R_{AP}, through magnetizing directions of the free layer, those cells on selected and unselected bit lines do not lose stored data. Thus, unlike DRAM, it is not necessary to have the sense amplifier and its write-back mechanism on each bit line pair. Or it rather has to say that in the first place, the STT-MRAM sensing scheme cannot have such a sense amplifier in each bit line pair. The sensing bit line draws DC currents, and the bit line pair during sensing is not true and complement bit lines like DRAM, but lines to apply a read bias voltage, as shown in Figure C.8b. The pair is necessary to write data into the cell, as Figure C.8c shows. The sense amplifier compares the cell current, which flows from W0BL to W1BL, and the reference current, which is shared by many bit lines and located far away from the bit line under the read access. This is possible due to the robust DC current sensing, and the sense amplifier generally consists of current-mirror circuits, which cannot fit the tight bit line pitch. Thus, one sense amplifier is also shared by many bit lines through bit switches. Nondestructive read of STT-MRAM and no necessity of sense amplifier at each bit line are good features. However, unlike DRAM, this means it is impossible to have page-based many latched read data.

C.4.2 On-Chip Error-Correcting Code (ECC)

Memory products have to meet the stringent reliability requirements that customers demand. One hundred FITs are typical criteria. Converting FIT into bit error rate (BER) gives clear views of how it is severe. FIT is a failed product count in one billion device hours, typically during 1000 operating hours of one million chips. For a 1 Gb-chip chip case, DDR4 data rate is 3.2 GB/s, 25.6 Mb/s. During 1000 hours, there are 25.6 Gb/s $\times$ 3.6 $\times$ 10^6s = about 10^{17} cell accesses per chip, and there are one million chips for FIT calculation. Thus, one billion device hours are equal to $10^{17} \times 10^6 = 10^{23}$ cell accesses, and having 100 FITs means $100/10^{23} = 10^{-21}$ BER, because one cell error is counted as one product failure. To satisfy the typical product failure rate of 100 FITs, BER needs to be about 10^{-21} in 1 Gb chips.

Regarding BER, STT-MRAM faces a difficult situation compared to the customer requirements. Its cell has magnetics-unique characteristics no other memory cell possesses. As Chapter 7 shows the detailed description of the stochastic write properties, there is a finite probability that write access with sufficiently large current density cannot switch the cell state, a write error rate (WER). Yet, there is a finite probability that read access with small current density flips the cell state, a read disturb rate (RDR). Furthermore, there is a finite probability that cell data changes even during no operation at all, a data retention error rate (RER). These are not due to defective cells nor due to variations of write condition. Pick up a perfectly healthy cell, and write 1 and 0 alternately at the exact same condition for a million to a billion times, write fails a few times. All cells are alike, but there are variations in cell parameters naturally. Thus, chip-level BER is even worse than cell-level. WER, RDR, and RER are all BERs, and it is impossible to meet such tight BER requirements of other memory products, 10^{-21}. For example, suppose the cell is designed to have RDR and RER of 10^{-21}, yet to have the same 10^{-21} for WER, the write current density must be enormously high. In realistic designs, STT-MRAM BERs are in a range of ppm – ppb (10^{-6}~10^{-9}).

To satisfy the typical reliability requirements of memory products, high-performance on-chip error-correcting code (ECC) is an absolute necessity for STT-MRAM. ECC makes the product failure rate to about 100FITs, even though BER of STT-MRAM is so high as 10^{-6}~10^{-9}. Figure C.9 shows an example of the relationship between product failure rate and BER regarding two ECC architectures [4]. Since 10^{12} random trials are about 10^9 device hours, the 100pdm line is identical to 100 FITs. To meet the product reliability of 100 FITs, BERs of all errors are less than about 10^{-7} using such a powerful on-chip ECC, three-error correction with 128 data bits per codeword.

The consequent meaning of having on-chip ECC is that every access of the chip has to read/write many cells simultaneously internally, although the chip has only 16 I/Os. The number of simultaneous activating bits is determined to compromise ECC performance and area overhead by check bits, but data bits of the codeword are in a range of 64–512. The other outcome of the on-chip ECC is to have write access after read. When a read access has an error, it is corrected by ECC before the data appear on I/Os, but the cell itself keeps the original wrong data. Such error remains until the cell is written, but there are many more reads than writes generally. To avoid such error accumulation, if error occurs in reading, the cell gets a corrective write. Thus, read access allocates a write timing spot after read. This access mode, read followed by write, is also useful to save write currents. In write access, the chip reads the date first and compares it with the ECC-encoded write data and then writes only cells that need to be altered. Therefore, generally, STT-MRAM has a single access mode, read first and write, for both read and write operations.

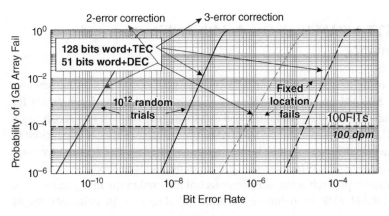

Figure C.9 Product failure rate and BER regarding two ECC architectures [4].

C.4.3 High-Bandwidth Resource

STT-MRAM needs to activate many cells simultaneously because of ECC codewords, but it cannot fire so many sense amplifiers as 16 K of SDRAM at a time. Such data in a large number of latched sense amplifiers are the significant resource of high bandwidth capability of SDRAM. In STT-MRAM, nondestructive read and no necessity of sense amplifier in each bit line are advantages in terms of circuit area and access power. However, at the same time, it means that the page-based column access mode for high-bandwidth designs is not available for STT-MRAM. Because of differences in the nature of cells, page-like access modes are incompatible to the STT-MRAM array.

On the other hand, STT-MRAM has some advantages compared to SDRAM. Because of no leak current constraints, it can choose logic process technologies for CMOS, where transistors are much faster than DRAM process technologies, two to three generations ahead at the same design rule. Recent progress in write speed is remarkable; a write pulse width well shorter than 10 ns is reported [5]. Selecting MTJ resistance by using lower resistance-area product, R_a, allows faster read speeds. Because of the nondestructive read, there is no need of the restore operation. Those suggest that STT-MRAM possess a fast-random access capability. SDRAM has high-bandwidth resource by the many data latched in sense amplifiers to realize the maximum data rates through the page-based column accesses, and this is a significant strength. However, its weakness is the much inferior random row access capability. Quite contrary to SDRAM, many data in the latched sense amplifiers are not available to STT-MRAM, and its high-bandwidth design resource is the better random row access capability relying on fast cycle time rather than the page-based column accesses.

C.5 STT-MRAM High-Bandwidth Architecture and Performance

Because of the inherent differences in cell characteristics and sensing schemes between SDRAM and STT-MRAM, each has its own strategic focus points for high-bandwidth designs. Exploiting and enhancing the random row access capability is the way to pursue for STT-MRAM. It means to make BCR large, close to one; namely, reducing cycle time is the most important task. Besides the advantage of using fast logic CMOS transistors, there are some circuit methods to reduce cycle time. A full-bit prefetch is the old circuit technique used in the first generation of 16 Mb SDRAM SDR to reduce cycle time [6]. Figure C.10 demonstrates its operation.

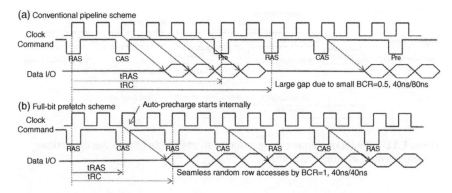

Figure C.10 (a) Conventional pipeline SDRAM SDR operating mechanism. The memory array has to be activated until the last burst data leaves it to result in a long cycle time, 80 ns. (b) The full-bit prefetch scheme transfers all burst data right after the sense amplifiers are activated, which makes the array to precharge slightly after CAS. Thus, cycle time can be shorter, 40 ns, to make the BCR one resulting in the seamless operation for random row accesses.

In a conventional pipeline SDRAM SDR, the burst mode transfers each column data from latched sense amplifiers by each clock after CAS. Since the array has to wait to start restore until the last burst data leaves it, the possible precharge timing is at one clock before the last burst data, as shown in Figure C.10a. Thus, tRAS, RAS-to-precharge timing, is 60 ns, the chip needs 20 ns to restore, and tRC, cycle time, becomes 80 ns for the 4-bit burst case resulting in BCR = 0.5, and the averaged data rate is half of the maximum. On the other hand, the full-bit prefetch scheme shown in (b) transfers full 4-bit burst data to small latches outside of the array as soon as sense amplifiers are activated. It allows the array to begin precharge as early as slightly after CAS, which results in 20 ns tRAS, and after 20 ns of tRP, the array becomes ready for a next row access. Cycle time, therefore, is shortened to 40 ns, which is the same as the burst time, resulting in BCR = 1 to make the seamless accesses.

Figure C.11 illustrates the other circuit arrangement, an array-locking system, to shorten cycle time further. In a read access, after a row access begins, when the address is decoded to select a word line, the array is locked to isolate from the address circuit and keeps the selected word line activated by a precoder or a decoder circuit step. After sense amplifiers are fired, the full-bit prefetch circuit transfers all burst data to latches outside of the array, an ECC-corrected bit writing proceeds, while burst data appear at I/Os. Because the array is locked and isolated from the peripheral circuits, the next address can be inserted without waiting for the write completion. As soon as the write operation ends, the chip releases the lock, and the array accepts the new address to activate the next word line. In a

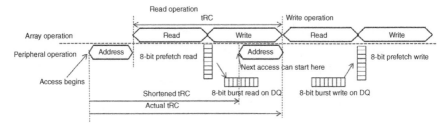

Figure C.11 The locking array configuration to shorten cycle time to the theoretical minimum length as array cycle time.

write operation, the chip reads the data first. In parallel to this, burst write data are asserted, and when burst write inputs end, the read data are compared with the ECC-encoded write data. The full-bit prefetch write stores all burst data simultaneously to cells, although actual bits to be written are limited to those that are different from the read bits.

Using the full-bit prefetch scheme and the array-locking system, the array and the peripheral circuits such as address and data path can run in parallel without waiting for each task's completion. This makes the chip operate at the shortest possible cycle time, simply only reading and writing portions of the array. Namely, the chip cycle time can be shortened to the array cycle time, which is the theoretically minimum.

With the minimum possible cycle time, it is ready to consider the high-bandwidth STT-MRAM chip based on the better random row access capability. The architecture implements the full 8-bit prefetch, in which the chip transfers 128 data bits to small latches from the sense amplifiers at a time for reads, and stores 128 data into cells simultaneously for writes in the 16-I/O configuration. The total 128 bits are just about good size for data bits of the ECC codeword. There are some options for selecting ECC circuits, two- or three-error correction per 128 data, or two sets ECC for upper and lower bytes, each has 64 data bits for the codewords.

The most important parameters are read and write times. For recent STT-MRAM technologies, 10 ns for both read and write times including ECC decoding/encoding are reasonable. Setting the minimum cycle time at 20 ns, Table C.3 lists possible high-bandwidth product lines, and Figure C.12 shows their timing charts. It covers a wide range of performance grades, from DDR to DDR4, depending on the sorting of cycle time speeds and the clock frequencies. All cases assume 8-bit burst operations. In a low-speed DDR class, 100–200 MHz, the cycle time of each speed sort is the same as burst time, 8 times 1-bit data time; thus, BCR is 1, and all random accesses become seamless without any bank. Since the clock speed doubles from DDR to DDR2, and from DDR2 to DDR3–4, BCR drops to 0.5

Table C.3 STT-MRAM high-bandwidth chip offering depending on sorting of cycle time speeds and the clock frequencies.

Category	Speed sort, tRC (ns)	40	36	30	24	20
DDR class	1-bit data time (ns)	5.0	4.5	3.75	3.0	2.5
BCR = 1 (8-bit burst)	Clock (MHz)	100	111	133	167	200
No bank necessary	Clock for data (MHz)	200	222	267	333	400
Always seamless	Data rate (MB/s) for x16	400	444	533	667	800
DDR2 class	1-bit data time (ns)	2.5	2.25	1.875	1.5	1.25
BCR = 0.5 (8-bit burst)	Clock (MHz)	200	222	267	333	400
Seamless by 2 banks	Clock for data (MHz)	400	444	533	667	800
	Data rate (MB/s) for x16	800	889	1,067	1,333	1,600
DDR3-4 class	1-bit data time (ns)	1.25	1.125	0.9375	0.75	0.625
BCR = 0.25 (8-bit burst)	Clock (MHz)	400	444	533	667	800
Seamless by 4 banks	Clock for data (MHz)	800	889	1,067	1,333	1,600
	Data rate (MB/s) for x16	1,600	1,778	2,133	2,667	3,200

and 0.25, respectively; thus, DDR2 and DDR3–4 need two and four banks, respectively, to make seamless operations.

In the STT-MRAM high-bandwidth designs, there is no fast column access mode, because it is not compatible to the page-based architecture. However, to have the seamless data I/O operations, the DDR2-DDR4 chips use the bank-interleaved modes, as shown in later parts of Figure C.12. This achieves the same maximum data rates SDRAM delivers by page-based column accesses. Besides the same maximum data rate capability, STT-MRAM has better random row access performance than SDRAM because of the faster cycle times.

The data rates of the DDR class always stay at the maximum levels for any random row address regardless of the clock frequency, from 100 to 200 MHz. There is no need for a bank. The BCR of the DDR2 class with 200–400 MHz clock speeds is

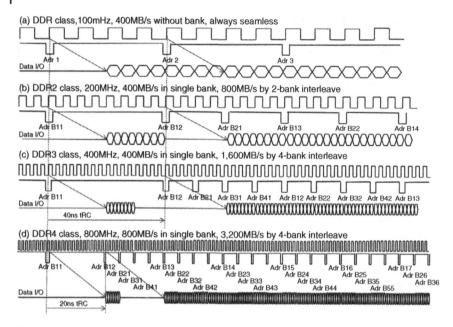

Figure C.12 STT-MRAM high-bandwidth chip timing charts.

0.5, and the averaged data rate of random row accesses becomes 1/2 of the maximum, 400 MB/s at 200 MHz. With two banks, it can make seamless operations to have the maximum data rate, 800 MB/s. Both DDR3 and DDR4 classes have 0.25 in the BCR, and the averaged data rates for random row accesses are 1/4 of the maximum. To have seamless operations, the number of banks has to be increased to four. Unlike SDRAM DDR3 and DDR4, they do not need eight banks because of the short cycle time. The timing charts of DDR, DDR2, and DDR3 classes shown in Figure C.12 are based on a 40 ns cycle time. In speed sorting of each class, since the cycle time and the clock frequency change at the same rate, the timing relations between the clock and the burst data remain the same. For example, in a 20 ns cycle time case, just tRC has to be labeled to 20 ns instead of 40 ns of Figure C.12a–c. Thus, the BCRs of each class are the same for all speed sorts from 40 ns to 20 ns cycle times, as shown in Table C.3.

C.6   Competitiveness Analysis

Data rates: Table C.4 lists the data rate comparison between JEDEC SDRAM and STT-MRAM high-bandwidth design. Since both architectures can deliver the maximum data rate in different ways, page-based column accesses in SDRAM and

Table C.4 Random row access data rate comparison between SDRAM and STT MRAM high-bandwidth design.

						Random row access performance							
						JEDEC DDRx				STT MRAM			
		8-bit burst time	Maxi-um data rate	tRC	#of		Data rate (MB/s)		tRC	#of		Data rate (MB/s)	
Category	Clock (MHz)	(ns)	(Mb/s)	(ns)	bank	BCR	1-bank	Mul-bank	(ns)	bank	BCR	1-bank	Mul-bank
	100	40	400			0.67	267	400	40			400	NA
	133	30	533			0.50	267	533	30			533	NA
DDR	200	20	800	60	2	0.33	267	533	20	0	1.0	800	NA
	200	20	800			0.33	267	800	40			400	800
	267	15	1,067			0.25	267	1,067	30			534	1,067
DDR2	400	10	1,600	60	4	0.17	267	1,067	20	2	0.5	800	1,600
	400	10	1,600			0.20	320	1,280	40			400	1,600
DDR3	533	7.5	2,133	50	8*	0.15	320	1,280	30	4	0.25	533	2,133
	533	7.5	2,133			0.15	320	1,280	30			533	2,133
DDR4	800	5	3,200	50	8*	0.10	320	1,280	20	4	0.25	800	3,200

Note 8*: there are 8 banks, but only 4 banks are allowed to activate simultaneously.

multiple bank-interleaved mode by STT-MRAM, the table shows differences in random row access performance. The worst-case happens when continuous accesses come on different rows in the same bank, in which data rates are listed in column marked as "1-bank." SDRAM always has inferior data throughput than STT-MRAM because of the small BCR. In each category, data rates stay constant, with no improvement at higher clock speeds. Even with DDR4 800 MHz, the data rate is very low, only 320 Mb/s, which is 10% of the maximum.

On the other hand, because BCR is 1 in the STT-MRAM DDR class, it always delivers the maximum data rates. Even 100 MHz operation has a better data throughput, 4000 Mb/s, than 320 MB/s of JEDEC DDR4 at 800 MHz. However, in DDR2 to DDR4 classes of STT-MRAM, data rates of random row accesses in the single bank remain the same as the DDR class. With multiple banks, data rates become better in both JEDEC SDRAM and STT-MRAM cases, which are listed in the column marked by "mul-bank." However, improvements of JEDEC SDRAM is much smaller than STT-MRAM, which always achieves the maximum data rates using two-four banks. In JEDEC DDR3 and DDR4, data rates never reach the maximum, 8–40% of it, because only four banks are allowed to activate simultaneously even though the chips have eight banks. Overall, STT-MRAM high-bandwidth design has significantly better performance than JEDEC SDRAM.

Active currents: To estimate active currents, STT-MRAM has to assume a certain MTJ design. Table C.1 of Appendix B lists various MTJ designs, and using a 40 nm MTJ seems to be the optimum selection, small switching current yet not too high in read disturb rate. However, the table is based on a rather slow read speed, and for 10 ns read with ECC decoding, it had better assume a 7 ns signal development time. To meet this, R_p and R_{Ap} are designed to 1.7 KΩ and 4.0 KΩ, respectively. In this case, the switching current with 0.01 ppm WER is slightly higher, 200 µA, for the 40 nm MTJ. For full 8-bit prefetch, the chip has to activate about 160 bits simultaneously, which are 128 data bit and ECC check bits. At a 20 ns cycle time and a 10 ns write pulse width, the averaged write current is 16 mA. During the 4-bank interleaved operations of DDR4, the total write-cell current becomes 64 mA. STT-MRAM has to use the bank-interleaved mode instead of the page-based column access mode of SDRAM. Because of the smaller number of simultaneous activating cells, the total bank-interleaved active current, which is 64 mA plus others due to peripheral circuits, may not be a serious concern.

Timing specification: Because there are differences in cell characteristics and chip architectures, STT-MRAM is not package-pin compatible with JEDEC SDRAM. Unlike SDRAM, there is no reason for the column address to wait after tRCD from a row address. STT-MRAM can insert all addresses instantaneously, and an SRAM-like nonmultiplexed addressing scheme is rather preferable to save time for avoiding the two-step address insertion. Although address pin count increases, the memory controller becomes simpler, because tRC and tRRD are the only timings to be controlled. Table C.5 shows a timing specification of JEDEC SDRAM and STT-MRAM for DDR3 and DDR4. Because of the nonmultiplexed address scheme, tRCD and tAA are not applicable for SATT MRAM. It also does not need the restore operation due to nondestructive read; thus, tRAS and tRP are not necessary to specify. For SDRAM, tFAW is required to limit the simultaneous activations to four banks out of eight, but STT-MRAM has only four banks to make seamless I/O operations in DDR3 and DDR4, thus, it is also not necessary

C.7 Summary

Because of the destructive read, SDRAM has to have sense amplifiers in each bit line pair to latch the read data and write them back to the cells. In a typical 8-bank 1Gb chip, a single row access needs to fire 16 K sense amplifiers at a time. This is a

Table C.5 Timing specification of JEDEC SDRAM and STT-MRAM for DDR3 and DDR4.

DDR3, DDR4 common timing specification			
Parameter	Symbol	JEDEC	STT-MRAM
Active to read or write delay	tRCD	14	NA
First read data from column	tAA	14	NA
Active to first read data		28	24
Precliard command period	tRP	14	NA
Active to active command period, cycle time	tRC	49	20–40
Active to precharge command period	tRAS	35	NA
ACT to ACT command period, different banks	tRRD	5–10	5–10
Four active window	tFAW	25–50	NA
		Unit : ns	

significant disadvantage because of large currents and slow row access speed. However, SDRAM turns this disadvantage into the resource for page-based high-bandwidth operations. Since there are so many data in sense amplifiers, applying column address with the burst mode enables chip I/Os to process read/write data at high-speed clocks. All SDRAM evolution from 100 MB/s SDR to 3.2GB/s DDR4 rely on this single concept, page-based many data in latched sense amplifiers. As far as access goes on the same row, data I/Os keep the seamless operations. However, when an address changes to a different row, the chip has to go through precharge and another row activation, which takes rather long cycle time, tRC, and the seamless operation halts resulting in lower data throughput. Implementing multiple banks recovers data rates, but its improvement is small. Thus, random row access performance remains the weak point of SDRAM, because of the small BCR, burst time to cycle time, ratio.

STT-MRAM is nondestructive read; thus, it does not need a sense amplifier at each bit line. Or it cannot have such a configuration in the first place because of the DC current sensing scheme. Sense amplifiers, which consist of current mirror circuits, are shared by many bit lines. This means page-based fast column accesses are not available as the high-bandwidth resource for STT-MRAM. On the other hand, the nondestructive read does not need the restore operation, so no tRP is necessary. DC current sensing has robust signals with fast read time. Write pulse width less than 10 ns is now common. STT-MRAM can use fast transistors of logic CMOS technologies because of no leak current constraint. All those facts imply that infrastructures for better random row access performance are available. Thus, the fast row access capability is the resource for high-bandwidth designs of STT-MRAM. The focused task is to make BCR large, close to one, namely, to shorten cycle time. The full-bit prefetch architecture and the array-locking system isolate the memory array operations from all peripheral circuit actions, and cycle time can be shortened to the theoretical minimum array cycle time, which includes only the read and write portions of array operation. Setting this minimum cycle time to 20 ns and relaxing it to 40 ns by speed sort, STT-MRAM high-bandwidth designs offer the wide variety of product lines from 200 Mb/s SDR class to 3.2Gb/s DDR4 class. The bank-interleaved mode delivers the same maximum data rates as SDRAM column accesses. STT-MRAM has substantially better random row access performances than SDRAM in all classes.

References

1 Everspin datasheet: EMD3D256M08BS1/EMD3D256M16BS1, https://www.everspin.com/spin-transfer-torque-mram-products, access on 2020, July 23rd.

2 Everspin datasheet: EMD4E001GAS2, https://www.everspin.com/spin-transfer-torque-mram-products, access on 2020, July 23rd.

3 Micron datasheet: MT41J128M8/ MT41J64M16, https://www.micron.com/products/dram, access on 2020, July 23rd.

4 Alzate, J.G. et al. (2019). 2 MB Array-Level Demonstration of STT-MRAM Process and Performance Towards L4 Cache Applications. *IEEE digest of IEDM*: 2.4.1–2.4.4.

5 Jan, G. et al. (2014). Demonstration of fully functional 8Mb perpendicular STT-MRAM chips with sub-5ns writing for non-volatile embedded memories. *IEEE digest of VLSI Tech.*: 50–51.

6 Sunaga, T. et al. (January 1997). An Eight-Bit Prefetch Circuit for High-Bandwidth DRAMs. *IEEE J of Solid-State Circuits* 32 (1): 105–110.

Index

a

abnormal switching
  backhopping 164–165
  ballooning (*.cf* Bifurcation
    switching) 165
Ampère's molecular current model 20
angular momentum 20–26, 29, 32, 43,
  56–57
anisotropic magnetoresistance (AMR)
  25, 77–79
annealing 57, 83–86
antiferromagnetic/antiferromagnetism
  28, 37–40

b

Bethe-Slater curve 44
binary device 243, 246, 263
Bohr magneton 23, 27, 37, 61
Brillouin function 31, 35–36, 38
burst mode access 302, 315,
  321–315
  bifurcation switch 165

c

CMOS transistor
  cf (FET, transistor) 151–152, 156–163,
  174–176, 178, 182–183, 186–166

colossal magnetoresistance (CMR) 78
compensation temperature 40–41
Coulomb interaction 25, 42, 56
crystal field effect 26, 56–57
cubic anisotropy 53–54
Curie's law 32
Curie temperature 34–35, 37, 45
Curie-Weiss' law 34
current-in-plane (CIP) 81
current-in-plane tunneling
  (CIPT) 87–89
current-perpendicular-to-plane
  (CPP) 81
cyber security 261

d

damping 22, 51, 115–120, 123, 126–127,
  129, 136
damping constant 117, 119, 121–122,
  129, 136, 199
  MTJ capping 177
data retention
  at chip level 173
  magnetic decay model 170
  retention bake 171
density of states 30, 32–33
demagnetization field 10, 116, 119,
  123, 126

Magnetic Memory Technology: Spin-Transfer-Torque MRAM and Beyond,
First Edition. Denny D. Tang and Chi-Feng Pai.
© 2021 The Institute of Electrical and Electronics Engineers, Inc.
Published 2021 by John Wiley & Sons, Inc.

demagnetization field (*cont'd*)
demag factor   10, 152–154, 184
dielectric breakdown, MgO dielectric
breakdown   153–164,
167–168, 195
diamagnetic, diamagnetism   28–30
discrete memory (MRAM,
DRAM)   241–242, 246, 248, 250,
254, 258–259, 261, 267
domain   51, 57–60
domain wall motion   48, 51, 57–59, 135,
205–206, 220
current-induced domain
wall motion (CIDM) cell,
single bit   205–209, 230–231
Racetrack   201–211, 231–232
threshold current density   137,
206–208, 211, 213, 228, 232
dual MgO MTJ   185–187
dynamic memory (DRAM)   241–249,
251, 254–26, 267–269, 299
synchronous DRAM   299, 301–310,
314–315, 317–322
Dzyaloshinsky-Moriya interaction
(DMI)   42, 48, 59

e
easy axis   52–55
ECC
error correcting code   312–316, 320
Einstein-de Haas experiment   20
electron spin   21–23, 29
embedded memory (MRAM)   241–242,
248–251, 264, 267, 271
endurance
MgO endurance   151, 173, 177–178,
187, 191–192, 194–195
MgO degradation model   167
energy barrier
switching energy barrier   152–154,
161–163, 170–171, 176, 184

thermal energy barrier   15, 151,
169–171, 176, 185, 187–188,
193, 195
exchange coupling constant   44–48
exchange, direct   42–45
exchange field   34
exchange, indirect   42, 45–48
exchange interactions   28–29, 42
exchange of spin moment (torque) from
carrier to local moment   120, 125,
127, 134, 135
of magnetic tunnel junction   128, 131
of spin valve   127

f
Faraday effect   76
Faraday's Law   5
Fermi energy   30, 33, 37, 50
Fermi wave vector   46
ferrimagnetic/ferrimagnetism   28,
40–41
ferromagnetic/ferromagnetism   28, 34
ferromagnetic resonance   118
resonance half- line-width   199
field MRAM
half-select bit disturbance   101
read / write operation   100
FIT   302, 312–313

g
garnets   35, 45, 49, 51
giant magnetoresistance (GMR)   77–81
gyromagnetic ratio   20, 23

h
Hall effect
Anomalous Hall effect   105–108, 140
inverse spin Hall effect   113
Ordinary Hall effect   105, 107, 108
spin Hall effect   106
hard axis   54–55

hard-disk drive (HDD)   79, 81
Heisenberg model   44
Hund's rules   25–27, 37, 57, 61

i
Internet of things (IOT)   261
Ion-beam etching (IBE)   255–256

j
Julliere's formula   82, 89

k
Knudsen cell   71

l
Landau diamagnetism   30
Landau-Lifshitz equation   114
Landau-Lifshitz-Gilbert (LLG) equation   114, 116–118, 121, 125, 133
Landé g-factor   23, 26, 163
Larmor precession frequency   21, 23

m
Macrospin model   125
magnetic anisotropy   51–58, 63
   origins of   54–57
magnetic field from current
   Ampère's circuital law   6
   Biot-Savart's Law   7
magnetic field from material   11
magnetic flux   5
magnetic induction   5
magnetic moment   3
   of current loop   4
   origin of   19–27
magnetic poles   1
   monopole   3
magnetic precession   21–22, 60
magnetic torque   4, 5
magnetic thin film
   characterization of   72–76
   growth of   67–72

magnetic tunnel junction (MTJ)
   constant thermal barrier scaling   188
   design, consideration   177
   functional-based scaling   279
   scaling   173, 181–183, 187–188
magnetic tunnel junction (MTJ)   81–89
magnetization   6
   irreversible switching   95
   origin of   19–27
   reversible rotation   94
   saturated magnetization   6
   under two orthogonal eternal
      fields   96
magnetocrystalline anisotropy (MCA)   25, 56–57
magneto-optical Kerr effect (MOKE)   74–76
magnetoresistance (MR)   77–78
magnetostatic energy   4
mean field   45
memristor   232
molecular beam epitaxy (MBE)   71–72
moment exchange
   direct   42
   Dzyaloshinskii-Moriya-interaction
      (DMI)   48, 221
   indirect   42, 44–46
   RKKY exchange   47, 123
MRAM
   cell size   250–251, 254–257
   discrete MRAM   241,
      246, 254, 259, 261, 265,
      299, 304
   embedded MRAM   177, 241–242, 245,
      248–251, 267, 271
   production ecosystem   266

n
Néel temperature   37–39
non-volatile Dual-in-line memory
   module (nvDIMM)   242, 260–261
non-volatile memory (nvM)   241

o

orbital angular momentum   21–23, 25,
    29, 56

p

page mode access   299, 301–303
paramagnetic, paramagnetism
    28–33, 35
Pauli exclusion principle   25, 42, 45
Pauli paramagnetism   33, 37
PCM (Phase change memory)   242–246
perpendicular anisotropy
    (pMA)   176, 184–187
perpendicular magnetic anisotropy
    (PMA)   57, 59, 62–63, 74–76, 85
physically unclonable function
    (PUF)   262–264, 274
physical vapor deposition (PVD)   68, 70
precession
    cone angle   153, 155–157, 193
    of magnetic moment   117
    precession cycle   155, 157–158
    precession cycle period   155, 157
    precession toggle   205–206, 229–231
precharge, bit line   300–303, 305–309,
    315, 321
prefetch   299, 303, 305–306, 314–316,
    320–322

r

rare-earth elements   25, 51, 57
read operation
    read current   173, 178, 193
    read disturb   161, 163, 172 173, 178,
        188–189, 192
    read marginal bits   192
read to write voltage margin   179
reflection high-energy electron
    diffraction (RHEED)   71
resistance area (RA)   87–88
resistance-area product
    RA   155, 178
resistance RAM (ReRAM)   242–245

retention
    data retention   151–152, 169–173,
        176–178, 183–185, 187,
        191–193, 196
rigid band model   5051
RKKY interaction   42, 46–48, 86

s

SAF under external field   99
    saturate   99
    spin flop   99
saturated (saturation) magnetization
    25, 31, 34–36, 73
sense amplifier   300–305, 311–312,
    315–316, 320–321
    covalent reference sense
        amplifier   252–253
shape anisotropy   55, 85
Slater-Pauling curve   49–50
SPICE model   189
    embedded Macrospin calculator   189
spin angular momentum   22–26, 43
spin current   109, 111–113, 116, 120,
    123, 132
    drift-diffusion equation   114
    nonlocal spin valve   119
spin dice   263
spin diffusion   110–111, 114, 120
    spin accumulation   108, 113, 115,
        120, 121
    spin diffusion length   110, 120
spin Hall angle   115
spin Hall effect (SHE)   25, 78
spin Hall magnetoresistance (SMR)   78
spin-orbit interaction   24–26, 42, 48, 56,
    74, 78
spin orbit torque transfer
    (SOT)   205–206, 211–212, 215,
        222–224, 227–220, 229
spin polarization   109–110, 113–116,
    124–125, 128, 131–132
spin pumping   119, 122, 123, 129, 200
spin quantum number   22, 25

spin valve   81–82, 111, 113, 126–128,
   131–133, 137
sputter   48, 57, 68–70, 82
Stoner criterion   37
Stoner-Walfarth Astroid (Asteroid)   97
stray field
   of perpendicular MTJ   179–180, 188
superexchange   42, 45–46
superlattice   81
superparamagnetism   60
susceptibility   26, 28, 30–33, 38–39, 62
switching efficiency
   switching energy efficiency   152–153,
      166, 196
switching threshold current density   128
synthetic anti ferromagnetic (SAF)
   film stack   97
   layer   47, 82, 85–86

t
test chip   191
thermal stability   17, 152–157,
   162, 172, 178, 183–187,
   193–195, 197
   factor   193–194
Thomas precession   25
time dependent dielectric
   breakdown (TDDB)   161, 194
   bipolar stress   167–168, 195, 281
   unipolar stress   168, 195, 281
topological insulator material   219–221
torkance   133
tunneling magnetoresistance
   (TMR)   77–78, 81–89
twin cell   253
two-channel model   80
two-way flip data retention
   model   277–278

u
uniaxial anisotropy   52–53

v
vibrating sample magnetometer
   (VSM)   73–74
voltage control magnetic anisotropy
   (VCMA)   206, 224–230, 233, 270

w
weak bit screening   191
   low endurance bits   194
   margin bits   193
   short retention bits   193
Weiss molecular field   34, 38, 45
   constant   35, 37, 45
write operation
   critical current, current density   155,
      158, 208, 213, 223
   precession regime   154
   stochastic switching   151,
      156–157, 164
   thermal regime   154–155, 158
   WER of SOT cell   229–230
   write current (.cf switching current,
      switching threshold current)   13,
      151, 153–155, 160–161,
      165, 169, 175–176, 181–183,
      185, 187
   write error rate (WER) of STT
      cell   151–152, 156–163,
      165–166, 173–174, 178–179,
      182, 188–190, 193, 229, 230
   write margin bits   193
   write to breakdown voltage
      margin   164, 167

y
yttrium iron garnet (YIG)   35, 41

z
Zeeman effect   22
Zeeman energy   22, 29, 31, 33, 58,
   109, 117, 118